# ISO AND IEC SELECTED STANDARDS FOR THE PLASTICS INDUSTRY

## 1996

ASTM Publication Code Number (PCN): 29-420096-19

ASTM
100 Barr Harbor Drive
West Conshohocken, PA 19428-2959

Printed in Philadelphia , Pa.,USA
March 1996

# CONTENTS

# International Standard

**ISO** 62

INTERNATIONAL ORGANIZATION FOR STANDARDIZATION•МЕЖДУНАРОДНАЯ ОРГАНИЗАЦИЯ ПО СТАНДАРТИЗАЦИИ•ORGANISATION INTERNATIONALE DE NORMALISATION

# Plastics — Determination of water absorption

*Plastiques — Détermination de l'absorption d'eau*

**First edition — 1980-09-15**
Corrected and reprinted 1990-10-01

UDC 678.5/.8 : 678.01 : 532.696.52

Descriptors : plastics, tests, water absorption tests.

Ref. No. ISO 62-1980 (E)

# Foreword

ISO (the International Organization for Standardization) is a worldwide federation of national standards institutes (ISO member bodies). The work of developing International Standards is carried out through ISO technical committees. Every member body interested in a subject for which a technical committee has been set up has the right to be represented on that committee. International organizations, governmental and non-governmental, in liaison with ISO, also take part in the work.

Draft International Standards adopted by the technical committees are circulated to the member bodies for approval before their acceptance as International Standards by the ISO Council.

International Standard ISO 62 was developed by Technical Committee ISO/TC 61, *Plastics*, and was circulated to the member bodies in May 1978.

It has been approved by the member bodies of the following countries :

| | | |
|---|---|---|
| Australia | India | Romania |
| Austria | Iran | South Africa, Rep. of |
| Belgium | Israel | Spain |
| Bulgaria | Italy | Sweden |
| Canada | Japan | Switzerland |
| Czechoslovakia | Korea, Rep. of | Turkey |
| Finland | Mexico | United Kingdom |
| France | Netherlands | USA |
| Greece | New Zealand | USSR |
| Hungary | Poland | Yugoslavia |

The member body of the following country expressed disapproval of the document on technical grounds :

Germany, F. R.

This International Standard cancels and replaces ISO Recommendations R 62-1958 and R 117-1959, of which it constitutes a technical revision.

# Plastics — Determination of water absorption

## 0  Introduction

Under the action of water, a plastic may be subjected to two phenomena at more or less the same time :

— the absorption of water, sometimes accompanied by swelling;

— the extraction of any water-soluble constituents.

The action of the water may also result in changes in the dimensions and/or in the physical properties of the material.

The water absorption values of different plastics may be used for comparing their behaviour under moist conditions, but they do not represent the maximum amount of water that may be absorbed.

Comparison of the water absorption of various plastics is possible only if the test specimens are of identical dimensions and, as nearly as possible, in the same physical state.

Correspondence between the methods specified in this International Standard and the previous Recommendations ISO/R 62 and ISO/R 117 is indicated in the standard itself.

## 1  Scope and field of application

**1.1**  This International Standard specifies four conventional methods for the determination of the mass of water absorbed by a plastic test specimen of defined dimensions that is immersed in water for a specified time and at a specified temperature. Two of the methods are for use when it is desired to take account of the possible presence of water-soluble matter.

**1.2**  The four conventional methods specified are applicable to all plastics except cellular plastics and plastics that soften at the temperature of boiling water to an extent that causes them to lose their shape entirely. In the latter case, only tests at 23 °C are applicable.

**1.3**  The standard immersion periods are 24 h at 23 °C and 30 min at the boiling point of water, but provision is made for longer test durations, by agreement between the interested parties.

NOTE — Except for the variations in mass covered in this International Standard, the influence of water absorption on the properties of plastics should be determined in accordance with ISO 175.

## 2  References

ISO 175, *Plastics — Determination of the effects of liquid chemicals including water.*[1]

ISO 293, *Plastics — Compression moulding test specimens of thermoplastic materials.*[2]

ISO 294, *Plastics — Injection moulding test specimens of thermoplastic materials.*

ISO 295, *Plastics — Compression moulding test specimens of thermosetting materials.*

ISO 2508, *Unplasticized polyvinyl chloride (PVC) pipes — Determination of water absorption.*

ISO 2818, *Plastics — Preparation of test specimens by machining.*

## 3  Principle

Complete immersion of test specimens of the plastic material in water for a specified period of time and at a specified temperature. Determination of changes in the mass of the test specimens after immersion in water and if required after elimination of the water by drying.

The water absorption may be expressed in the following ways :

a)  as the mass of water absorbed;

b)  as the mass of water absorbed per unit of surface area;

---

[1]  At present at the stage of draft. (Revision of ISO/R 175 and ISO/R 462.)

[2]  At present at the stage of draft. (Revision of ISO 293-1974.)

c) as a percentage by mass of water absorbed with respect to the mass of the test specimen.

IMPORTANT NOTE — Comparisons between various plastics on the basis of this test are valid only if the test specimens are of identical dimensions and, as nearly as possible, in the same physical state (surface, internal stress, etc.).

# 4 Apparatus

**4.1 Balance,** with an accuracy of 1 mg.

**4.2 Oven,** capable of being controlled at 50 $\pm$ 2 °C or at any other agreed temperature (see clause 6, note 2).

**4.3 Containers,** containing distilled water, or water of equivalent purity, equipped with a means of heating and capable of being controlled at the temperature specified.

**4.4 Desiccator.**

**4.5 Means** of measuring dimensions of specimens, if required (see 7.2.1).

# 5 Test specimens

Three specimens shall be tested. They may be obtained directly by moulding or by machining. In the latter case, the cut surface shall be smooth and shall not show any trace of charring that may be due to the method of preparation.

## 5.1 Moulding material

The test specimen shall have a diameter of 50 $\pm$ 1 mm and a thickness of 3 $\pm$ 0,2 mm. It shall be moulded under the conditions given in the relevant specification for the material (or under the conditions prescribed by the supplier of the material).

NOTES

1 The general principles for preparing moulded test specimens are the subject of the following International Standards : ISO 293, ISO 294 and ISO 295.

2 In certain specific cases, a square test specimen of side 50 $\pm$ 1 mm and a thickness of 4 $\pm$ 0,2 mm may be used by agreement between the interested parties.

## 5.2 Extrusion compounds

The test specimen shall have a diameter of 50 $\pm$ 1 mm and a thickness of 3 $\pm$ 0,2 mm. It shall be cut from a sheet of this thickness, prepared under the conditions given in the relevant specification for the material (or under the conditions prescribed by the supplier of the material).

## 5.3 Sheet

The specimen shall be 50 $\pm$ 1 mm square and shall be machined from the sheet under test, for example in accordance with ISO 2818.

The thickness of the test specimen shall be the same as that of the sheet under test if the nominal thickness of the sheet is equal to or less than 25 mm.

If the nominal thickness is greater than 25 mm and in the absence of special provisions in the relevant specification, the thickness of the test specimen shall be reduced to 25 mm by machining on one surface only.

## 5.4 Tubes and rods

### 5.4.1 Tubes

When possible, reference shall be made to the relevant International Standards for the material under test.[1] In the absence of specific International Standards, the test specimen shall be a piece of tube, of length 50 $\pm$ 1 mm, obtained by cutting it at right angles to its longitudinal axis.

For tubes of outside diameter larger than 50 mm, a length of 50 $\pm$ 1 mm shall be cut and the test specimen prepared from this length by making a cut along each of two planes containing the longitudinal axis of the tube, so as to give a developed width of 50 $\pm$ 1 mm when measured on the outer surface.

### 5.4.2 Rods

For rods of diameter below or equal to 50 mm, the test specimen shall be a piece of the rod, of length 50 $\pm$ 1mm, obtained by cutting it at right angles to its longitudinal axis.

For rods of diameter greater than 50 mm, in the absence of any specification agreed between the interested parties, the test specimen shall be a 50 $\pm$ 1 mm length of the rod with its diameter reduced to 50 $\pm$ 1 mm by machining concentrically.

## 5.5 Profiles

In the absence of specific International Standards, a piece of the profile of length 50 $\pm$ 1 mm shall be cut and the test specimen shall be either :

a) this piece of the profile, or

b) this piece after machining so as to reduce one or more of the dimensions of the cross-section of the profile in such a way that the thickness in particular is as near as possible to 3 $\pm$ 0,2 mm. In this case, the dimensions to be obtained and the machining conditions shall be the subject of an agreement between the interested parties.

---

1) For example, for rigid unplasticized PVC pipes, see ISO 2508.

# 6 Procedure

NOTES

1 With certain materials it may be necessary to weigh the test specimens in a weighing vessel.

2 Other drying procedures than those described in 6.2 to 6.5 may be used by agreement between the interested parties.

## 6.1 General conditions

**6.1.1** The volume of water used shall be at least 8 ml per square centimetre of the total surface of the test specimens, so as to avoid any extraction product becoming excessively concentrated in the water during the test.

**6.1.2** In general, place each set of three test specimens in a separate container (4.3) with the specimens immersed completely in the water.

However, when several samples of the same composition have to be tested, it is permissible to place several sets of test specimens in the same container.

In no case shall any significant area of the surface of a test specimen come into contact with the surface of other test specimens, or with the walls of the container.

**6.1.3** The times for immersion in the water are stated in 6.2 and 6.4. However, longer times may be used by agreement between the interested parties. In such cases the following precautions shall be taken :

— If testing in water at 23 °C, agitate the water at least once daily, for example by rotating the containers.

— If testing in boiling water, add boiling water from time to time, as required to maintain the volume.

## 6.2 Method 1 [Formerly procedure A in ISO/R 62-1958]

Dry three test specimens for 24 $\pm$ 1 h in the oven (4.2) controlled at 50 $\pm$ 2 °C, allow to cool to ambient temperature in the desiccator (4.4) and weigh each specimen to the nearest 1 mg (mass $m_1$). Then place the specimens in a container (4.3) containing distilled water, controlled at 23 °C with a tolerance of $\pm$ 0,5 °C or $\pm$ 2 °C according to the relevant specification. In the absence of a specification, the tolerance shall be $\pm$ 0,5 °C.

After immersion for 24 $\pm$ 1 h, take the specimens from the water and remove all surface water with a clean, dry cloth or with filter paper. Re-weigh the specimens to the nearest 1 mg within 1 min of taking them from the water (mass $m_2$).

## 6.3 Method 2 [Formerly procedure B in ISO/R 62-1958]

If it is desired to allow for the presence of water-soluble matter, dry the test specimens again for 24 $\pm$ 1 h in the oven (4.2), controlled at 50 $\pm$ 2 °C, after completion of method 1 (6.2).

Allow the specimens to cool to ambient temperature in the desiccator (4.4) and reweigh to the nearest 1 mg (mass $m_3$).

## 6.4 Method 3 [Formerly procedure A in ISO/R 117-1959]

Dry three test specimens for 24 $\pm$ 1 h in the oven (4.2), controlled at 50 $\pm$ 2 °C, allow to cool to ambient temperature in the desiccator (4.4) and weigh each specimen to the nearest 1 mg (mass $m_1$). Place the specimens in a container (4.3) containing boiling distilled water.

After immersion for 30 $\pm$ 1 min, take the specimens from the boiling water and allow them to cool for 15 $\pm$ 1 min in distilled water at ambient temperature. Take the specimens from the water and remove all surface water with a clean, dry cloth or with filter paper. Re-weigh the specimens to the nearest 1 mg within 1 min of taking them from the water (mass $m_2$).

## 6.5 Method 4 [Formerly procedure B in ISO/R 117-1959]

If it is desired to allow for the presence of water-soluble matter dry the test specimens again for 24 $\pm$ 1 h in the oven (4.2), controlled at 50 $\pm$ 2 °C, after completion of method 3 (6.4). Allow the specimens to cool to ambient temperature in the desiccator (4.4) and re-weigh to the nearest 1 mg (mass $m_3$).

# 7 Expression of results

## 7.1 Standard way of expressing results (as the mass of water absorbed)

**7.1.1** For methods 1 and 3, calculate for each test specimen the mass, in milligrams, of water absorbed, according to the formula

$$m_2 - m_1$$

where

$m_1$ is the mass, in milligrams, of the test specimen before immersion;

$m_2$ is the mass, in milligrams, of the test specimen after immersion.

**7.1.2** For methods 2 and 4, calculate for each test specimen the mass, in milligrams, of water absorbed, according to the formula

$$m_2 - m_3$$

where

$m_2$ is as defined in 7.1.1;

$m_3$ is the mass, in milligrams, of the test specimen after immersion and drying.

**7.1.3** For all four methods, express the result as the arithmetic mean of the three values obtained.

## 7.2 Other possible ways of expressing results

If the material specification so demands, or by agreement between the interested parties, the results may be expressed in one of the following ways.

**7.2.1** As the mass of water absorbed per unit of surface area

Calculate for each test specimen the water absorption, in milligrams per square centimetre, by the following formulae, as appropriate :

$$\frac{m_2 - m_1}{A} \quad \text{or} \quad \frac{m_2 - m_3}{A}$$

where

$m_1$, $m_2$ and $m_3$ are as defined in 7.1.1 and 7.1.2;

$A$ is the initial total surface area, in square centimetres, of the test specimen.

**7.2.2** As a percentage by mass of water absorbed

Calculate for each test specimen the water absorption as a percentage by mass of the initial mass, by the following formulae, as appropriate :

$$\frac{m_2 - m_1}{m_1} \times 100 \quad \text{or} \quad \frac{m_2 - m_3}{m_1} \times 100$$

NOTE — If it is required to express the water absorption as a percentage of the mass of the test specimen after drying, use the formula

$$\frac{m_2 - m_3}{m_3} \times 100$$

where $m_1$, $m_2$ and $m_3$ are as defined in 7.1.1 and 7.1.2.

**7.2.3** For the calculations described in 7.2.1 and 7.2.2, express the result as the arithmetic mean of the three values obtained.

## 8 Test report

The test report shall include the following particulars :

a) reference to this International Standard;

b) complete identification of the material or the product tested;

c) the type of test specimens used, the method of preparing them, their dimensions, their initial mass, and, if applicable, their initial surface area and their surface condition (for example whether machined or not);

d) the method (1, 2, 3 or 4) used and the immersion period if this is different from that specified in the method;

e) the water absorption calculated by one or more of the ways of expressing results given in clause 7;

NOTE — If the calculations described in 7.1 and 7.2 give a negative value for water absorption, this fact should be stated clearly in the test report.

f) any incidents likely to have affected the results.

# INTERNATIONAL STANDARD

**ISO
75-1**

First edition
1993-09-15

# Plastics — Determination of temperature of deflection under load —

## Part 1:
## General test method

*Plastiques — Détermination de la température de fléchissement sous charge —*

*Partie 1: Méthode générale d'essai*

Reference number
ISO 75-1:1993(E)

# Foreword

ISO (the International Organization for Standardization) is a worldwide federation of national standards bodies (ISO member bodies). The work of preparing International Standards is normally carried out through ISO technical committees. Each member body interested in a subject for which a technical committee has been established has the right to be represented on that committee. International organizations, governmental and non-governmental, in liaison with ISO, also take part in the work. ISO collaborates closely with the International Electrotechnical Commission (IEC) on all matters of electrotechnical standardization.

Draft International Standards adopted by the technical committees are circulated to the member bodies for voting. Publication as an International Standard requires approval by at least 75 % of the member bodies casting a vote.

International Standard ISO 75-1 was prepared by Technical Committee ISO/TC 61, *Plastics*, Sub-Committee SC 2, *Mechanical properties*.

Together with the other parts, it cancels and replaces the second edition of ISO 75 (ISO 75:1987), which has been technically revised.

ISO 75 consists of the following parts, under the general title *Plastics — Determination of temperature of deflection under load*:

— *Part 1: General test method*

— *Part 2: Plastics and ebonite*

— *Part 3: High-strength thermosetting laminates and long-fibre-reinforced plastics*

International Organization for Standardization
Case Postale 56 • CH-1211 Genève 20 • Switzerland

Printed in Switzerland

# Plastics — Determination of temperature of deflection under load —

# Part 1:
## General test method

## 1 Scope

**1.1** ISO 75 specifies methods for the determination of the temperature of deflection under load (bending stress under three-point loading) of different types of material.

**1.2** Part 1 of ISO 75 gives a general test method, part 2 gives specific requirements for plastics and ebonite and part 3 gives specific requirements for high-strength thermosetting laminates and long-fibre-reinforced plastics.

**1.3** The methods specified are suitable for assessing the behaviour of the different types of material at elevated temperature under load at a specified rate of temperature increase. The results obtained do not necessarily represent maximum use temperatures, because in practice essential factors such as time, loading conditions and nominal surface stress may differ from the test conditions.

## 2 Normative reference

The following standard contains provisions which, through reference in this text, constitute provisions of this part of ISO 75. At the time of publication, the edition indicated was valid. All standards are subject to revision, and parties to agreements based on this part of ISO 75 are encouraged to investigate the possibility of applying the most recent edition of the standard indicated below. Members of IEC and ISO maintain registers of currently valid International Standards.

ISO 291:1977, *Plastics — Standard atmospheres for conditioning and testing.*

## 3 Definitions

For the purposes of this and the other parts of ISO 75, the following definitions apply.

**3.1 deflection:** The distance, at mid-span, over which the top or bottom surface of a test specimen deviates, during flexure, from its original position. It is expressed in millimetres.

**3.2 standard deflection $s$:** The deflection which will result in the flexural strain, at the surface of the test specimen, which is specified in the relevant part of this International Standard. The standard deflection depends on the dimensions and position of the test specimen and the span between the supports. It is expressed in millimetres.

**3.3 temperature of deflection under load, $T_f$:** The temperature at which the deflection of the test specimen reaches the standard deflection as the temperature is increased. It is expressed in degrees Celsius.

**3.4 flexural strain:** The nominal fractional change in length of an element at the surface of the test specimen at mid-span. It is expressed as a dimensionless quantity.

**3.5 flexural stress, $\sigma$:** The nominal stress at the surface of the test specimen at mid-span. It is expressed in megapascals.

## 4 Principle

A standard test specimen is subjected to a bending stress to produce one of the nominal surface stresses given in the relevant part of this International Standard. The temperature is raised at a uniform rate, and the temperature at which a specified deflection occurs is measured.

## 5 Apparatus

### 5.1 Means of applying a bending stress

The apparatus shall be constructed essentially as shown in figure 1. It consists of a rigid metal frame in which a rod can move freely in the vertical direction. The rod is fitted with a weight-carrying plate and a loading edge. The base of the frame is fitted with test-specimen supports; these and the vertical members of the frame are made of a metal having the same coefficient of linear expansion as the rod.

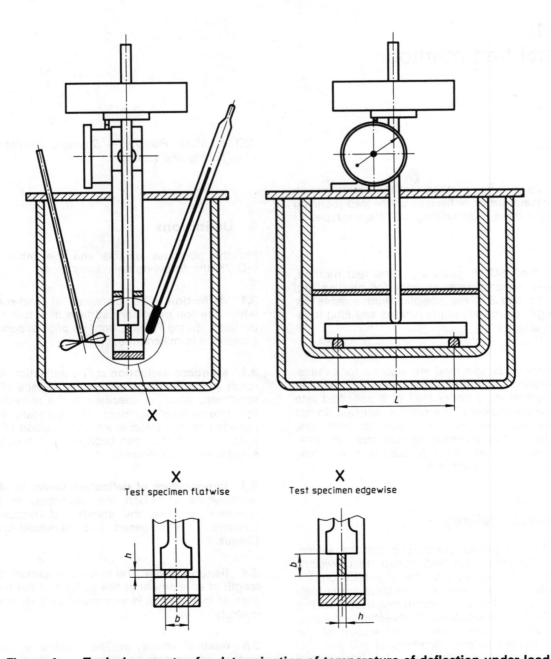

X

Test specimen flatwise

X

Test specimen edgewise

**Figure 1 — Typical apparatus for determination of temperature of deflection under load**

The test-specimen supports consist of cylindrical metal pieces at a distance apart defined in the relevant part of this International Standard and with their lines of contact with the specimen in a horizontal plane. The supports are fitted to the base of the frame in such a way that the vertical force applied to the test specimen by the loading edge is midway between them. The contact edges of the supports are parallel to the loading edge and at right angles to the length direction of a test specimen placed symmetrically across them. The contact edges of the supports and loading edge are rounded to a radius of 3,0 mm ± 0,2 mm and shall be longer than the width of the test specimen.

Unless vertical parts of the apparatus have the same coefficient of linear expansion, the differential change in the length of these parts introduces an error in the reading of the apparent deflection of the test specimen. A blank test shall be made on each apparatus using a test specimen made of rigid material having a low coefficient of expansion. The blank test shall cover the temperature ranges to be used in the actual determination, and a correction term shall be determined for each temperature. If the correction term is 0,010 mm or greater, its value and algebraic sign shall be noted and the term applied to each test by adding it algebraically to the reading of the apparent deflection of the test specimen.

NOTE 1    Invar and borosilicate glass have been found suitable as materials for the test specimen in the blank test.

## 5.2  Heating equipment

The heating equipment may be a heating bath containing a suitable liquid, a fluidized bed or an oven with forced circulation of air or nitrogen.

The heating equipment shall be provided with a control unit so that the temperature can be raised at a uniform rate of 120 °C/h ± 10 °C/h. This heating rate shall be considered to be met if, over every 6-min interval during the test, the temperature change is 12 °C ± 1 °C.

The difference in the temperature of the heat-transfer medium between the ends and the centre of the test specimen shall not exceed 1 °C.

The apparatus may be designed to stop heating automatically and sound an alarm when the specified deflection has been reached.

NOTE 2    The heat-transfer media given above vary in their thermal conductivity. To prevent significant errors in the results, the thermometer or other temperature-measuring instrument used must be carefully adjusted in accordance with 8.2.

### 5.2.1  Heating bath, containing a suitable liquid in which the test specimen can be immersed to a depth of at least 50 mm. An efficient stirrer shall be provided. It shall be established that the liquid chosen is

stable over the temperature range used and does not affect the material under test, for example causing it to swell or crack.

NOTE 3    Liquid paraffin, transformer oil, glycerol and silicone oils are suitable liquid heat-transfer media, but other liquids may be used.

### 5.2.2  Oven, with forced air or nitrogen circulation of about 60 times per minute; with a volume of not less than 10 litres per specimen holder (see 5.1), and with the air or nitrogen flow directed horizontally, perpendicular to the length of the specimen, at a speed of 1,5 m/s to 2 m/s.

NOTE 4    Commercially available ovens are often fitted with suitable air or nitrogen circulation equipment. If not, the necessary heat-transfer conditions may be ensured by fitting guide-vanes to direct the air or nitrogen flow horizontally across the specimen.

## 5.3  Weights

A set of weights shall be provided so that the test specimen can be loaded to the required nominal surface stress, calculated as specified in 8.1.

NOTE 5    It may be necessary to adjust these weights in 1 g increments.

## 5.4  Temperature-measuring instrument

This may be a mercury-in-glass thermometer of the partial-immersion type, or any other suitable temperature-measuring instrument, with an appropriate range and reading to 0,5 °C or less. Mercury-in-glass thermometers shall be calibrated at the depth of immersion required by 8.2.

## 5.5  Deflection-measuring instrument

This may be a calibrated micrometer dial gauge, or any other suitable measuring instrument, capable of measuring to 0,01 mm the deflection at the midpoint between the test-specimen supports.

NOTE 6    In certain types of apparatus, the force $F_s$ exerted by the dial gauge spring acts upwards and therefore reduces the downward force exerted by the weighted rod, while, in other types, $F_s$ acts downwards and augments that exerted by the weighted rod. In such cases, it is necessary to determine the magnitude and direction of $F_s$ so as to be able to compensate for it (see 8.1).

Since, in certain dial gauges, $F_s$ varies considerably over the range of the instrument, it should be measured in that part of the range in which the instrument is to be used.

## 5.6  Micrometers and gauges

These are used to measure the main dimensions of the test specimens. They shall be accurate to 0,01 mm.

11

# 6 Test specimens

At least two test specimens shall be used for each sample. Each test specimen shall be a bar of rectangular cross-section (length $l$, width $b$, thickness $h$).

The dimensions of the test specimens shall be as specified in the relevant part of this International Standard, the thickness always being smaller than the width (i.e. $h < b$). All test specimens shall be free of warpage.

In each test specimen, the thickness and width over the central one-third of the length shall nowhere deviate by more than 2 % from the mean value.

# 7 Conditioning

Unless otherwise required by the specification for the material being tested, the atmosphere used for conditioning and testing shall be in accordance with ISO 291.

# 8 Procedure

## 8.1 Calculation of force to be applied

In the three-point loading method employed in this International Standard, the force applied to the specimen is given, in newtons, as a function of the flexural stress by one of the following equations:

$$F = \frac{2\sigma \cdot b \cdot h^2}{3L}$$

when the test specimen is in the flatwise position:

and

$$F = \frac{2\sigma \cdot h \cdot b^2}{3L}$$

when the test specimen is in the edgewise position:

where

$\sigma$ is the maximum nominal surface stress, in megapascals, in the test specimen;

$b$ is the width, in millimetres, of the test specimen;

$h$ is the thickness, in millimetres, of the test specimen;

$L$ is the span, in millimetres, between the test-specimen supports.

Measure dimensions $b$ and $h$ to the nearest 0,1 mm and dimension $L$ to the nearest 0,5 mm.

The span between the test specimen supports and the nominal surface stress shall be as specified in the relevant part of this International Standard.

The effect of the mass $m_r$ of the rod that applies the test force $F$ shall be taken into account as contributing to the test force. If a spring-loaded instrument such as a dial gauge is used, the magnitude and direction of the force $F_s$ exerted by the spring shall also be taken into account as a positive or negative contribution to the force $F$ (see 5.5, note 6).

The mass $m_w$ of any additional weights which need to be placed on the rod to produce the required total force $F$ is given by the equation

$$F = 9,81(m_w + m_r) + F_s$$

from which

$$m_w = \frac{F - F_s}{9,81} - m_r$$

where

$m_r$ is the mass, in kilograms, of the rod assembly that applies the test force;

$m_w$ is the mass, in kilograms, of the additional weights;

$F$ is the total force, in newtons, applied to the test specimen;

$F_s$ is the force, in newtons, exerted by any spring-loaded instrument used; the value of the force is positive if the thrust of the spring is directed towards the test specimen (i.e. downwards), negative if the thrust of the spring is in the opposite direction (i.e. opposing the descent of the rod) or zero if no such instrument is used.

The actual force applied shall be the calculated force $F \pm 2,5$ %.

## 8.2 Initial temperature of the heating equipment

The thermometer bulb or the sensor element of the temperature-measuring instrument (5.4) shall be at the same level as, and as close as possible to, the test specimen (within 10 mm), but not in contact with it. The temperature of the heating equipment (5.2) shall be between 20 °C and 23 °C at the start of each test, unless previous tests have shown that, for the particular material under test, no error is introduced by starting at some other temperature.

## 8.3 Measurement

Check and, if necessary, adjust the span between the test-specimen supports (see 5.1) to the appropriate value. Measure this distance to the nearest 0,5 mm and record for use in the calculations in 8.1.

Place a test specimen on the supports so that the longitudinal axis of the specimen is perpendicular to the supports. Place the loading assembly (5.1) in the heat-transfer medium. Apply the force calculated in 8.1 to give a nominal surface stress in the test specimen as specified in the relevant part of this International Standard. Allow the force to act for 5 min (see note 7). Record the reading of the deflection-measuring instrument (5.5) or set it to zero.

Raise the temperature of the heating equipment at a uniform rate of 120 °C/h ± 10 °C/h. Note the temperature at which the bar reaches the standard deflection, i.e. the temperature of deflection under load at the nominal surface stress specified in the relevant part of this International Standard. The standard deflection is a function of the height ($h$ or $b$, depending on the orientation of the test specimen, see 8.1) and is given in the relevant part of this International Standard.

NOTES

7  The 5-min waiting period is provided to compensate partially for the creep exhibited by some materials at room temperature when subjected to the specified nominal surface stress. The creep which occurs in the first 5 min is usually a significant fraction of that which occurs in the first 30 min. This waiting period may be omitted when testing materials that show no appreciable creep during the first 5 min at the initial temperature used.

8  It is frequently helpful in the interpretation of test results if specimen deflection is known as a function of specimen temperature. It is thus recommended that, where possible, the specimen deflection be monitored continuously during the waiting and heating periods.

The test shall be carried out at least in duplicate. Each test specimen shall be used only once.

## 9  Expression of results

Unless otherwise specified in the relevant part of this International Standard, express the temperature of deflection under load of the material under test as the arithmetic mean of the temperatures of deflection under load of the specimens tested.

## 10  Precision

The precision of this test method is not known because inter-laboratory data are not available. When such data are obtained, a precision statement will be added at the next revision.

## 11  Test report

The test report shall include the following information:

a)  a reference to the relevant part of this International Standard;

b)  all details necessary for the identification of the material tested;

c)  the method of preparation of the test specimens;

d)  the heat-transfer medium used;

e)  the conditioning and annealing procedures used, if any;

f)  the temperature of deflection under load, in degrees Celsius (if the individual results of two measurements differ by more than the limit given in the relevant part of this International Standard, all individual results shall be reported);

g)  the dimensions of the test specimens used;

h)  the orientation of the test specimen (flatwise or edgewise);

i)  the nominal surface stress used;

j)  the span between the test-specimen supports;

k)  any unusual behaviour of the test specimen noted during the test or after removal from the apparatus.

# INTERNATIONAL STANDARD

**ISO 75-2**

First edition
1993-09-15

## Plastics — Determination of temperature of deflection under load —

### Part 2:
Plastics and ebonite

*Plastiques — Détermination de la température de fléchissement sous charge —*

*Partie 2: Plastiques et ébonite*

Reference number
ISO 75-2:1993(E)

14

# Foreword

ISO (the International Organization for Standardization) is a worldwide federation of national standards bodies (ISO member bodies). The work of preparing International Standards is normally carried out through ISO technical committees. Each member body interested in a subject for which a technical committee has been established has the right to be represented on that committee. International organizations, governmental and non-governmental, in liaison with ISO, also take part in the work. ISO collaborates closely with the International Electrotechnical Commission (IEC) on all matters of electrotechnical standardization.

Draft International Standards adopted by the technical committees are circulated to the member bodies for voting. Publication as an International Standard requires approval by at least 75 % of the member bodies casting a vote.

International Standard ISO 75-2 was prepared by Technical Committee ISO/TC 61, *Plastics*, Sub-Committee SC 2, *Mechanical properties*.

Together with the other parts, it cancels and replaces the second edition of ISO 75 (ISO 75:1987), which has been technically revised.

ISO 75 consists of the following parts, under the general title *Plastics — Determination of temperature of deflection under load*:

— *Part 1: General test method*

— *Part 2: Plastics and ebonite*

— *Part 3: High-strength thermosetting laminates and long-fibre-reinforced plastics*

International Organization for Standardization
Case Postale 56 • CH-1211 Genève 20 • Switzerland

Printed in Switzerland

15

# Plastics — Determination of temperature of deflection under load —

## Part 2:
Plastics and ebonite

## 1 Scope

**1.1** This part of ISO 75 specifies three methods for the determination of the temperature of deflection under load (bending stress) of plastics and ebonite:

— method A, using a nominal surface stress of 1,80 MPa;

— method B, using a nominal surface stress of 0,45 MPa;

— method C, using a nominal surface stress of 8,00 MPa.

**1.2** The test specimens are tested in one of two positions, flatwise or edgewise, the requirements on test-specimen dimensions being different in each case (see clause 6).

**1.3** See ISO 75-1:1993, subclause 1.3.

NOTE 1  The methods give better reproducibility with amorphous plastics than with semi-crystalline ones. With some materials, it may be necessary to anneal the test specimens to obtain reliable results. Annealing procedures, if applied, generally result in an increase in the temperature of deflection under load (see 6.2 and 6.3).

## 2 Normative references

The following standards contain provisions which, through reference in this text, constitute provisions of this part of ISO 75. At the time of publication, the editions indicated were valid. All standards are subject to revision, and parties to agreements based on this part of ISO 75 are encouraged to investigate the possibility of applying the most recent editions of the standards indicated below. Members of IEC and ISO maintain registers of currently valid International Standards.

ISO 75-1:1993, *Plastics — Determination of temperature of deflection under load — Part 1: General test method.*

ISO 293:1986, *Plastics — Compression moulding test specimens of thermoplastic materials.*

ISO 294:—[1], *Plastics — Injection moulding of test specimens of thermoplastic materials.*

ISO 2818:—[2], *Plastics — Preparation of test specimens by machining.*

ISO 3167:1993, *Plastics — Multipurpose test specimens.*

## 3 Definitions

See ISO 75-1:1993, clause 3.

## 4 Principle

A standard test specimen made of plastic or ebonite is subjected to a bending stress to produce one of the nominal surface stresses given in 1.1. The temperature is raised at a uniform rate, and the temperature at which a specified deflection occurs is measured.

---

1) To be published. (Revision of ISO 294:1975)

2) To be published. (Revision of ISO 2818:1980)

# 5 Apparatus

## 5.1 Means of applying a bending stress

See ISO 75-1:1993, subclause 5.1.

The span between the test-specimen supports shall be 64 mm $\pm$ 1 mm if the specimen is tested in the flatwise position and 100 mm $\pm$ 2 mm if the specimen is tested in the edgewise position.

## 5.2 Heating equipment

See ISO 75-1:1993, subclause 5.2.

## 5.3 Weights

See ISO 75-1:1993, subclause 5.3.

## 5.4 Temperature-measuring instrument

See ISO 75-1:1993, subclause 5.4.

## 5.5 Deflection-measuring instrument

See ISO 75-1:1993, subclause 5.5.

# 6 Test specimens

See ISO 75-1:1993, clause 6.

**6.1** One of two different types of test specimen shall be used, depending on the orientation of the specimen in the test apparatus.

If the specimen is tested in the flatwise position, its dimensions shall be

| | |
|---|---|
| length, $l$ : | 80 mm $\pm$ 2,0 mm |
| width, $b$ : | 10 mm $\pm$ 0,2 mm |
| thickness, $h$ : | 4 mm $\pm$ 0,2 mm |

If the specimen is tested in the edgewise position, its dimensions shall be

| | |
|---|---|
| length, $l$ : | 120,0 mm $\pm$ 10,0 mm |
| width, $b$ : | 9,8 mm to 15,0 mm |
| thickness, $h$ : | 3,0 mm to 4,2 mm |

The test specimen shall be produced in accordance with ISO 293 and ISO 2818, or ISO 294, or as agreed upon by the interested parties. In the case of compression-moulded specimens, the thickness shall be in the direction of the moulding force. For materials in sheet form, the thickness of the test specimen (this dimension is usually the thickness of the sheet) shall be in the range 3 mm to 13 mm, preferably between 4 mm and 6 mm.

NOTES

2 The test results obtained on specimens approaching 13 mm thick may be 2 °C to 4 °C above those obtained from thin test specimens because of poorer heat transfer.

3 The possibility of carrying out the test with a smaller (80 mm × 10 mm × 4 mm) specimen in the flatwise position has been introduced because it gives the following advantages:

— the specimen can be taken from the narrow central part of the multipurpose test specimen specified in ISO 3167;

— it is more stable on the supports;

— it does not tend to stand on one edge like the edgewise test specimen.

**6.2** The test results obtained on moulded test specimens depend on the moulding conditions used in their preparation. Moulding conditions shall be in accordance with the standard for the material, or shall be agreed upon by the interested parties.

**6.3** Discrepancies in test results due to variations in moulding conditions may be minimized by annealing the test specimens before testing them. Since different materials require different annealing conditions, annealing procedures shall be employed only if required by the materials standard or if agreed upon by the interested parties.

# 7 Conditioning

See ISO 75-1:1993, clause 7.

# 8 Procedure

## 8.1 Calculation of force to be applied

See ISO 75-1:1993, subclause 8.1.

The maximum nominal surface stress applied shall be one of the following:

1,80 MPa, in which case the method is designated method A;

0,45 MPa, in which case the method is designated method B;

8,00 MPa, in which case the method is designated method C.

The dimensions of the test specimen are given in 6.1. The span between the test-specimen supports is given in 5.1.

## 8.2 Initial temperature of the heating equipment

See ISO 75-1:1993, subclause 8.2.

## 8.3 Measurement

See ISO 75-1:1993, subclause 8.3.

Apply one of the nominal surface stresses specified in 8.1 of this part of ISO 75.

Note the temperature at which the bar reaches the standard deflection given in table 1 or 2 for the test-specimen height concerned (thickness $h$ for specimens tested in the flatwise position and width $b$ for specimens tested in the edgewise position). This temperature is the temperature of deflection under load.

NOTES

4 The initial flexural strain due to the loading of the specimen at room temperature is neither specified nor measured in these methods. The specified quantity, the standard deflection $s$, is essentially a deflection difference, corresponding to a flexural-strain difference. The ratio of this flexural-strain difference to the initial flexural strain depends on the modulus of elasticity, at room temperature, of the material under test. This method is not suitable, therefore, for comparing the temperatures of deflection under load of materials with widely differing elastic properties.

5 The standard deflections given in tables 1 and 2 correspond to a flexural strain of 0,2 % at the surface of the test specimen.

## 9 Expression of results

See ISO 75-1:1993, clause 9.

If the individual results for amorphous plastics or ebonite differ by more than 2 °C, or those for semi-crystalline materials by more than 5 °C, repeat tests shall be carried out.

**Table 1 — Standard deflection for different test-specimen heights — 80 mm × 10 mm × 4 mm specimen tested in the flatwise postion**

| Test-specimen height (thickness $h$ of specimen) mm | Standard deflection mm |
|---|---|
| 3,8 | 0,36 |
| 3,9 | 0,35 |
| 4,0 | 0,34 |
| 4,1 | 0,33 |
| 4,2 | 0,32 |

**Table 2 — Standard deflection for different test-specimen heights — 120 mm × (3,0 to 4,2) mm × (9,8 to 15,0) mm specimen tested in the edgewise position**

| Test-specimen height (width $b$ of specimen) mm | Standard deflection mm |
|---|---|
| 9,8 to 9,9 | 0,33 |
| 10,0 to 10,3 | 0,32 |
| 10,4 to 10,6 | 0,31 |
| 10,7 to 10,9 | 0,30 |
| 11,0 to 11,4 | 0,29 |
| 11,5 to 11,9 | 0,28 |
| 12,0 to 12,3 | 0,27 |
| 12,4 to 12,7 | 0,26 |
| 12,8 to 13,2 | 0,25 |
| 13,3 to 13,7 | 0,24 |
| 13,8 to 14,1 | 0,23 |
| 14,2 to 14,6 | 0,22 |
| 14,7 to 15,0 | 0,21 |

## 10 Precision

See ISO 75-1:1993, clause 10.

## 11 Test report

See ISO 75-1:1993, clause 11.

The information on

h) the orientation of the test specimen (flatwise or edgewise);

i) the nominal surface stress;

may be given as follows:

Use method A, B or C to designate the nominal surface stress and the letters "e" and "f" to designate the test-specimen orientation.

Thus a test using a nominal surface stress of 1,80 MPa and flatwise test-specimen orientation would be referred to as "method Af". Similarly, a test using a nominal surface stress of 0,45 MPa and edgewise orientation would be called "method Be".

# INTERNATIONAL STANDARD

## ISO
## 75-3

First edition
1993-09-15

# Plastics — Determination of temperature of deflection under load —

## Part 3:
High-strength thermosetting laminates and long-fibre-reinforced plastics

*Plastiques — Détermination de la température de fléchissement sous charge —*

*Partie 3: Stratifiés thermodurcissables à haute résistance et plastiques renforcés de fibres longues*

Reference number
ISO 75-3:1993(E)

# Foreword

ISO (the International Organization for Standardization) is a worldwide federation of national standards bodies (ISO member bodies). The work of preparing International Standards is normally carried out through ISO technical committees. Each member body interested in a subject for which a technical committee has been established has the right to be represented on that committee. International organizations, governmental and non-governmental, in liaison with ISO, also take part in the work. ISO collaborates closely with the International Electrotechnical Commission (IEC) on all matters of electrotechnical standardization.

Draft International Standards adopted by the technical committees are circulated to the member bodies for voting. Publication as an International Standard requires approval by at least 75 % of the member bodies casting a vote.

International Standard ISO 75-3 was prepared by Technical Committee ISO/TC 61, *Plastics*, Sub-Committee SC 2, *Mechanical properties*.

Together with the other parts, it cancels and replaces the second edition of ISO 75 (ISO 75:1987), which has been technically revised.

ISO 75 consists of the following parts, under the general title *Plastics — Determination of temperature of deflection under load*:

— *Part 1: General test method*

— *Part 2: Plastics and ebonite*

— *Part 3: High-strength thermosetting laminates and long-fibre-reinforced plastics*

International Organization for Standardization
Case Postale 56 • CH-1211 Genève 20 • Switzerland

Printed in Switzerland

# Plastics — Determination of temperature of deflection under load —

# Part 3:
High-strength thermosetting laminates and long-fibre-reinforced plastics

## 1 Scope

**1.1** This part of ISO 75 specifies a method for the determination of the temperature of deflection under load (bending stress) of high-strength thermosetting laminates and compression-moulded long-fibre-reinforced plastics. The test load used is not a fixed load, as in part 2 of this International Standard, but a function (1/10) of the ultimate or specified load. This allows the method to be applied to materials with a wide range of strengths and bending moduli.

**1.2** The test specimen is tested in a flatwise position.

**1.3** See ISO 75:1993, subclause 1.3.

## 2 Normative references

The following standards contain provisions which, through reference in this text, constitute provisions of this part of ISO 75. At the time of publication, the editions indicated were valid. All standards are subject to revision, and parties to agreements based on this part of ISO 75 are encouraged to investigate the possibility of applying the most recent editions of the standards indicated below. Members of IEC and ISO maintain registers of currently valid International Standards.

ISO 75-1:1993, *Plastics — Determination of temperature of deflection under load — Part 1: General test method.*

ISO 178:1993, *Plastics — Determination of flexural properties.*

ISO 295:1991, *Plastics — Compression moulding of test specimens of thermosetting materials.*

ISO 2818:—[1], *Plastics — Preparation of test specimens by machining.*

ISO 3167:1993, *Plastics — Multipurpose test specimens.*

## 3 Definitions

See ISO 75-1:1993, clause 3.

## 4 Principle

A standard test specimen made of thermosetting laminate or long-fibre-reinforced plastic is subjected to a bending stress equal to 1/10 of a specified or measured flexural strength. The temperature is raised at a uniform rate, and the temperature at which a specified deflection occurs is measured.

NOTE 1   It is recommended that, to facilitate comparison between materials, whenever this property is stated in product literature the test stress should also be given.

## 5 Apparatus

### 5.1 Means of applying a bending stress

See ISO 75-1:1993, subclause 5.1.

---

1) To be published. (Revision of ISO 2818:1980)

If apparatus in which the distance between the test-specimen supports can be varied is used, this distance shall be variable from 60 mm to 210 mm.

## 5.2 Heating equipment

See ISO 75-1:1993, subclause 5.2.

## 5.3 Weights

See ISO 75-1:1993, subclause 5.3.

## 5.4 Temperature-measuring instrument

See ISO 75-1:1993, subclause 5.4.

## 5.5 Deflection-measuring instrument

See ISO 75-1:1993, subclause 5.5.

## 6 Test specimens

See ISO 75-1:1993, clause 6.

**6.1** The test specimen shall have the following dimensions:

length, $l$: at least 10 mm longer than the distance chosen for the span between the test-specimen supports

width, $b$: 9,8 mm to 12,8 mm

thickness, $h$: 2,0 mm to 7,0 mm

The test specimen shall be produced in accordance with ISO 295 (and ISO 2818, if applicable), or as agreed upon by the interested parties. In the case of compression-moulded test specimens, the width shall be perpendicular to the direction of the moulding force. For materials in sheet form, the thickness of the test specimens (this dimension is usually the thickness of the sheet) shall be in the range 2 mm to 7 mm. For samples over 7 mm thick, reduce the thickness to 7 mm by machining one face. If the faces of the test specimen are dissimilar, report the face machined in the test report.

In view of the requirement for the distance between the test-specimen supports to be 30 times the test-specimen thickness (see 8.3), the distance between the supports may be anywhere between 60 mm and 210 mm. Some test machines have a fixed span of 100 mm, however, and can therefore only be used with test specimens up to 3 mm thick. Such a machine may be used but, if the test-specimen thickness is greater than 3 mm, it will have to be reduced by machining. As before, machine only one face and, if the faces are dissimilar, report which face was machined in the test report.

NOTE 2 Most reinforced thermoset laminates are anisotropic, and machining may significantly alter their properties.

Ensure that all cut surfaces are as smooth as possible, and that any unavoidable machining marks are in the lengthwise direction.

**6.2** The test results obtained on moulded test specimens depend on the moulding conditions used in their preparation. Moulding conditions shall be in accordance with the standard for the material, or shall be agreed upon by the interested parties.

**6.3** Discrepancies in test results due to variations in moulding conditions may be minimized by annealing the test specimens before testing them. Since different materials require different annealing conditions, annealing procedures shall be employed only if required by the materials standard or if agreed upon by the interested parties.

## 7 Conditioning

See ISO 75-1:1993, clause 7.

## 8 Procedure

## 8.1 Calculation of force to be applied

See ISO 75-1:1993, subclause 8.1.

The force applied shall be such as to generate a bending stress $\sigma$ equal to 1/10 of the requirement for the flexural strength quoted in the relevant standard for the material. If there is no such requirement, the bending stress $\sigma$ shall be 1/10 of the flexural strength determined in accordance with ISO 178.

The dimensions of the test specimen are given in 6.1, The span between the test-specimen supports is given in 8.3.

## 8.2 Initial temperature of the heating equipment

See ISO 75-1:1993, subclause 8.2.

## 8.3 Measurement

See ISO 75-1:1993, subclause 8.3.

Place the test specimen on the supports in the flatwise position.

Adjust the span $L$ between the supports to $(30 \pm 2)$ times the thickness $h$ of the test specimen.

Apply the calculated force (see 8.1) to give the required nominal surface stress.

Note the temperature at which the bar reaches the standard deflection calculated from the test-specimen height (the thickness $h$ as the specimen is in the flatwise position) using the formula

$$\frac{L^2 \times 10^{-3}}{6h}$$

where

$L$     is the span, in millimetres, between the test-specimen supports;

$h$     is the thickness, in millimetres, of the test specimen.

This temperature is the temperature of deflection under load.

NOTE 3    The standard deflection corresponds to a flexural strain of 0,1 % at the surface of the test specimen.

## 9 Expression of results

See ISO 75-1:1993, clause 9.

If the individual results differ by more than 5 °C, repeat tests shall be carried out.

## 10 Precision

See ISO 75-1:1993, clause 10.

## 11 Test report

See ISO 75-1:1993, clause 11.

In addition, identify

l)    the face machined, if it was necessary the reduce the test-specimen thickness by machining.

# INTERNATIONAL STANDARD

**ISO**
**178**

Third edition
1993-05-15

# Plastics — Determination of flexural properties

*Plastiques — Détermination des propriétés en flexion*

Reference number
ISO 178:1993(E)

# Foreword

ISO (the International Organization for Standardization) is a worldwide federation of national standards bodies (ISO member bodies). The work of preparing International Standards is normally carried out through ISO technical committees. Each member body interested in a subject for which a technical committee has been established has the right to be represented on that committee. International organizations, governmental and non-governmental, in liaison with ISO, also take part in the work. ISO collaborates closely with the International Electrotechnical Commission (IEC) on all matters of electrotechnical standardization.

Draft International Standards adopted by the technical committees are circulated to the member bodies for voting. Publication as an International Standard requires approval by at least 75 % of the member bodies casting a vote.

International Standard ISO 178 was prepared by Technical Committee ISO/TC 61, *Plastics*, Sub-Committee SC 2, *Mechanical properties*.

This third edition cancels and replaces the second edition (ISO 178:1975), which has been improved in the following ways:

— normative references have been added especially for specimen preparation and the use of multipurpose test specimens complying with ISO 3167;

— a definition of modulus is given;

— one strain rate only is recommended;

— designation of quantities has been harmonized with those of other International Standards for testing plastics, in accordance with ISO 31.

International Organization for Standardization
Case Postale 56 • CH-1211 Genève 20 • Switzerland
Printed in Switzerland

# Plastics — Determination of flexural properties

## 1 Scope

**1.1** This International Standard specifies a method for determining the flexural properties of plastics under defined conditions. A standard test specimen is defined, but parameters are included for alternative specimen sizes for use where appropriate. A range of testing speeds is included.

**1.2** The method is used to investigate the flexural behaviour of the test specimens and for determining the flexural strength, flexural modulus and other aspects of the flexural stress/strain relationship under the conditions defined. It applies to a freely supported beam, loaded at midspan (three-point loading test).

**1.3** The method is suitable for use with the following range of materials:

— thermoplastics moulding and extrusion materials, including filled and reinforced compounds in addition to unfilled types; rigid thermoplastics sheets;

— thermosetting moulding materials, including filled and reinforced compounds; thermosetting sheets, including laminates;

— fibre-reinforced thermoset and thermoplastics composites, incorporating unidirectional or non-unidirectional reinforcements such as mat, woven fabrics, woven rovings, chopped strands, combination and hybrid reinforcements, rovings and milled fibres; sheets made from pre-impregnated materials (prepregs);

— thermotropic liquid-crystal polymers.

The method is not normally suitable for use with rigid cellular materials and sandwich structures containing cellular material.

NOTE 1   For certain types of textile-fibre-reinforced plastics, a four-point bending test is preferred. This is currently under consideration in ISO.

**1.4** The method is performed using specimens which may be either moulded to the chosen dimensions, machined from the central portion of a standard multi-purpose test specimen (see ISO 3167) or machined from finished and semi-finished products such as mouldings, laminates and extruded or cast sheet.

**1.5** The method specifies preferred dimensions for the test specimen. Tests which are carried out on specimens of different dimensions, or on specimens which are prepared under different conditions, may produce results which are not comparable. Other factors, such as the speed of testing and the conditioning of the specimens, can also influence the results. Consequently, when comparative data are required, these factors must be carefully controlled and recorded.

**1.6** Flexural properties can only be used for engineering design purposes for materials with linear stress/strain behaviour. For non-linear material behaviour the flexural properties are only nominal. The bending test should preferentially be used with brittle materials, for which tensile tests are difficult.

## 2 Normative references

The following standards contain provisions which, through reference in this text, constitute provisions of this International Standard. At the time of publication, the editions indicated were valid. All standards are subject to revision, and parties to agreements based on this International Standard are encouraged to investigate the possibility of applying the most recent editions of the standards indicated below. Members of IEC and ISO maintain registers of currently valid International Standards.

ISO 291:1977, *Plastics — Standard atmospheres for conditioning and testing.*

ISO 293:1986, *Plastics — Compression moulding test specimens of thermoplastic materials.*

ISO 294:—[1], *Plastics — Injection moulding of test specimens of thermoplastic materials.*

ISO 295:1991, *Plastics — Compression moulding of test specimens of thermosetting materials.*

ISO 1209-1:1990, *Cellular plastics, rigid — Flexural tests — Part 1: Bending test.*

ISO 1209-2:1990, *Cellular plastics, rigid — Flexural tests — Part 2: Determination of flexural properties.*

ISO 1268:1974, *Plastics — Preparation of glass fibre reinforced, resin bonded, low-pressure laminated plates or panels for test purposes.*

ISO 2557-1:1989, *Plastics — Amorphous thermoplastics — Preparation of test specimens with a specified maximum reversion — Part 1: Bars.*

ISO 2557-2:1986, *Plastics — Amorphous thermoplastics — Preparation of test specimens with a specified reversion — Part 2: Plates.*

ISO 2602:1980, *Statistical interpretation of test results — Estimation of the mean — Confidence interval.*

ISO 2818:—[2], *Plastics — Preparation of test specimens by machining.*

ISO 3167:—[3], *Plastics — Multipurpose test specimens.*

ISO 5893:1985, *Rubber and plastics test equipment — Tensile, flexural and compression types (constant rate of traverse) — Description.*

## 3  Definitions

For the purposes of this International Standard, the following definitions apply.

**3.1  speed of testing,** $v$: Rate of relative movement between the supports and the striking edge, expressed in millimetres per minute (mm/min).

**3.2  flexural stress,** $\sigma_f$: Nominal stress of the outer surface of the test specimen at midspan.

It is calculated according to the relationship given in 9.1, equation (3), and is expressed in megapascals (MPa).

**3.3  flexural stress at break,** $\sigma_{fB}$: Flexural stress at break of the test specimen (see figure 1, curves a and b).

It is expressed in megapascals (MPa).

**3.4  flexural strength,** $\sigma_{fM}$: Maximum flexural stress sustained by the test specimen during a bending test (see figure 1, curves a and b).

It is expressed in megapascals (MPa).

**3.5  flexural stress at conventional deflection,** $\sigma_{fc}$: Flexural stress at the conventional deflection $s_C$ according to 3.7 (see figure 1, curve c).

It is expressed in megapascals (MPa).

**3.6  deflection,** $s$: Distance over which the top or bottom surface of the test specimen at midspan has deviated during flexure from its original position.

It is expressed in millimetres (mm).

**3.7  conventional deflection,** $s_C$: Deflection equal to 1,5 times the thickness $h$ of the test specimen.

It is expressed in millimetres (mm).

Using the span $L = 16h$, the conventional deflection corresponds to a flexural strain of 3,5 % (see 3.8).

**3.8  flexural strain,** $\varepsilon_f$: Nominal fractional change in length of an element of the outer surface of the test specimen at midspan.

It is expressed as a dimensionless ratio or a percentage (%).

It is calculated according to the relationship given in 9.2, equation (4).

**3.9  flexural strain at break,** $\varepsilon_{fB}$: Flexural strain at break of the test specimen (see figure 1, curves a and b).

It is expressed as a dimensionless ratio or a percentage (%).

**3.10  flexural strain at flexural strength,** $\varepsilon_{fM}$: Flexural strain at maximum flexural stress (see figure 1, curves a and b).

It is expressed as a dimensionless ratio or a percentage (%).

**3.11  modulus of elasticity in flexure; flexural modulus,** $E_f$: Ratio of the stress difference $\sigma_{f2} - \sigma_{f1}$ to the corresponding strain difference values $(\varepsilon_{f2} = 0,002\ 5) - (\varepsilon_{f1} = 0,000\ 5)$ [see 9.2, equation (5)].

It is expressed in megapascals (MPa).

---

1) To be published. (Revision of ISO 294:1975)

2) To be published. (Revision of ISO 2818:1980)

3) To be published. (Revision of ISO 3167:1983)

NOTES

2   The flexural modulus is only an approximate value of Young's modulus of elasticity.

3   With computer-aided equipment, the determination of the modulus $E_f$ using two distinct stress/strain points can be replaced by a linear regression procedure applied on the part of the curve between these two points.

## 4   Principle

The test specimen, supported as a beam, is deflected at constant rate at the midspan until the specimen fractures or until the deformation reaches some pre-

determined value. During this procedure the force applied to the specimen is measured.

## 5   Apparatus

### 5.1   Testing machine

#### 5.1.1   General

The machine shall comply with ISO 5893 and the requirements given in 5.1.2 to 5.1.4, as follows.

#### 5.1.2   Speed of testing

The testing machine shall be capable of maintaining the speed of testing (see 3.1), as specified in table 1.

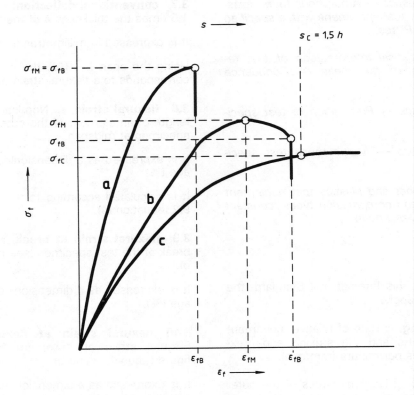

Curve a      Specimen that breaks before yielding

Curve b      Specimen that shows a maximum and then breaks before the conventional deflection $s_C$

Curve c      Specimen that neither has a yield point nor breaks before the conventional deflection $s_C$

**Figure 1 — Typical curves of flexural stress $\sigma_f$ versus flexural strain $\varepsilon_f$ and deflection $s$**

**Table 1 — Recommended values for the speed of testing**

| Speed mm/min | Tolerance % |
|---|---|
| 1 [1] | ± 20 [2] |
| 2 | ± 20 [2] |
| 5 | ± 20 |
| 10 | ± 20 |
| 20 | ± 10 |
| 50 | ± 10 |
| 100 | ± 10 |
| 200 | ± 10 |
| 500 | ± 10 |

1) The lowest speed is used for specimens with thicknesses between 1 mm and 3,5 mm, see 8.3.

2) The tolerances on the speeds 1 mm/min and 2 mm/min are lower than indicated in ISO 5893.

### 5.1.3 Supports and striking edge

Two supports and a central striking edge are arranged according to figure 2. The parallel alignment of the supports and the striking edge shall lie within ± 0,02 mm.

The radius $R_1$ of the striking edge and the radius $R_2$ of the supports shall be as follows:

$R_1 = 5,0$ mm ± 0,1 mm

$R_2 = 2,0$ mm ± 0,2 mm for thicknesses of the test specimen $\leqslant 3$ mm, and

$R_2 = 5,0$ mm ± 0,2 mm for thicknesses of the test specimen > 3 mm.

The span $L$ shall be adjustable.

### 5.1.4 Indicators for load and deflection

The error for the indicated force shall not exceed 1 %, and the error for the indicated deflection shall not exceed 1 % of full scale (see ISO 5893).

## 5.2 Micrometers and gauges

**5.2.1 Micrometer**, or equivalent, reading to $\leqslant 0,01$ mm (for measuring the thickness $h$ and width $b$ of the test specimen, see figure 3).

**5.2.2 Vernier caliper**, or equivalent, accurate to within ± 0,1 % of the span $L$, for determining the span (see 8.2 and figure 2).

## 6 Test specimens

## 6.1 Shape and dimensions

### 6.1.1 General

The dimensions of the test specimens shall comply with the relevant material standard and, as applicable, with 6.1.2 or 6.1.3. Otherwise, the type of specimen shall be agreed between the interested parties.

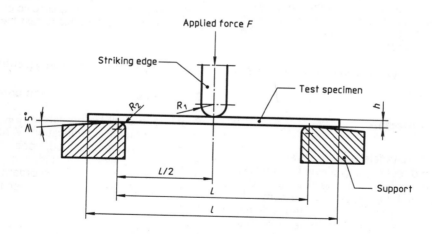

**Figure 2 — Position of test specimen at the start of the test**

### 6.1.2 Preferred specimen type

The preferred dimensions, in millimetres, shall be

length:       $l = 80 \pm 2$
width:        $b = 10{,}0 \pm 0{,}2$
thickness:    $h = 4{,}0 \pm 0{,}2$

In any one test specimen, the thickness within the central third of the length shall nowhere deviate by more than 2 % from its mean value. The corresponding maximum deviation for width is 3 %. The cross-section shall be rectangular, with no rounded edges.

NOTE 4   The preferred specimen may be machined from the central part of a multi-purpose test specimen complying with ISO 3167.

### 6.1.3 Other test specimens

When it is not possible or desirable to use the preferred test specimen, the following limits shall apply.

The length and thickness of the test specimen shall be in the same ratio as in the preferred test specimen, i.e.:

$$\frac{l}{h} = 20 \pm 1 \qquad \ldots (1)$$

unless affected by the provisions of 8.2 a), 8.2 b) or 8.2 c).

NOTE 5   Certain specifications require that test specimens from sheets of thickness greater than a specified upper limit shall be reduced to a standard thickness by machining one face only. In such cases, it is conventional practice to place the test specimen such that the original surface of the specimen is in contact with the two supports and the force is applied by the central striking edge to the machined surface of the specimen.

The values of applicable width given in table 2 shall be used.

### 6.2 Anisotropic materials

**6.2.1**   In the case of materials having physical properties that depend on direction, e.g. elasticity, the test specimens shall be chosen so that the flexural stress in the test procedure will be applied in the same direction as that to which products (moulded articles, sheets, tubes, etc.) will be subjected in service. The relationship of the test specimen to the application will determine the possibility of using standard test specimens (see 6.1 and 8.2).

#### Table 2 — Values for the width, $b$, in relation to the thickness $h$

Dimensions in millimetres

| Nominal thickness, $h$ | Width $b \pm 0{,}5$ [1] | |
|---|---|---|
| | Moulding and extrusion compounds, thermoplastic and thermosetting sheets | Textile and long-fibre-reinforced plastics materials |
| $1 < h \leqslant 3$ | 25,0 | } 15,0 |
| $3 < h \leqslant 5$ | 10,0 | |
| $5 < h \leqslant 10$ | 15,0 | |
| $10 < h \leqslant 20$ | 20,0 | 30,0 |
| $20 < h \leqslant 35$ | 35,0 | 50,0 |
| $35 < h \leqslant 50$ | 50,0 | 80,0 |

1)   For materials with very coarse fillers, the minimum width shall be 20 mm to 50 mm.

NOTE 6   The position or orientation and the dimensions of the test specimens sometimes have a very significant influence on the test results. This is particularly true for laminates.

**6.2.2**   When the material shows a significant difference in flexural properties in two principal directions, it shall be tested in these two directions. The orientation of the test specimens relative to the principal directions shall be recorded (see figure 3).

NOTE 7   If, because of the application, this material is subjected to stress at some specific orientation to the principal direction, it is desirable to test the material in that orientation.

### 6.3 Preparation of test specimens

#### 6.3.1 Moulding or extrusion compounds

Specimens shall be prepared in accordance with the relevant material specification. When none exists, or unless otherwise specified, specimens shall be either directly compression moulded or injection moulded from the material in accordance with ISO 293, ISO 294, ISO 295, ISO 2557-1 or ISO 2557-2, as appropriate.

#### 6.3.2 Sheets

Specimens shall be machined from sheets in accordance with ISO 2818.

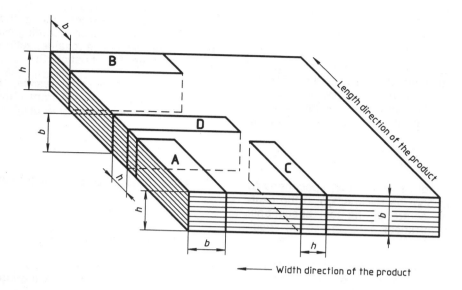

| Position of specimen | Direction of product | Direction of force |
|:---:|:---:|:---:|
| A<br>B | length<br>width | } normal |
| C<br>D | length<br>width | } parallel |

**Figure 3 — Position of test specimen in relation to direction of product and direction of force**

### 6.3.3 Long-fibre-reinforced plastics materials

A panel of the material shall be prepared in accordance with ISO 1268 or another specified or agreed-upon preparation procedure. Specimens shall be machined in accordance with ISO 2818.

### 6.4 Checking

The specimens shall be free of twist and shall have mutually perpendicular parallel surfaces. The surfaces and edges shall be free from scratches, pits, sink marks and flash.

The specimens shall be checked for conformity with these requirements by visual observation against straightedges, squares and flat plates, and by measuring with micrometer calipers.

Specimens showing measurable or observable departure from one or more of these requirements shall be rejected or machined to proper size and shape before testing.

### 6.5 Number of test specimens

**6.5.1** At least five test specimens shall be tested in each direction of test, see figure 3. The number of measurements may be more than five if greater precision of the mean value is required. It is possible to evaluate this by means of the confidence interval (95 % probability, see ISO 2602).

**6.5.2** The results from test specimens that rupture outside the central third of their span length shall be discarded and new test specimens tested in their place.

### 7 Conditioning

The test specimens shall be conditioned as specified in the standard for the material tested. In the absence of this information, select the most appropriate conditions from ISO 291, unless otherwise agreed upon by the interested parties, e.g. for testing at high or low temperatures.

# 8 Procedure

**8.1** Conduct the test in the atmosphere specified in the standard for the material tested. In the absence of this information, select the most appropriate conditions from ISO 291, unless otherwise agreed upon by the interested parties, e.g. for testing at high or low temperatures.

**8.2** Measure the width $b$ of the test specimens to the nearest 0,1 mm and the thickness $h$ to the nearest 0,01 mm in the centre of the test specimens. Calculate the mean thickness $\bar{h}$ for the set of specimens.

Discard any specimen(s) with a thickness exceeding the tolerance of $\pm$ 0,5 % of the mean value and replace it by another specimen chosen at random.

Adjust the span $L$ to comply with the following equation:

$$L = (16 \pm 1)\bar{h} \qquad \qquad \ldots (2)$$

and measure the resulting span to the nearest 0,5 %.

Equation (2) shall be used except in the following cases.

a) For very thick and unidirectional fibre-reinforced test specimens, if necessary to avoid delamination in shear, use a distance between supports calculated on a higher ratio of $L/\bar{h}$.

b) For very thin test specimens, if necessary to enable measurements to be made within the load capacity of the testing machine, use a span length calculated on a lower ratio of $L/\bar{h}$.

c) For soft thermoplastics, if necessary to prevent indentation of the supports into the test specimen, use a larger ratio for $L/\bar{h}$.

**8.3** Set the speed of testing according to the standard for the material being tested. In the absence of this information, select a value from table 1 that gives a strain rate as near as possible to 1 % per minute. This gives a testing speed that produces the deflection closest to 0,4 times the specimen thickness in 1 min, e.g. 2 mm/min for the preferred specimen complying with 6.1.2.

**8.4** Place the test specimen symmetrically on the two supports and apply the force at midspan (see figure 2).

**8.5** Record the force and the corresponding deflection of the specimen during the test, using, if practicable, an automatic recording system that yields a complete flexural stress/deflection curve for this operation [see 9.1, equation (3)].

Determine all relevant stresses, deflections and strains defined in clause 4 from a force/deflection or a stress/deflection curve or equivalent data.

# 9 Calculation and expression of results

## 9.1 Flexural stress

Calculate the flexural stress $\sigma_f$, expressed in megapascals, using the following equation:

$$\sigma_f = \frac{3FL}{2bh^2} \qquad \qquad \ldots (3)$$

where

$F$     is the applied force, in newtons;

$L$     is the span, in millimetres;

$b$     is the width, in millimetres, of the specimen;

$h$     is the thickness, in millimetres, of the specimen.

## 9.2 Flexural modulus

For the measurement of the flexural modulus, calculate the deflections $s_1$ and $s_2$, which correspond to the given values of flexural strain $\varepsilon_{f1} = 0,000\ 5$ and $\varepsilon_{f2} = 0,002\ 5$, by the following equation:

$$s_i = \frac{\varepsilon_{fi}L^2}{6h} \quad (i = 1;2) \qquad \qquad \ldots (4)$$

where

$s_i$     is an individual deflection, in millimetres;

$\varepsilon_{fi}$     is the corresponding flexural strain, whose values $\varepsilon_{f1}$ and $\varepsilon_{f2}$ are given above;

$L$     is the span, in millimetres;

$h$     is the thickness, in millimetres, of the specimen.

Calculate the flexural modulus $E_f$, expressed in megapascals, using the following equation:

$$E_f = \frac{\sigma_{f2} - \sigma_{f1}}{\varepsilon_{f2} - \varepsilon_{f1}} \qquad \qquad \ldots (5)$$

where

$\sigma_{f1}$    is the flexural stress, expressed in megapascals, measured at the deflection $s_1$;

$\sigma_{f2}$    is the flexural stress, expressed in megapascals, measured at the deflection $s_2$.

For computer-aided equipment, see note 3 in 3.11.

NOTE 8    All equations referring to flexural properties hold exactly for linear stress/strain behaviour only (see 1.6); thus for most plastics they are accurate at small deflections only.

## 9.3 Statistical parameters

Calculate the arithmetic mean of the test results and, if required, the standard deviation and the 95 % confidence interval of the mean value using the procedure given in ISO 2602.

## 9.4 Significant figures

Calculate the stresses and the modulus to three significant figures. Calculate the deflections to two significant figures.

## 10 Precision

The precision of this test method is not known because interlaboratory data are not available. When interlaboratory data are obtained, a precision statement will be added in the next revision.

## 11 Test report

The test report shall include the following information:

a) a reference to this International Standard;

b) all the information necessary for identification of the material tested, including type, source, manufacturer's code number, form and previous history, where these are known;

c) for sheets, the thickness of the sheet and, if applicable, the direction of the major axes of the specimens in relation to some feature of the sheet;

d) the shape and dimensions of the test specimens;

e) the method of preparing the specimens;

f) the test conditions and conditioning procedures, if applicable;

g) the number of specimens tested;

h) the nominal length of span used;

i) the speed of testing;

j) the accuracy grading of the test machine (see ISO 5893);

k) the surface of the force application;

l) the individual test results, if required;

m) the mean values of the individual results;

n) the standard deviations and the 95 % confidence intervals of these mean values, if required;

o) the date of the test.

33

# INTERNATIONAL STANDARD

**ISO**

**179**

Second edition
1993-05-15

Corrected and reprinted
1993-08-15

# Plastics — Determination of Charpy impact strength

*Plastiques — Détermination de la résistance au choc Charpy*

Reference number
ISO 179:1993(E)

# Foreword

ISO (the International Organization for Standardization) is a worldwide federation of national standards bodies (ISO member bodies). The work of preparing International Standards is normally carried out through ISO technical committees. Each member body interested in a subject for which a technical committee has been established has the right to be represented on that committee. International organizations, governmental and non-governmental, in liaison with ISO, also take part in the work. ISO collaborates closely with the International Electrotechnical Commission (IEC) on all matters of electrotechnical standardization.

Draft International Standards adopted by the technical committees are circulated to the member bodies for voting. Publication as an International Standard requires approval by at least 75 % of the member bodies casting a vote.

International Standard ISO 179 was prepared by Technical Committee ISO/TC 61, *Plastics*, Sub-Committee SC 2, *Mechanical properties*.

This second edition cancels and replaces the first edition (ISO 179:1982), which has been revised in the following ways.

— The recommended specimen types for testing moulding materials are reduced to one only, which can be taken from the multipurpose test specimen complying with ISO 3167 by simple machining.

— The notch types are reduced to three only: V-type, 45°, with different notch base radii.

— The preferred direction of blow is changed from "flatwise" (parallel to the dimension thickness) to "edgewise" (parallel to the dimension width) in order to align the test method with that of the Izod test according to ISO 180.

— Special test methods are included for laminated sheets in order to respect the thicknesses of semifinished products and interlaminar shear failure.

— The designation of sizes are harmonized with those of a great number of other International Standards for testing plastics, in accordance with ISO 31.

International Organization for Standardization
Case Postale 56 • CH-1211 Genève 20 • Switzerland
Printed in Switzerland

— The method designations are harmonized with the modifications described above (direction of blow and types of notch).

Annex A of this International Standard is for information only.

# Plastics — Determination of Charpy impact strength

## 1 Scope

**1.1** This International Standard specifies a method for determining the Charpy impact strength of plastics under defined conditions. A number of different types of specimen and test configurations are defined. Different test parameters are specified according to the type of material, the type of test specimen and the type of notch.

**1.2** The method is used to investigate the behaviour of specified types of specimen under the impact conditions defined and for estimating the brittleness or toughness of specimens within the limitations inherent in the test conditions.

The method has a greater range of applicability than that given in ISO 180 (Izod)[1] and is more suitable for the testing of materials showing interlaminar shear fracture or of materials exhibiting surface effects due to environmental factors.

**1.3** The method is suitable for use with the following range of materials:

— rigid thermoplastics moulding and extrusion materials, including filled and reinforced compounds in addition to unfilled types; rigid thermoplastics sheets;

— rigid thermosetting moulding materials, including filled and reinforced compounds; rigid thermosetting sheets, including laminates;

— fibre-reinforced thermoset and thermoplastics composites incorporating unidirectional or non-unidirectional reinforcements such as mat, woven fabrics, woven rovings, chopped strands, combination and hybrid reinforcements, rovings and milled fibres; sheets made from pre-impregnated materials (prepregs);

— thermotropic liquid-crystal polymers.

The method is not normally suitable for use with rigid cellular materials and sandwich structures containing cellular material. Also, notched specimens are not normally used for long-fibre-reinforced composites or for thermotropic liquid-crystal polymers.

**1.4** The method is adapted to the use of specimens which may be either moulded to the chosen dimensions, machined from the central portion of a standard multipurpose test specimen (see ISO 3167) or machined from finished and semifinished products such as mouldings, laminates and extruded or cast sheet.

**1.5** The method specifies preferred dimensions for the test specimen. Tests which are carried out on specimens of different dimensions and notches, or on specimens which are prepared under different conditions, may produce results which are not comparable. Other factors, such as the energy capacity of the pendulum, its impact velocity and the conditioning of the specimens can also influence the results. Consequently, when comparative data are required, these factors must be carefully controlled and recorded.

**1.6** The method should not be used as a source of data for design calculations of components. Information on the typical behaviour of a material can be obtained, however, by testing at different temperatures, by varying the notch radius and/or the thickness and by testing specimens prepared under different conditions.

## 2 Normative references

The following standards contain provisions which, through reference in this text, constitute provisions of this International Standard. At the time of publication, the editions indicated were valid. All standards are subject to revision, and parties to agreements based on this International Standard are encouraged to investigate the possibility of applying the most recent editions of the standards indicated below.

---

1) ISO 180:1993, *Plastics — Determination of Izod impact strength.*

Members of IEC and ISO maintain registers of currently valid International Standards.

ISO 291:1977, *Plastics — Standard atmospheres for conditioning and testing.*

ISO 293:1986, *Plastics — Compression moulding test specimens of thermoplastic materials.*

ISO 294:—[2], *Plastics — Injection moulding of test specimens of thermoplastic materials.*

ISO 295:1991, *Plastics — Compression moulding of test specimens of thermosetting materials.*

ISO 1268:1974, *Plastics — Preparation of glass fibre reinforced, resin bonded, low-pressure laminated plates or panels for test purposes.*

ISO 2557-1:1989, *Plastics — Amorphous thermoplastics — Preparation of test specimens with a specified maximum reversion — Part 1: Bars.*

ISO 2557-2:1986, *Plastics — Amorphous thermoplastics — Preparation of test specimens with a specified reversion — Part 2: Plates.*

ISO 2602:1980, *Statistical interpretation of test results — Estimation of the mean — Confidence interval.*

ISO 2818:—[3], *Plastics — Preparation of test specimens by machining.*

ISO 3167:1993, *Plastics — Multipurpose test specimens.*

## 3 Definitions

For the purposes of this International Standard, the following definitions apply.

**3.1 Charpy impact strength of unnotched specimens,** $a_{cU}$: Impact energy absorbed in breaking an unnotched specimen, referred to the original cross-sectional area of the specimen.

It is expressed in kilojoules per square metre (kJ/m²).

**3.2 Charpy impact strength of notched specimens,** $a_{cN}$: Impact energy absorbed in breaking a notched specimen, referred to the original cross-sectional area of the specimen at the notch, where N = A, B or C depending on the notch type (see 6.3.1.1.2).

It is expressed in kilojoules per square metre (kJ/m²).

**3.3 edgewise impact** (e): Direction of blow parallel to the dimension $b$, with impact on the narrow longitudinal surface $h \times l$ of the specimen (see figure 1, left, and figures 2 and 5).

**3.4 flatwise impact** (f): Direction of blow parallel to the dimension $h$, with impact on the broad longitudinal surface $b \times l$ of the specimen (see figure 1, right, and figures 3 and 5).

**3.5 normal impact** (n): Direction of blow normal to the plane of reinforcement (see figure 5).

It is used for laminar-type reinforced plastics.

**3.6 parallel impact** (p): Direction of blow parallel to the plane of reinforcement (see figure 5).

## 4 Principle

The test specimen, supported as a horizontal beam, is broken by a single swing of a pendulum, with the line of impact midway between the supports.

In the case of edgewise impact with notched specimens, the line of impact is directly opposite the single notch (see figure 1, left, and figure 2).

## 5 Apparatus

### 5.1 Testing machine

**5.1.1** The testing machine shall be of the pendulum type and shall be of rigid construction. It shall be capable of measuring the impact energy, $W$, absorbed in breaking a test specimen. The value of this energy is defined as the difference between the initial energy, $E$, of the pendulum and the energy remaining in the pendulum after breaking the test specimen. The energy shall be accurately corrected for losses due to friction and air resistance (see table 1 and 7.4).

**5.1.2** The machine shall have the characteristics shown in table 1.

In order to apply the test to the full range of materials specified in 1.3, it is necessary to use more than one machine or to use a set of interchangeable pendulums (see 7.3). It is not advisable to compare results obtained with different pendulums. The frictional losses shall be periodically checked.

2) To be published. (Revision of ISO 294:1975)

3) To be published. (Revision of ISO 2818:1980)

## Table 1 — Characteristics of pendulum impact testing machines

| Energy E (nominal) J | Velocity of impact $v_0$ m/s | Maximum permissible frictional loss without specimen J | Permissible error[1] after correction with specimen J |
|---|---|---|---|
| 0,5<br>1,0<br>2,0<br>4,0<br>5,0 | 2,9 (± 10 %) | 0,02 | 0,01<br>0,01<br>0,01<br>0,02<br>0,02 |
| 7,5<br>15,0<br>25,0<br>50,0 | 3,8 (± 10 %) | 0,04<br>0,05<br>0,10<br>0,20 | 0,05<br>0,05<br>0,10<br>0,10 |

1) The permissible error shall not be exceeded within the 10 % to 80 % range of the pendulum capacity.

**5.1.3** The machine shall be securely fixed to a foundation having a mass at least 40 times that of the heaviest pendulum in use. The foundation shall be capable of being adjusted so that the striker and supports are as specified in 5.1.4 and 5.1.6.

**5.1.4** The striking edge of the pendulum shall be hardened steel tapered to an included angle of $30° ± 1°$ and shall be rounded to a radius $R_1 = 2$ mm $± 0,5$ mm. It shall pass midway, to within $± 0,2$ mm, between the test specimen supports, and shall be aligned so that it contacts the full width or thickness of rectangular test specimens. The line of contact shall be perpendicular, within $± 2°$, to the longitudinal axis of the test specimen.

**5.1.5** The distance between the axis of rotation and the point of impact at the centre of the specimen shall be within $± 1$ % of the pendulum length $L_p$.

NOTE 1    The pendulum length $L_p$, in metres, may be determined experimentally from the period of small amplitude oscillations of the pendulum by means of the following equation:

$$L_p = \frac{g_n}{4\pi^2} \times T^2 \qquad \ldots (1)$$

where

$g_n$    is the standard acceleration of free fall, in metres per second squared (9,81 m/s²);

$T$    is the period, in seconds, of a single complete swing (to and fro) determined from at least 50 consecutive and uninterrupted swings (known to an accuracy of 1 part in 2 000). The angle of swing shall be less than 5° to each side of the centre.

**5.1.6** The test specimen supports shall be two rigidly mounted smooth blocks, arranged so that the longitudinal axis of a perfectly rectangular test specimen is horizontal to within 1 part in 200, and the striking face of such a test specimen is parallel to the striking edge of the pendulum to within 1 part in 200 at the moment of impact. The specimen supports shall not inhibit the movement of the specimen.

The shape of the supports shall be as shown in figure 1. The span, $L$, is the distance between the contact lines of the specimen on the supports and shall be as specified in table 2. Means shall be provided to centre test specimens, in relation to the striker, to within $± 0,5$ mm. Separate support blocks may be required for each type of test specimen.

## 5.2    Micrometers and gauges

Micrometers and gauges suitable for measuring the essential dimensions of test specimens to an accuracy of 0,02 mm are required. For measuring the dimension $b_N$ of notched specimens, the micrometer shall be fitted with an anvil of width 2 mm to 3 mm and of suitable profile to fit the shape of the notch.

## 6    Test specimens

### 6.1    Preparation

#### 6.1.1    Moulding or extrusion compounds

Specimens shall be prepared in accordance with the relevant material specification. When none exists, or unless otherwise specified, specimens shall be either directly compression moulded or injection moulded from the material in accordance with ISO 293, ISO 294, ISO 295, ISO 2557-1 or ISO 2557-2 as appropriate, or machined in accordance with ISO 2818 from sheet that has been compression or injection moulded from the compound.

NOTE 2    Type 1 specimens may be cut from multi-purpose test specimens complying with ISO 3167 type A.

**Table 2 — Specimen types, dimensions and span** (see figure 1)

Dimensions in millimetres

| Specimen type[1] | Length[2] $l$ | Width[2] $b$ | Thickness[2] $h$ | Span $L$ |
|---|---|---|---|---|
| 1 | $80 \pm 2$ | $10,0 \pm 0,2$ | $4,0$[3] $\pm 0,2$ | $62 {}^{+0,5}_{0}$ |
| 2[4] 3[4] | $25\,h$ $(11 \text{ or } 13)\,h$ | 10 or 15[5] | $3$[3] | $20\,h$ $(6 \text{ or } 8)\,h$ |

1) Attention is drawn to the changes in the specimen type numbers from those used in ISO 179:1982.

2) The specimen dimensions (thickness $h$, width $b$ and length $l$) are defined according to: $h \leqslant b < l$.

3) Preferred thickness. If the specimen is cut from a sheet or a piece, $h$ shall be equal to the thickness of the sheet or piece, up to 10,2 mm (see 6.3.1.2).

4) Specimen types 2 and 3 shall be used only for materials described in 6.3.2.

5) 10 mm for materials reinforced with a fine structure, 15 mm with a large stitch structure (see 6.3.2.2).

### 6.1.2  Sheets

Specimens shall be machined from sheets in accordance with ISO 2818.

### 6.1.3  Long-fibre-reinforced polymers

A panel shall be prepared in accordance with ISO 1268 or another specified or agreed upon preparation procedure. Specimens shall be machined in accordance with ISO 2818.

### 6.1.4  Checking

The specimens shall be free of twist and shall have mutually perpendicular parallel surfaces. The surfaces and edges shall be free from scratches, pits, sink marks and flash.

The specimens shall be checked for conformity with these requirements by visual observation against straightedges, squares and flat plates, and by measuring with micrometer calipers.

Specimens showing measurable or observable departure from one or more of these requirements shall be rejected or machined to proper size and shape before testing.

### 6.1.5  Notching

**6.1.5.1**  Machined notches shall be prepared in accordance with ISO 2818. The profile of the cutting tooth shall be such as to produce in the specimen a notch of the contour and depth shown in figure 4, at right angles to its principal axes.

**6.1.5.2**  Specimens with moulded-in notches may be used if specified for the material being tested. Specimens with moulded-in notches do not give results comparable to those obtained from specimens with machined notches.

### 6.2  Anisotropy

Certain types of sheet or panel materials may show different impact properties according to the direction in the plane of the sheet or panel. In such cases, it is customary to cut groups of test specimens with their major axes respectively parallel and perpendicular to the direction of some feature of the sheet or panel which is either visible or inferred from knowledge of the method of its manufacture.

### 6.3  Shape and dimensions

### 6.3.1  Materials not exhibiting interlaminar shear fracture

### 6.3.1.1  Moulding and extrusion compounds

**6.3.1.1.1**  Type 1 test specimens with three different types of notch shall be used as specified in tables 2 and 3, and shown in figures 2 and 4. The notch shall be located at the centre of the specimen.

NOTE 3   Type 1 specimens (see table 2) may be taken from the central part of the multi-purpose test specimen type A complying with ISO 3167.

**Table 3 — Method designations, specimen types, notch types and notch dimensions — Materials not exhibiting interlaminar shear fracture**

Dimensions in millimetres

| Method designation[1] [2] | Specimen type[1] | Blow direction | Notch type[1] | | Notch base radius $r_N$ | Remaining width, $b_N$, at notch base |
|---|---|---|---|---|---|---|
| ISO 179/1eU [3] | 1 | edgewise | unnotched | | | |
| | | | single notch | | | |
| ISO 179/1eA [3] ISO 179/1eB ISO 179/1eC | | | | A B C | $0,25 \pm 0,05$ $1,00 \pm 0,05$ $0,10 \pm 0,02$ | $8,0 \pm 0,2$ $8,0 \pm 0,2$ $8,0 \pm 0,2$ |
| ISO 179/1fU [4] | 1 | flatwise | unnotched | | | |

1) Attention is drawn to the changes in the specimen type number, notch type letter designations and method designation number from those used in ISO 179:1982.

2) If specimens are taken from sheet or products, the thickness of the sheet or product shall be added to the designation, and unreinforced specimens shall not be tested with their machined surface under tension.

3) Preferred method.

4) Especially for study of surface effects (see 1.2 and 6.3.1.1.3).

**6.3.1.1.2** The preferred type of notch is type A (see table 3 and figure 4). For most materials, unnotched specimens or specimens with a single notch of type A tested according to 3.3 (edgewise impact) are suitable. If specimens with notch type A do not break during the test, specimens with notch type C shall be used. If information on the notch sensitivity of the material is desired, specimens with notch type A, B and C shall be tested.

NOTE 4   Notch type C replaces the former U notch, which in some cases gives test results that are not comparable.

**6.3.1.1.3** Unnotched or double-notched specimens tested according to 3.4 (flatwise impact) can be used to study surface effects (see 1.2 and annex A).

### 6.3.1.2   Sheet materials

The recommended thickness $h$ is 4 mm. If the specimen is cut from a sheet or a piece taken from a structure, the thickness of the specimen, up to 10,2 mm, shall be the same as the thickness of the sheet or the structure.

Specimens taken from pieces thicker than 10,2 mm shall be machined to 10 mm $\pm$ 0,2 mm from one surface, providing that the sheet is homogeneous in its thickness and contains only one type of reinforcement regularly distributed. If unnotched or double-notched specimens are tested according to 3.4 (flatwise impact), the original surface shall be tested under tension, in order to avoid surface effects.

### 6.3.2   Materials exhibiting interlaminar shear fracture (e.g. long-fibre-reinforced materials)

**6.3.2.1** Unnotched specimens of type 2 or 3 are used. There are no specified specimen sizes. The only important parameter is the ratio of the span, $L$, to the specimen dimension in the direction of blow (see table 2).

Usually specimens are tested in the normal direction (see figure 5).

**6.3.2.2** "Flatwise normal" testing (see figure 5): the width of the specimen shall be 10 mm for materials reinforced with a fine structure (thin fabrics and parallel yarns) and 15 mm for materials reinforced with a large stitch structure (roving fabrics) or an irregularly manufactured structure.

**6.3.2.3** "Edgewise parallel" testing (see figure 5): when testing specimens in the parallel direction, the specimen dimension perpendicular to the blow direction shall be the thickness of the sheet from which the specimen was cut.

**6.3.2.4** The length, $l$, of the specimen shall be chosen according to the span to thickness ratio $L/h$ of 20 (for type 2 specimens) and 6 (for type 3 specimens) as indicated in table 2.

If the apparatus does not allow a ratio $L/h = 6$, a ratio $L/h = 8$ may be used, especially for thin sheets.

**6.3.2.5** With type 2 specimens, tensile-type failure occurs; with type 3 specimens, interlaminar shear failure of the sheet can occur. The different types of failure that can occur are summarized in table 4.

NOTE 5   In some cases (thin-fabric reinforcement) shear failure does not occur. In the case of type 3 specimens, the fracture initiates as a single- or multiple-shear failure and continues as a tensile failure.

## 6.4   Number of test specimens

**6.4.1**   Unless otherwise specified in the standard for the material being tested, a set consisting of a minimum of ten specimens shall be tested. When the coefficient of variation (see ISO 2602) has a value of less than 5 %, a minimum number of five test specimens is sufficient.

**6.4.2**   If laminates are tested in the normal and parallel directions, ten specimens shall be used for each direction.

## 6.5   Conditioning

Unless otherwise specified in the standard for the material under test, the specimens shall be conditioned for at least 16 h at 23 °C and 50 % relative humidity according to ISO 291, unless other conditions are agreed upon by the interested parties.

**Table 4 — Method designations, specimen types, notch types and notch dimensions — Materials exhibiting interlaminar shear fracture**

| Method designation | Specimen type | $L/h$ | Type of failure | | Schematic |
|---|---|---|---|---|---|
| ISO 179/2 | 2 | 20 | tension | t | |
| n or p [1] | | | compression | c | |
| | | | buckling | b | |
| ISO 179/3 | 3 | 6 or 8 | shear | s | |
| n or p [1] | | | multiple shear | ms | |
| | | | shear followed by a tensile failure | st | |
| 1)  n is the normal direction and p is the parallel direction with respect to the sheet plane (see figure 5). | | | | | |

42

# 7 Procedure

**7.1** Conduct the test in the same atmosphere as that used for conditioning, unless otherwise agreed upon by the interested parties, e.g. for testing at high or low temperatures.

**7.2** Measure the thickness, $h$, and the width, $b$, of each test specimen, in the centre, to the nearest 0,02 mm. In the case of notched specimens, carefully measure the remaining width $b_N$ to the nearest 0,02 mm.

NOTE 6    In the case of injection-moulded specimens, it is not necessary to measure the dimensions of each specimen. It is sufficient to measure one specimen from a set to make sure that the dimensions correspond to those in table 2.

With multiple cavity moulds, ensure that the dimensions of the specimens are the same for each cavity.

In the case of specimens type 2 or 3, adjust the span $L$ according to table 2.

**7.3** Check that the pendulum machine has the specified velocity of impact (see table 1) and that it is in the correct range of absorbed energy, $W$, which shall be between 10 % and 80 % of the pendulum energy, $E$. If more than one of the pendulums described in table 1 meet these requirements, the pendulum having the highest energy shall be used.

**7.4** Carry out a blank test (i.e. without a specimen in place) and record the frictional energy loss. Ensure that this energy loss does not exceed the appropriate value given in table 1.

If frictional losses are equal to or less than the values indicated in table 1, they may be used in the calculations of corrected energy absorbed. If frictional losses exceed the values indicated in table 1, care should be taken to evaluate the cause of any excess frictional losses and corrections made as necessary to the equipment.

**7.5** Lift and support the pendulum. Place the specimen on the supports of the machine in such a manner that the striking edge will hit the centre of the specimen. Carefully align notched specimens so that the centre of the notch is located directly in the plane of impact (see figure 1, left).

**7.6** Release the pendulum. Record the impact energy absorbed by the specimen and apply any necessary corrections for frictional losses etc. (see table 1 and 7.4).

**7.7** For moulding and extrusion compounds, four types of failure according to the following letter codes may occur:

C    complete break; a break in which the specimen separates into two or more pieces

H    hinge break; an incomplete break such that both parts of the specimen are held together only by a thin peripheral layer in the form of a hinge having no residual stiffness

P    partial break; an incomplete break that does not meet the definition for a hinge break

NB    non-break; in the case where there is no break, and the specimen is only bent and pushed through the support blocks, possibly combined with stress whitening

The measured values of complete and hinged breaks can be used for a common mean value without remark. If in the case of partial breaks a value is required, it shall be designated with the letter P. In the case of non-break, NB, no values shall be reported.

For materials with interlaminar shear fracture, the types of failure and their codes are shown in table 4.

**7.8** If, within one sample, the test specimens show both P and C (or H) failures, the mean value for each failure type shall be reported.

# 8 Calculation and expression of results

## 8.1 Unnotched specimens

Calculate the Charpy impact strength of unnotched specimens, $a_{cU}$, expressed in kilojoules per square metre, using the formula

$$a_{cU} = \frac{W}{h \cdot b} \times 10^3 \qquad \qquad \dots (2)$$

where

$W$    is the corrected energy, in joules, absorbed by breaking the test specimen;

$h$    is the thickness, in millimetres, of the test specimen;

$b$    is the width, in millimetres, of the test specimen.

## 8.2 Notched specimens

Calculate the Charpy impact strength of notched specimens, $a_{cN}$, expressed in kilojoules per square metre, with notches N = A, B or C, using the formula

$$a_{cN} = \frac{W}{h \cdot b_N} \times 10^3 \qquad \qquad \dots (3)$$

where

$W$    is the corrected energy, in joules, absorbed by breaking the test specimen;

$h$  is the thickness, in millimetres, of the test specimen;

$b_N$  is the remaining width, in millimetres, at the notch base of the test specimen.

## 8.3  Statistical parameters

Calculate the arithmetic mean of test results and, if required, the standard deviation and the 95 % confidence interval of the mean value using the procedure given in ISO 2602. For different types of failure within one sample, the relevant numbers of specimens shall be given and mean values shall be calculated.

## 8.4  Significant figures

Report all calculated mean values to two significant figures.

## 9  Precision

The precision of this method is not known because interlaboratory data are not available. When interlaboratory data are obtained, a precision statement will be added in the next revision.

## 10  Test report

The test report shall include the following information:

a) a reference to this International Standard;

b) the method designation according to table 3, e.g.:

```
Charpy impact test              ISO 179/1  e  A
Specimen type (see table 2)
Direction of blow (see figure 5)
Type of notch (see figure 4)
```

or according to table 4, e.g.:

```
Charpy impact test              ISO 179/2  n
Specimen type (see table 2)
Direction of blow (see figure 5)
```

c) all the information necessary for identification of the material tested, including type, source, manufacturer's code, grade and form, history, where these are known;

d) description of the nature and form of the material, i.e. whether a product, semifinished product, test plaque or specimen, including principal dimensions, shape, method of manufacture, etc. where these are known;

e) the velocity of impact;

f) the nominal pendulum energy;

g) method of test specimen preparation;

h) if the material is in the form of a product, or a semifinished product, the orientation of the test specimen in relation to the product or semifinished product from which it is cut;

i) number of tested specimens;

j) the standard atmosphere for conditioning and for testing, plus any special conditioning treatment if required by the standard for the material or product;

k) the type(s) of failure observed;

l) the individual test results;

m) the impact strength of the material, reported as the arithmetic mean value of the results for each specimen type, and type of failure where appropriate (see 7.7);

n) the standard deviations and the 95 % confidence intervals of these mean values, if required;

o) the date(s) of the test.

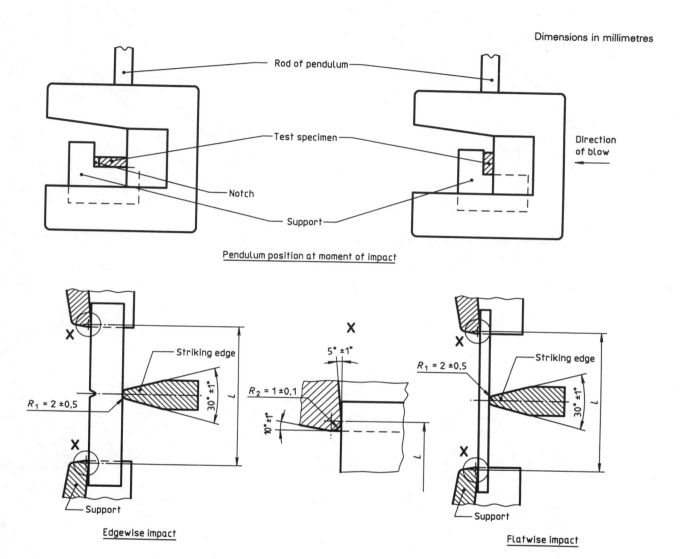

Figure 1 — Striking edge and support blocks for type 1 test specimens

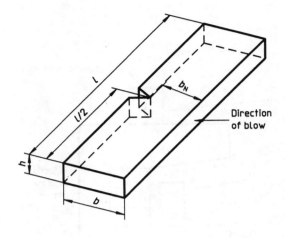

**Figure 2 — Charpy edgewise impact (e), with single-notched specimen**

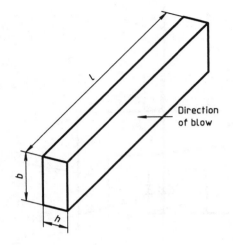

**Figure 3 — Charpy flatwise impact (f)**

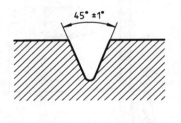

Radius of notch base
$r_N = 0,25$ mm $\pm 0,05$ mm

Type A notch

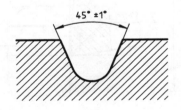

Radius of notch base
$r_N = 1$ mm $\pm 0,05$ mm

Type B notch

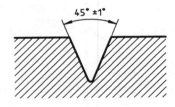

Radius of notch base
$r_N = 0,1$ mm $\pm 0,02$ mm

Type C notch

**Figure 4 — Notch types**

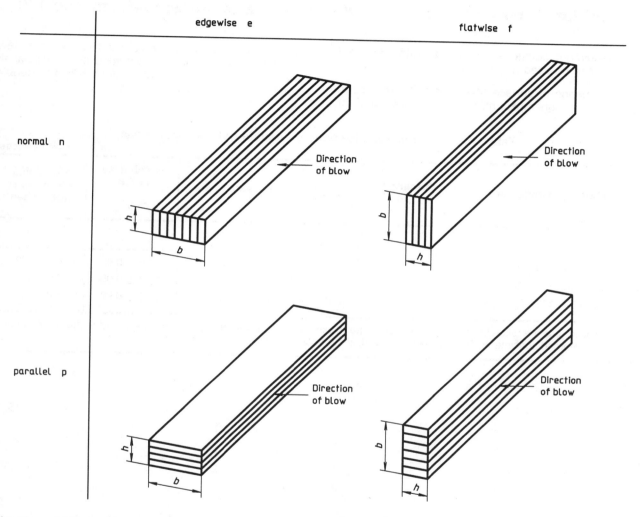

Direction of blow with respect to specimen thickness, $h$, and specimen width, $b$: edgewise (e) and flatwise (f); with respect to the laminate plane: normal (n) and parallel (p).

The Charpy "fn" and "ep" tests are used for laminates, and the Charpy "e" test is used for all other materials; the Charpy "f" test is used for testing surface effects.

**Figure 5 — Scheme of designations describing the direction of blow**

# Annex A

## (informative)

## Additional methods for testing the influence of surface effects (see 1.2)

The following additional methods with double-V notches can be used for materials described in 6.3.1.

If the influence of surface effects is to be measured moderate- or high-impact materials, the method of for flatwise impact may be applied using double-V notches. Two notches are provided perpendicular to the line of impact. The length of each of the double notches is $h$, as shown in figure A.1.

### Table A.1 — Parameters for tests on double-notched specimens

Dimensions in millimetres

| Method designation[1] | Specimen type | Blow direction | Notch type | Notch base radius $r_N$ | Remaining width, $b_N$, at notch base |
|---|---|---|---|---|---|
| | | | double notch | | |
| ISO 179/1fA | 1 | flatwise | A | 0,25 ± 0,05 | 6,0 ± 0,2 |
| ISO 179/1fB | 1 | flatwise | B | 1,00 ± 0,05 | 6,0 ± 0,2 |
| ISO 179/1fC | 1 | flatwise | C | 0,10 ± 0,02 | 6,0 ± 0,2 |

1) If specimens are taken from sheet or products, the thickness of the sheet or product shall be added to the designation, and unreinforced specimens shall not be tested with their machined surface under tension.

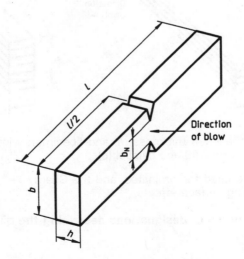

Figure A.1 — Charpy flatwise impact (f) with double-notched specimen

# INTERNATIONAL STANDARD

**ISO
180**

Second edition
1993-05-15

# Plastics — Determination of Izod impact strength

*Plastiques — Détermination de la résistance au choc Izod*

Reference number
ISO 180:1993(E)

# Foreword

ISO (the International Organization for Standardization) is a worldwide federation of national standards bodies (ISO member bodies). The work of preparing International Standards is normally carried out through ISO technical committees. Each member body interested in a subject for which a technical committee has been established has the right to be represented on that committee. International organizations, governmental and non-governmental, in liaison with ISO, also take part in the work. ISO collaborates closely with the International Electrotechnical Commission (IEC) on all matters of electrotechnical standardization.

Draft International Standards adopted by the technical committees are circulated to the member bodies for voting. Publication as an International Standard requires approval by at least 75 % of the member bodies casting a vote.

International Standard ISO 180 was prepared by Technical Committee ISO/TC 61, *Plastics*, Sub-Committee SC 2, *Mechanical properties*.

This second edition cancels and replaces the first edition (ISO 180:1982), which has been revised in the following ways.

— The recommended specimen types for testing moulding materials are reduced to one only, which can be taken from the central part of the multipurpose test specimen complying with ISO 3167 by simple machining.

— Instead of testing in a "reversed-notch" configuration, the use of unnotched specimens is recommended.

— The designations of sizes are harmonized to those of a great number of other International Standards for testing plastics, in accordance with ISO 31.

— The method designations are changed and fitted to the modifications described above.

Annex A forms an integral part of this International Standard.

International Organization for Standardization
Case Postale 56 • CH-1211 Genève 20 • Switzerland

Printed in Switzerland

# Plastics — Determination of Izod impact strength

## 1  Scope

**1.1**  This International Standard specifies a method for determining the Izod impact strength of plastics under defined conditions. A number of different types of specimen and test configurations are defined. Different test parameters are specified according to the type of material, the type of test specimen and the type of notch.

**1.2**  The method is used to investigate the behaviour of specified types of specimen under the impact conditions defined and for estimating the brittleness or toughness of specimens within the limitations inherent in the test conditions.

**1.3**  The method is suitable for use with the following range of materials:

— rigid thermoplastics moulding and extrusion materials, including filled and reinforced compounds in addition to unfilled types; rigid thermoplastics sheet;

— rigid thermosetting moulding materials, including filled and reinforced compounds; rigid thermosetting sheet, including laminates;

— fibre-reinforced thermoset and thermoplastics composites incorporating unidirectional or non-unidirectional reinforcements such as mat, woven fabrics, woven rovings, chopped strands, combination and hybrid reinforcements, rovings and milled fibres; sheet made from pre-impregnated materials (prepregs);

— thermotropic liquid-crystal polymers.

The method is not normally suitable for use with rigid cellular materials and sandwich structures containing cellular material. Also, notched specimens are not normally used for long-fibre-reinforced composites or for thermotropic liquid-crystal polymers.

**1.4**  The method is adapted to the use of specimens which may be either moulded to the chosen dimensions, machined from the central portion of a standard multipurpose test specimen (see ISO 3167) or machined from finished and semifinished products such as mouldings, laminates and extruded or cast sheet.

**1.5**  The method specifies preferred dimensions for the test specimen. Tests which are carried out on specimens of different dimensions and notches, or on specimens which are prepared under different conditions may produce results which are not comparable. Other factors, such as the energy capacity of the pendulum, its impact velocity and the conditioning of the specimens can also influence the results. Consequently, when comparative data are required, these factors must be carefully controlled and recorded.

**1.6**  The method should not be used as a source of data for design calculations of components. Information on the typical behaviour of a material can be obtained, however, by testing at different temperatures, by varying the notch radius and/or the thickness and by testing specimens prepared under different conditions.

## 2  Normative references

The following standards contain provisions which, through reference in this text, constitute provisions of this International Standard. At the time of publication, the editions indicated were valid. All standards are subject to revision, and parties to agreements based on this International Standard are encouraged to investigate the possibility of applying the most recent editions of the standards indicated below. Members of IEC and ISO maintain registers of currently valid International Standards.

ISO 291:1977, *Plastics — Standard atmospheres for conditioning and testing.*

ISO 293:1986, *Plastics — Compression moulding test specimens of thermoplastic materials.*

ISO 294:—[1], *Plastics — Injection moulding of test specimens of thermoplastic materials.*

ISO 295:1991, *Plastics — Compression moulding of test specimens of thermosetting materials.*

ISO 1268:1974, *Plastics — Preparation of glass fibre reinforced, resin bonded, low-pressure laminated plates or panels for test purposes.*

ISO 2557-1:1989, *Plastics — Amorphous thermoplastics — Preparation of test specimens with a specified maximum reversion — Part 1: Bars.*

ISO 2557-2:1986, *Plastics — Amorphous thermoplastics — Preparation of test specimens with a specified reversion — Part 2: Plates.*

ISO 2602:1980, *Statistical interpretation of test results — Estimation of the mean — Confidence interval.*

ISO 2818:—[2], *Plastics — Preparation of test specimens by machining.*

ISO 3167:—[3], *Plastics — Multipurpose test specimens.*

## 3  Definitions

For the purposes of this International Standard, the following definitions apply.

**3.1  Izod impact strength of unnotched specimens,** $a_{iU}$: Impact energy absorbed in breaking an unnotched specimen, referred to the original cross-sectional area of the specimen.

It is expressed in kilojoules per square metre $(kJ/m^2)$.

**3.2  Izod impact strength of notched specimens,** $a_{iN}$: Impact energy absorbed in breaking a notched specimen, referred to the original cross-sectional area of the specimen at the notch, the pendulum striking the face containing the notch.

It is expressed in kilojoules per square metre $(kJ/m^2)$.

**3.3  Izod impact strength of reversed-notch specimens,** $a_{iR}$: Impact energy absorbed in breaking a reversed-notch specimen, referred to the original cross-sectional area of the specimen at the notch, the pendulum striking the face opposite the notch.

It is expressed in kilojoules per square metre $(kJ/m^2)$.

**3.4  parallel impact** (p) (for laminar reinforced plastics): Direction of blow parallel to the laminate plane of sheet materials. The blow direction in the Izod test is "edgewise" (e) (see figure 1, "edgewise parallel").

**3.5  normal impact** (n) (for laminar reinforced plastics): Direction of blow normal to the laminate plane of sheet materials (see figure 1, "edgewise normal").

NOTE 1    This kind of impact is not used with the Izod test, but is indicated only for clarifying the designation system.

## 4  Principle

The test specimen, supported as a vertical cantilever beam, is broken by a single swing of a pendulum, with the line of impact at a fixed distance from the specimen clamp and, in the case of notched specimens, from the centreline of the notch (see figure 2).

---

1) To be published. (Revision of ISO 294:1975)

2) To be published. (Revision of ISO 2818:1980)

3) To be published. (Revision of ISO 3167:1983)

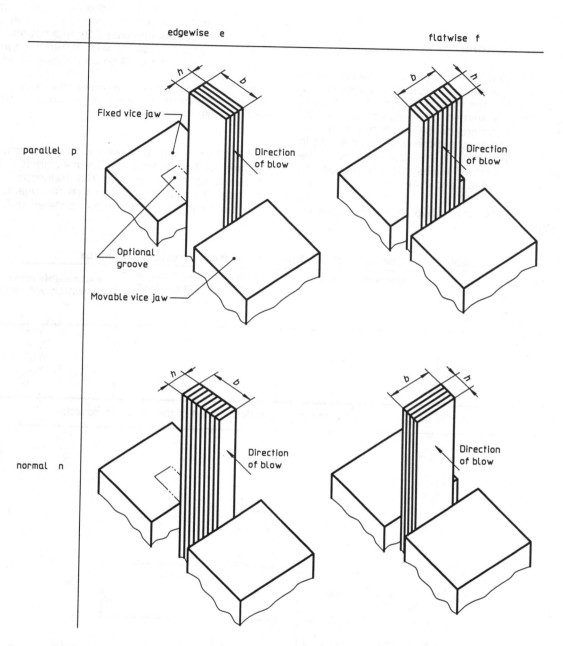

Direction of blow with respect to specimen thickness $h$ and specimen width $b$: edgewise (e) and flatwise (f); with respect to the laminate plane: parallel (p) and normal (n).

The usual Izod test is edgewise parallel. When $h = b$, then parallel as well as normal can be tested.

**Figure 1 — Scheme of designations describing the direction of blow**

# 5 Apparatus

## 5.1 Testing machine

**5.1.1** The testing machine shall be of the pendulum type and shall be of rigid construction. It shall be capable of measuring the impact energy, $W$, absorbed in breaking a test specimen. The value of this energy is defined as the difference between the initial energy, $E$, of the pendulum and the energy remaining in the pendulum after breaking the test specimen. The energy shall be corrected for losses due to friction and air resistance (see table 1 and 7.4).

**5.1.2** The machine shall have the characteristics shown in table 1.

In order to apply the test to the full range of materials specified in 1.3, it is necessary to use a set of interchangeable pendulums (see 7.3). It is not advisable to compare results obtained with different pendulums. The frictional losses shall be periodically checked.

NOTE 2    Pendulums with energies other than those given in table 1 are permitted, but it is planned to withdraw this option at the next revision.

**5.1.3** The machine shall be securely fixed to a foundation having a mass at least 40 times that of the heaviest pendulum in use. The foundation shall be capable of being adjusted so that the orientations of the pendulum and vice are as specified in 5.1.4 and 5.1.6.

### Table 1 — Characteristics of pendulum impact testing machines

| Energy $E$ (nominal) J | Velocity at impact $v_o$ m/s | Maximum permissible frictional loss without specimen J | Permissible error[1] after correction with specimen J |
|---|---|---|---|
| 1,0<br>2,75<br>5,5<br>11,0<br>22,0 | 3,5 (± 10 %) | 0,02<br>0,03<br>0,03<br>0,05<br>0,10 | 0,01<br>0,01<br>0,02<br>0,05<br>0,10 |
| 1)   The permissible error shall not be exceeded over the 10 % to 80 % range of the pendulum capacity. | | | |

Dimensions en millimètres

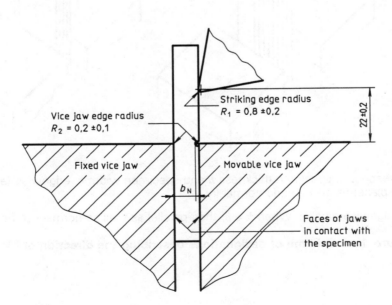

**Figure 2 — Vice support, test specimen and striking edge shown at impact of notched specimen**

**5.1.4** The striking edge of the pendulum shall be hardened steel with a cylindrical surface having a radius of curvature of $R_1 = 0,8$ mm $\pm$ 0,2 mm, with its axis horizontal and perpendicular to the plane of motion of the pendulum. It shall be aligned so that it contacts the full width or thickness of rectangular test specimens. The line of contact shall be perpendicular within $\pm$ 2° to the longitudinal axis of the test specimen.

**5.1.5** The distance between the axis of rotation and the point of impact shall be within $\pm$ 1 % of the pendulum length $L_P$.

NOTE 3    The pendulum length $L_P$, in metres, may be determined experimentally from the period of small amplitude oscillations of the pendulum by means of the following equation:

$$L_P = \frac{g_n}{4\pi^2} \times T^2$$

$$\ldots (1)$$

where

$g_n$    is the standard acceleration of free fall, in metres per second squared (9,81 m/s²);

$T$    is the period, in seconds, of a single complete swing (to and fro) determined from at least 50 consecutive and uninterrupted swings (known to an accuracy of one part in two thousand). The angle of swing shall be less than 5° to each side of the centre.

**5.1.6** The test specimen support shall comprise a vice consisting of a fixed and a moveable jaw. The clamping surfaces of the jaws shall be parallel to within 0,025 mm. The vice shall be arranged to hold the test specimen vertically with respect to its long axis and at right angles to the top plane of the vice (see figure 2). The top edges of the vice jaws shall have radii $R_2 = 0,2$ mm $\pm$ 0,1 mm.

Means shall be provided to ensure that, when a notched test specimen is clamped in the vice, the top plane of the vice is within 0,2 mm of the plane bisecting the angle of the notch.

The vice shall be positioned so that the test specimen is central, to within $\pm$ 0,05 mm, to the striking edge and so that the centre of the striking edge is 22,0 mm $\pm$ 0,2 mm above the top plane of the vice (see figure 2). The vice shall be designed to prevent the clamped portion of the test specimen from moving during the clamping or testing operations.

NOTE 4    The fixed vice jaw may be provided with a groove to improve positioning and handling of the test specimen (see figure 1).

**5.1.7** Some plastics are sensitive to clamping pressure. When testing such materials, a means of standardizing the clamping force shall be used and the clamping force shall be recorded in the test report. The clamping force can be controlled by using a calibrated torque wrench or a pneumatic or hydraulic device on the vice clamping screw.

## 5.2 Micrometers and gauges

Micrometers and gauges suitable for measuring the essential dimensions of the test specimens to an accuracy of 0,02 mm are required. For measuring the dimension $b_N$ of notched specimens, the micrometer shall be fitted with an anvil of width 2 mm to 3 mm and of suitable profile to fit the shape of the notch.

# 6  Test specimens

## 6.1  Preparation

### 6.1.1  Moulding or extrusion compounds

Specimens shall be prepared in accordance with the relevant material specification. When none exists, or unless otherwise specified, specimens shall be either directly compression moulded or injection moulded from the material in accordance with ISO 293, ISO 294, ISO 295, ISO 2557-1 or ISO 2557-2 as appropriate, or machined in accordance with ISO 2818 from sheets that have been compression or injection moulded from the compound.

NOTE 5    Type 1 specimens may be taken from the central part of the test specimen type A complying with ISO 3167 (see 6.3).

### 6.1.2  Sheets

Specimens shall be machined from sheets in accordance with ISO 2818. Whenever possible, specimens of type 1 with notch A shall be used. The machined surface of unnotched specimens shall not be tested under tension.

### 6.1.3  Long-fibre-reinforced polymers

A panel shall be prepared in accordance with ISO 1268 or another specified or agreed upon preparation procedure. Specimens shall be machined in accordance with ISO 2818.

### 6.1.4  Checking

The specimens shall be free of twist and shall have mutually perpendicular parallel surfaces. The surfaces and edges shall be free from scratches, pits, sink marks and flash.

The specimens shall be checked for conformity with these requirements by visual observation against straightedges, squares and flat plates, and by measuring with micrometer calipers.

Specimens showing measurable or observable departure from one or more of these requirements shall be rejected or machined to proper size and shape before testing.

## 6.1.5 Notching

**6.1.5.1** Machined notches shall be prepared in accordance with ISO 2818. The profile of the cutting tooth shall be such as to produce in the specimen a notch of the contour and depth shown in figure 3, at right angles to its principal axes.

**6.1.5.2** Specimens with moulded-in notches may be used if specified for the material being tested. Specimens with moulded-in notches do not give results comparable to those obtained from specimens with machined notches.

## 6.2 Anisotropy

Certain types of sheet or panel materials may show different impact properties according to the direction in the plane of the sheet or panel. In such cases, it is customary to cut groups of test specimens with their major axes respectively parallel and perpendicular to the direction of some feature of the sheet or panel which is either visible or inferred from knowledge of the method of its manufacture.

## 6.3 Shape and dimensions

The preferred specimen is of type 1 with the following dimensions, in millimetres:

| | |
|---|---|
| length: | $l = 80 \pm 2$ |
| width: | $b = 10{,}0 \pm 0{,}2$ |
| thickness: | $h = 4{,}0 \pm 0{,}2$ |

With respect to existing apparatus, the length may be shortened symmetrically to 63,5 mm.

Additional types 2, 3 and 4 are described in annex A.

The longitudinal direction of the notch is always parallel to the thickness $h$.

### 6.3.1 Moulding or extrusion compounds

Type 1 test specimens with two different types of notch shall be used as specified in table 2 and shown in figure 3. The notch shall be located at the centre of the specimen.

The preferred type of notch is type A. If information on the notch sensitivity of the material is desired, specimens with notch types A and B shall be tested.

### 6.3.2 Sheet materials, including long-fibre-reinforced polymers

The recommended thickness $h$ is 4 mm. If the specimen is cut from a sheet or a piece taken from a structure, the thickness of the specimen, up to 10,2 mm, shall be the same as the thickness of the sheet or the structure.

Specimens taken from pieces thicker than 10,2 mm shall be machined to 10 mm ± 0,2 mm from one surface, providing that the sheet is homogeneous in its thickness and contains only one type of reinforcement regularly distributed. If unnotched specimens are tested according to 3.1, the original surface shall be tested under tension, in order to avoid surface effects.

Specimens are tested edgewise in the parallel direction with the exception of specimens with $h = b = 10$ mm, which can be tested parallel and normal to the sheet plane (see figure 1).

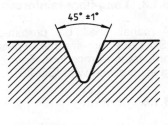

Radius of notch base
$r_N = 0{,}25$ mm ±0,05 mm

Type A notch

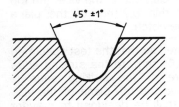

Radius of notch base
$r_N = 1$ mm ±0,05 mm

Type B notch

**Figure 3 — Notch types**

**Table 2 — Method designations, specimen types, notch types and notch dimensions**

Dimensions in millimetres

| Method designation[1] | Specimen type | Notch type[1] | Notch base radius $r_N$ | Remaining width, $b_N$, at notch base |
|---|---|---|---|---|
| ISO 180/1U [2] | 1 [3] | unnotched | — | — |
| ISO 180/1A | | A | $0,25 \pm 0,05$ | $8,0 \pm 0,2$ |
| ISO 180/1B | | B | $1,00 \pm 0,05$ | $8,0 \pm 0,2$ |

1) Attention is drawn to changes in method designations and notch type designations from those used in ISO 180:1982.

2) If specimens are taken from sheets or products, the thickness of the sheet or product shall be added to the designation and unreinforced specimens shall not be tested with their machined surface under tension.

3) If the sheet thickness $h$ equals the width $b$, the blow direction (normal n, or parallel p) should be added to the designation.

## 6.4 Number of test specimens

**6.4.1** Unless otherwise specified in the standard for the material being tested, a set consisting of a minimum of ten specimens shall be tested. When the coefficient of variation (see ISO 2602) has a value of less than 5 %, a minimum number of five test specimens is sufficient.

**6.4.2** If laminates are tested in the parallel and normal directions, ten specimens shall be used for each direction.

## 6.5 Conditioning

Unless otherwise specified in the standard for the material under test, the specimens shall be conditioned for at least 16 h at 23 °C and 50 % relative humidity according to ISO 291, unless other conditions are agreed upon by the interested parties.

## 7 Procedure

**7.1** Conduct the test in the same atmosphere as that used for conditioning, unless otherwise agreed upon by the interested parties, e.g. for testing at high or low temperatures.

**7.2** Measure the thickness, $h$, and the width, $b$, of each test specimen, in the centre, to the nearest 0,02 mm. In the case of notched specimens, carefully measure the remaining width $b_N$ to the nearest 0,02 mm.

NOTE 6   In the case of injection-moulded specimens, it is not necessary to measure the dimensions of each specimen. It is sufficient to measure one specimen from a set to make sure that the dimensions correspond to those in 6.3.

With multiple-cavity moulds, ensure that the dimensions of the specimens are the same for each cavity.

**7.3** Check that the pendulum machine has the specified velocity of impact (see table 1) and that it is in the correct range of absorbed energy $W$, which shall be between 10 % and 80 % of the pendulum energy $E$. If more than one of the pendulums described in table 1 meet these requirements, the pendulum having the highest energy shall be used.

**7.4** Carry out a blank test (i.e. without a specimen in place) and record the measured values of the total frictional loss. Ensure that this energy loss does not exceed the appropriate value given in table 1.

If frictional losses are equal to or less than the values indicated in table 1, they may be used in the calculations of corrected energy absorbed. If frictional losses exceed the values indicated in table 1, care should be taken to evaluate the cause of any excess frictional losses and corrections made as necessary to the equipment.

**7.5** Lift and support the pendulum. Place the specimen in the vice and clamp it in accordance with 5.1.6 and as shown in figure 2. For the determination of notched Izod impact strength, the notch shall be positioned on the side that is to be struck by the striking edge of the pendulum (see figure 2).

**7.6** Release the pendulum. Record the impact energy absorbed by the specimen and apply any necessary corrections for frictional losses etc. (see table 1 and 7.4).

**7.7** Four types of failure according to the following letter codes may occur:

C   complete break; a break in which the specimen separates into two or more pieces

H   hinge break; an incomplete break such that both parts of the specimen are held together only by a thin peripheral layer in the form of a hinge having no residual stiffness

P    partial break; an incomplete break that does not meet the definition for a hinge break

NB   non-break; in the case where there is no break, the specimen is only bent, possibly combined with stress whitening

The measured values of complete and hinged breaks can be used for a common mean value without remark. If in the case of partial breaks a value is required, it shall be designated with the letter P. In the case of non-break, NB, no values shall be reported.

**7.8** If, within one sample, the test specimens show both P and C (or H) failures, the mean value for each failure type shall be reported.

## 8 Calculation and expression of results

### 8.1 Unnotched specimens

Calculate the Izod impact strength of unnotched specimens, $a_{iU}$, expressed in kilojoules per square metre, using the formula

$$a_{iU} = \frac{W}{h \cdot b} \times 10^3 \qquad \ldots (2)$$

where

    $W$    is the corrected energy, in joules, absorbed by breaking the test specimen;

    $h$    is the thickness, in millimetres, of the test specimen;

    $b$    is the width, in millimetres, of the test specimen.

### 8.2 Notched specimens

Calculate the Izod impact strength of notched specimens, $a_{iN}$, expressed in kilojoules per square metre, with notches N = A or B, using the formula

$$a_{iN} = \frac{W}{h \cdot b_N} \times 10^3 \qquad \ldots (3)$$

where

    $W$    is the corrected energy, in joules, absorbed by breaking the test specimen;

    $h$    is the thickness, in millimetres, of the test specimen;

    $b_N$    is the remaining width, in millimetres, at the notch base of the test specimen.

### 8.3 Statistical parameters

Calculate the arithmetic mean of the test results and, if required, the standard deviation and the 95 % confidence interval of the mean value using the procedure given in ISO 2602. For different types of failure within one sample, the relevant numbers of specimens shall be given and mean values shall be calculated.

### 8.4 Significant figures

Report all calculated mean values to two significant figures.

## 9 Precision

The precision of this test method is not known because interlaboratory data are not available. When interlaboratory data are obtained, a precision statement will be added in the next revision.

## 10 Test report

The test report shall include the following information:

a) a reference to this International Standard;

b) the method designation according to table 2, e.g.:

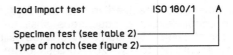

c) all the information necessary for identification of the material tested, including type, source, manufacturer's code, grade and form, history, where these are known;

d) description of the nature and form of the material, i.e. whether a product, semifinished product, test plaque or specimen, including principal dimensions, shape, method of manufacturer, etc. where these are known;

e) the velocity of impact;

f) the nominal pendulum energy;

g) clamping pressure, if applicable (see 5.1.7);

h) method of test specimen preparation;

i) if the material is in the form of a product, or a semifinished product, the orientation of the test specimen in relation to the product or semifinished product from which it is cut;

j)  number of tested specimens;

k)  the standard atmosphere for conditioning and for testing, plus any special conditioning treatment if required by the standard for the material or product;

l)  the type(s) of failure observed;

m) the individual test results;

n)  the impact strength of the material, reported as the arithmetic mean value of the results for each specimen type, and type of failure where appropriate (see 7.7);

o)  the standard deviations and the 95 % confidence intervals of these mean values, if required;

p)  the date(s) of the test.

# Annex A
## (normative)

# Izod impact test on specimen types 2, 3 and 4 of ISO 180:1982

## A.1 General

The Izod test with specimen types 2, 3 and 4 of ISO 180:1982 is reviewed in this annex to emphasize that their withdrawal is proposed for the next revision.

It is strongly recommended that type 1 specimens as described in 6.3 be used whenever possible.

## A.2 Test specimens (see table A.1)

For sheet materials (see 6.3.2), reversed-notch specimens (see table A.2) shall be used for specimen types 2, 3 and 4. The recommended thickness $h$ is 3,2 mm for type 4 if the specimen is cut from a piece or taken from a structure. The thickness of the specimen, up to 12,9 mm, shall be the same as the thickness of the sheet or the structure. Specimens taken from pieces thicker than 12,9 mm shall be machined from one side to achieve a thickness of 12,7 mm $\pm$ 0,2 mm providing that the sheet is homogeneous in its thickness and contains only one type of reinforcement, regularly distributed.

If reversed-notch specimens are tested, the original surface shall be loaded under tension in order to avoid surface effects.

Specimens are tested edgewise in the parallel direction, with the exception of specimens with $h = b = 12,7$ mm, which can be tested parallel and normal to the sheet plane (see figure 1).

## A.3 Calculation and expression of results

### A.3.1 Notched specimens

See 8.2:

### A.3.2 Reversed-notch specimens

Calculate the Izod impact strength of reversed-notch specimens $a_{iR}$, expressed in kilojoules per square metre, using the formula

$$a_{iR} = \frac{W}{h \cdot b_N} \times 10^3 \qquad \ldots (A.1)$$

where

$W$ is the corrected energy, in joules, absorbed by breaking the test specimen;

$h$ is the thickness, in millimetres, of the test specimen;

$b_N$ is the remaining width, in millimetres, at the notch base of the test specimen.

### Table A.1 — Specimen types and dimensions

Dimensions in millimetres

| Specimen type | Length $l$ | Width $b$ | Thickness $h$ |
|---|---|---|---|
| 2 | 63,5 $\pm$ 2 | 12,7 $\pm$ 0,2 | 12,7 $\pm$ 0,2 |
| 3 | 63,5 $\pm$ 2 | 12,7 $\pm$ 0,2 | 6,4 $\pm$ 0,2 |
| 4 | 63,5 $\pm$ 2 | 12,7 $\pm$ 0,2 | 3,2 $\pm$ 0,2 |

**Table A.2 — Method designations, specimen types, notch types and notch dimensions**

| Method designation[1] | Specimen type | Notch type[2] | Notch base radius $r_N$ | Remaining width, $b_N$, at notch base |
|---|---|---|---|---|
| ISO 180/2AR<br>ISO 180/2BR<br>ISO 180/2A<br>ISO 180/2B | 2 [3] | A reversed<br>B reversed<br>A<br>B | 0,25 ± 0,05<br>1,00 ± 0,05<br>0,25 ± 0,05<br>1,00 ± 0,05 | |
| ISO 180/3AR<br>ISO 180/3BR<br>ISO 180/3A<br>ISO 180/3B | 3 | A reversed<br>B reversed<br>A<br>B | 0,25 ± 0,05<br>1,00 ± 0,05<br>0,25 ± 0,05<br>1,00 ± 0,05 | 10,2 ± 0,2 |
| ISO 180/4AR<br>ISO 180/4BR<br>ISO 180/4A<br>ISO 180/4B | 4 | A reversed<br>B reversed<br>A<br>B | 0,25 ± 0,05<br>1,00 ± 0,05<br>0,25 ± 0,05<br>1,00 ± 0,05 | |

1)  If specimens are taken from sheets or products, the thickness of the sheet or product shall be added to the designation and unreinforced specimens shall not be tested with their machined surface under tension.

2)  For determination of the Izod impact strength of reversed-notch specimens, the pendulum strikes the face opposite that containing the notch.

3)  If the sheet thickness $h$ equals the width $b$, the blow direction (normal n or parallel p) should be added to the designation.

**INTERNATIONAL STANDARD ISO 180:1993**
**TECHNICAL CORRIGENDUM 1**

Published 1995-03-15

INTERNATIONAL ORGANIZATION FOR STANDARDIZATION • МЕЖДУНАРОДНАЯ ОРГАНИЗАЦИЯ ПО СТАНДАРТИЗАЦИИ • ORGANISATION INTERNATIONALE DE NORMALISATION

# Plastics — Determination of Izod impact strength

## TECHNICAL CORRIGENDUM 1

*Plastiques — Détermination de la résistance au choc Izod*

*RECTIFICATIF TECHNIQUE 1*

Technical corrigendum 1 to International Standard ISO 180:1993 was prepared by Technical Committee ISO/TC 61, *Plastics*, Subcommittee SC 2, *Mechanical properties.*

————————

Page 5

**Subclause 5.1.6**

In line 2 of the third paragraph, replace "± 0,05 mm" by "± 0,5 mm".

**UDC  678.077:620.178.746.24**

**Ref. No. ISO 180:1993/Cor.1:1995(E)**

**Descriptors:** plastics, rigid plastics, tests, Izod impact tests, test equipment, test specimens.

© ISO 1995

Printed in Switzerland

# INTERNATIONAL STANDARD

INTERNATIONAL ORGANIZATION FOR STANDARDIZATION •МЕЖДУНАРОДНАЯ ОРГАНИЗАЦИЯ ПО СТАНДАРТИЗАЦИИ •ORGANISATION INTERNATIONALE DE NORMALISATION

# Plastics — Standard atmospheres for conditioning and testing

*Plastiques — Atmosphères normales de conditionnement et d'essai*

**First edition — 1977-12-15**

UDC 678.01

**Ref. No. ISO 291-1977 (E)**

**Descriptors** : plastics, tests, testing conditions, standard atmospheres, test atmospheres, controlled atmospheres.

ISO 291-1977 (E)

## FOREWORD

ISO (the International Organization for Standardization) is a worldwide federation of national standards institutes (ISO member bodies). The work of developing International Standards is carried out through ISO technical committees. Every member body interested in a subject for which a technical committee has been set up has the right to be represented on that committee. International organizations, governmental and non-governmental, in liaison with ISO, also take part in the work.

Draft International Standards adopted by the technical committees are circulated to the member bodies for approval before their acceptance as International Standards by the ISO Council.

International Standard ISO 291 was developed by Technical Committee ISO/TC 61, *Plastics*, and was circulated to the member bodies in January 1976.

It has been approved by the member bodies of the following countries :

| | | |
|---|---|---|
| Australia | Hungary | Poland |
| Austria | India | Romania |
| Belgium | Iran | South Africa, Rep. of |
| Brazil | Ireland | Spain |
| Canada | Israel | Sweden |
| Chile | Italy | Switzerland |
| Czechoslovakia | Japan | Turkey |
| Finland | Korea, Rep. of | United Kingdom |
| France | Mexico | U.S.A. |
| Germany | New Zealand | U.S.S.R. |

The member body of the following country expressed disapproval of the document on technical grounds :

Netherlands

This International Standard cancels and replaces ISO Recommendation R 291-1963, of which it constitutes a technical revision.

# Plastics — Standard atmospheres for conditioning and testing

## 1 SCOPE AND FIELD OF APPLICATION

This International Standard sets out specifications relating to the conditioning and testing of all plastics and all types of test specimen under approximately normal ambient atmospheric conditions.

Special atmospheres applicable to a particular test or material or simulating a particular climatic environment are not included in this International Standard.

NOTE — This International Standard has been prepared taking into consideration ISO 554, *Standard atmospheres for conditioning and/or testing — Specifications.*

## 2 DEFINITIONS

**2.1 conditioning atmosphere :** The atmosphere in which a sample or test specimen is kept before being subjected to test.

**2.2 test atmosphere :** The atmosphere to which a sample or test specimen is exposed throughout the test.

NOTE — Both of these atmospheres are defined by specified values for temperature, relative humidity and pressure.

**2.3 conditioning :** The whole series of operations intended to bring a sample or test specimen into a state of equilibrium with regard to temperature and humidity.

The "conditioning procedure" is defined by the "conditioning atmosphere" and the "period of conditioning".

## 3 STANDARD ATMOSPHERES

| Designation | Temperature | Relative humidity | Pressure | Remarks |
|---|---|---|---|---|
| | °C | % | kPa | |
| Atmosphere 23/50 | 23 | 50 | 86 to 106 | Recommended atmosphere |
| Atmosphere 27/65 | 27 | 65 | | For tropical countries |

These atmospheres are to be used only when the properties of the samples or test specimens are influenced by both temperature and humidity.

If humidity has no influence on the properties being examined, the relative humidity may be uncontrolled. The two corresponding atmospheres are designated "atmosphere 23" and "atmosphere 27" respectively.

Similarly, if neither temperature nor humidity has any influence on the properties being examined, the temperature and relative humidity may be uncontrolled. In this case the atmosphere is termed the "ambient atmosphere".

NOTE — In the special case where the plastic material is, or has to be, used at the same time as other materials, an atmosphere of 20 °C/65 % relative humidity may be permitted to be used after agreement between the interested parties or if expressly specified.

## 4 TOLERANCES

| Tolerances | Temperature | Relative humidity |
|---|---|---|
| | °C | % |
| Normal | ± 2 | ± 5 |
| Close | ± 1 | ± 2 |

NOTES

1 Usually, the tolerances are coupled as pairs, i.e. normal tolerance for both temperature and relative humidity or close tolerance on both.

2 These tolerances apply to the working space of the test enclosure or conditioning enclosure and include both the variation with position in the enclosure and fluctuation with time.

## 5 CONDITIONING

The period of conditioning shall be stated in the relevant specifications for the materials.

NOTES

1 When the periods are not stated in the appropriate International Standard, the following shall be adopted :

 a) a minimum of 88 h for atmospheres 23/50 and 27/65;

 b) a minimum of 4 h for atmospheres 23 and 27.

2 For particular tests and for plastics or test specimens that are known to reach temperature and humidity equilibrium either very rapidly or very slowly, a shorter or longer time for the conditioning period may be specified in the appropriate International Standard (see the annex).

## 6 TESTING

Unless otherwise specified, the specimens shall be tested in the same atmosphere in which they have been conditioned.

In all cases, the test shall be made immediately after the removal of the test specimens from the conditioning enclosure.

ANNEX

## ATTAINMENT OF EQUILIBRIUM MOISTURE ABSORPTION BY PLASTICS IN A CONDITIONING ATMOSPHERE

The amount of moisture absorbed and its rate of absorption by a test specimen conditioned in a humid atmosphere vary significantly according to the nature of the plastic material being examined.

The normal conditions for conditioning established in this International Standard (see clause 5) are generally satisfactory, with the following exceptions :

— materials that are known to reach equilibrium with their conditioning atmosphere only after a long period of time (for example, certain polyamides);

— new materials or those of unknown structure, for which neither the capacity for absorbing moisture nor the time required for reaching equilibrium can be estimated beforehand.

In the two latter cases, one may either

a) dry the material at a high temperature, or

b) condition the test specimens at $23 \pm 2\ ^\circ C$ and $50 \pm 5\ \%$ relative humidity until equilibrium is reached.

Procedure a) has the disadvantage that certain property values, in particular mechanical ones, are different in the dry state from those obtained after conditioning in the atmosphere 23/50.

In the case of procedure b), one of the following criteria may be suitable :

1) constant mass is attained within 0,1 % for two determinations made at an interval of $d^2$ weeks ($d$ being the thickness, in millimetres, of the test specimen);

2) for certain polymers, it is sufficient to prepare a mass/time curve with intervals of time much less than $d^2$ weeks and to consider that a practical equilibrium is reached when the slope of the curve, expressed as a percentage, is equal to 0,1.

# International Standard

ISO 293

INTERNATIONAL ORGANIZATION FOR STANDARDIZATION•МЕЖДУНАРОДНАЯ ОРГАНИЗАЦИЯ ПО СТАНДАРТИЗАЦИИ•ORGANISATION INTERNATIONALE DE NORMALISATION

# Plastics — Compression moulding test specimens of thermoplastic materials

*Plastiques — Moulage par compression des éprouvettes en matières thermoplastiques*

**Second edition — 1986-10-01**

UDC  678.073 : 678.027.72 : 620.11

Ref. No.  ISO 293-1986 (E)

**Descriptors :**  plastics,  thermoplastic resins,  moulding materials,  compression moulding,  test specimens,  specimen preparation.

# Foreword

ISO (the International Organization for Standardization) is a worldwide federation of national standards bodies (ISO member bodies). The work of preparing International Standards is normally carried out through ISO technical committees. Each member body interested in a subject for which a technical committee has been established has the right to be represented on that committee. International organizations, governmental and non-governmental, in liaison with ISO, also take part in the work.

Draft International Standards adopted by the technical committees are circulated to the member bodies for approval before their acceptance as International Standards by the ISO Council. They are approved in accordance with ISO procedures requiring at least 75 % approval by the member bodies voting.

International Standard ISO 293 was prepared by Technical Committee ISO/TC 61, *Plastics*.

This second edition cancels and replaces the first edition (ISO 293-1974), of which it constitutes a technical revision.

Users should note that all International Standards undergo revision from time to time and that any reference made herein to any other International Standard implies its latest edition, unless otherwise stated.

# Plastics — Compression moulding test specimens of thermoplastic materials

## 0  Introduction

For reproducible test results, specimens with a defined state are required. In contrast to injection moulding, the aim of compression moulding is to produce test specimens and sheets for machining or stamping of test specimens that are homogeneous and isotropic.

In the process of compression moulding, mixing of material takes place on a negligible scale. Granules and powders fuse only at their surfaces and preforms (milled sheets) are only partially softened.

Isotropic and homogeneous specimens can, therefore, only be obtained when the moulding material is itself homogeneous and isotropic. This has to be considered when processing multiphase materials, such as ABS, which retain their internal structure.

## 1  Scope and field of application

This International Standard specifies the general principles and the procedures to be followed with thermoplastics in the preparation of compression-moulded test specimens and sheets from which test specimens may be machined or stamped.

In order to obtain mouldings in a reproducible state, the main steps of the procedure, including four different cooling methods, are standardized. For each material, the required moulding temperature and cooling methods shall be as specified in the appropriate International Standard for the material or as agreed between the interested parties.

NOTE — The procedure is not recommended for reinforced thermoplastics.

## 2  References

ISO/R 286, *ISO system of limits and fits — Part 1: General, tolerances and deviations.*

ISO 468, *Surface roughness — Parameters, their values and general rules for specifying requirements.*

## 3  Definitions

For the purpose of this International Standard, the following definitions apply.

**3.1  moulding temperature:** The temperature of the mould or the press platens during the preheating and moulding time, measured in the nearest vicinity to the moulded material.

**3.2  demoulding temperature:** The temperature of the mould or the press platens at the end of the cooling time, measured in the nearest vicinity to the moulded material.

NOTE — For positive moulds, holes are normally drilled in the mould for measuring the temperatures defined in 3.1 and 3.2.

**3.3  preheating time:** The time required to heat the material in the mould up to the moulding temperature while maintaining the contact pressure.

**3.4  moulding time:** The time during which full pressure is applied while maintaining the moulding temperature.

**3.5  average cooling rate** (non-linear): Rate of cooling by a constant flow of the cooling fluid, calculated by dividing the difference between moulding and demoulding temperatures by the time required to cool the mould to the demoulding temperature.

The average cooling rate is usually expressed in kelvins per minute.

**3.6  cooling rate:** Constant rate of cooling in a defined temperature range obtained by controlling the flow of the cooling fluid in such a way that over each 10 min interval the deviation from this specified cooling rate shall not exceed the specified tolerance.

The cooling rate is usually expressed in kelvins per hour.

## 4  Apparatus

### 4.1  Moulding press

The press shall have a clamping force capable of applying a pressure (conventionally given as the ratio of the clamping force to the area of the mould cavity) of at least 10 MPa.

The pressure shall be maintained to within 10 % of the specified pressure during the moulding cycle.

The platens shall be capable of

a) being heated to at least 240 °C;

b) being cooled at a rate given in the table.

The difference between the temperatures of any points of the mould surfaces shall not vary by more than ±2 K during heating and ±4 K during cooling.

When the heating and cooling system is incorporated in the mould, it shall comply with the same conditions.

The platens or mould shall be heated either by high-pressure steam, by a heat-conducting fluid in an appropriate channel system, or by using electric heating elements. The platens or mould are cooled by a heat-conducting fluid (usually cold water) in a channel system.

For quench cooling (see method C in the table), two presses shall be used, one for heating during moulding and the other for cooling.

NOTES

1  For a specified cooling method, the flow rate of the heat-conducting fluid should be predetermined in a test without any material in the mould.

2  The temperature may be constantly controlled in the centre between each upper and lower platen of the press.

## 4.2  Moulds

### 4.2.1  General

The characteristics of the test specimens prepared by using different types of mould are not the same. In particular, the mechanical properties depend on the pressure applied to the material during cooling.

In general, two types of moulds, "flash moulds" (see figure 1) and "positive moulds" (see figure 2), are used for compression moulding test specimens of thermoplastics.

Flash moulds permit excess moulding material to be squeezed out and do not exert moulding pressure on the moulding material during cooling. They are particularly convenient for preparing test specimens or panels of similar thickness or comparable levels of low internal stress.

With positive moulds, the full moulding pressure, neglecting friction, is exerted on the material during cooling. The thickness, stress and density of the resulting mouldings depend on mould construction, size of material charge and the moulding and cooling conditions. This type of mould produces consolidated test specimens with moulded surfaces and is therefore particularly suitable for obtaining flat surfaces or suppressing the formation of voids within test specimens.

### 4.2.2  Fabrication

The moulds shall be made of materials capable of withstanding the moulding temperature and pressure. The surfaces in contact with the material shall be polished to obtain a good surface condition on the specimens (recommended surface roughness 0,16 $R_a$, see ISO 468). Specimen removal can be made easier by chromium plating these surfaces. For specimens of small dimensions a 2° taper is strongly recommended.

Blind holes may be drilled in the mould so that temperature can be measured in the vicinity of the moulded material by using thermocouples or mercury thermometers.

NOTES

1  Depending on the performance of the press (see 4.1), the moulds may have built-in heating and/or cooling devices similar to those described for the press platens.

2  An alloy steel, resistant to mechanical shock and heat treated to provide a tensile strength of 2 200 MPa, will generally be satisfactory for the moulds. However, in the special case of PVC moulding materials, the use of martensite stainless steel treated to provide a tensile strength of 1 050 MPa is recommended.

### 4.2.3  Types

The type of mould used shall be capable of producing test specimens of the types and states specified in the appropriate International Standard for the material or shall be agreed upon between the interested parties.

#### 4.2.3.1  Flash ("picture frame") moulds

With this type of mould, the excess material is squeezed out and the moulding pressure during cooling is only exerted on the frame and not on the material. The thickness in the centre of the mouldings is slightly less than at the edges due to the shrinkage during cooling. Directly moulded test bars may also have sink marks or voids if the shrinkage is hindered by sticking of the plastic material to the mould.

To overcome these disadvantages, stamping or machining of test specimens from the central part of compression moulded sheets is preferred.

For moulding sheets, simple and economical flash moulds can be used, consisting of a frame covered with two plates (see figure 1). The lower and upper plates, having a thickness of about 1 to 2 mm, can be made from polished steel or chromium-plated brass to aid release. To avoid the plastic material sticking to the plates, they can be covered by a flexible foil, for example of aluminium or polyester.

Use of a release agent is not allowed.

The thickness of the chase shall be appropriate to the moulded sheet thickness.

The size of the moulding frame shall be such that specimens can be cut or machined without using the outer 20 mm perimeter of the sheet.

#### 4.2.3.2  Positive moulds

These moulds (see figure 2) are fitted with one or two male pistons and a female part. They allow known pressure, neglecting friction, to be applied to the material, and to be maintained during the moulding and cooling times.

The thickness of the moulding will depend on the quantity of material, its thermal expansion, and the loss of material due to clearances in the moulds. The losses will be a function of the flow of the material at the chosen moulding temperature, the applied pressure, the time over which the pressure is applied, mould construction, etc.

Correct guidance of the male part in the female part is facilitated by use of a round cavity. A fit between these parts of H7g6 (see ISO/R 286) is recommended, i.e. between 15 and 90 µm for a round cavity of diameter 200 mm. The mould may be fitted with one or several ejection pins to make part removal easier.

Shims may be used in positive moulds to aid in controlling thickness. These are removed at the start of the cooling phase.

# 5 Procedure

## 5.1 Preparation of moulding material

### 5.1.1 Drying of granular material

Dry the granular material as specified in the relevant International Standard or in accordance with the material supplier's instructions. If no instructions are given, dry for $24 \pm 1$ h at $70 \pm 2$ °C in an oven.

### 5.1.2 Preparation of preforms

Direct moulding of sheet from granules shall be the standard procedure, provided that a sufficiently homogeneous sheet is obtained. Normally this means that the sheet is free from surface irregularities and internal imperfections. Direct moulding from powder or granules may sometimes require melt homogenization using a hot melt milling or mixing procedure to achieve a satisfactory final sheet. Conditions shall be used that do not degrade the polymer. This can usually be achieved by not milling or mixing for more than 5 min after melting. The preform sheet obtained shall be thicker than the test sheet to be moulded and of sufficient size to enable the test sheet to be moulded.

NOTE — Storing the preforms in a dry, airtight container is recommended.

## 5.2 Moulding

Adjust the mould temperature to within $\pm 5$ K of the moulding temperature specified in the relevant International Standard or as agreed between the interested parties.

Place a weighed quantity of the material (granules or preforms) in the preheated mould. If granular material is used, make sure that it is evenly spread over the mould surface. The mass of the material shall be sufficient to fill the cavity volume when it is melted and allow about a 10 % loss for a flash mould and about a 3 % loss for a positive mould. With flash moulds, cover the mould with flexible foils (see 4.2.3.1) and then place the mould in the preheated press.

Close the press and preheat the material charge by applying a contact pressure for 5 min. Then apply full pressure for 2 min (moulding time, see 3.4) and then cool down (see 5.3)

NOTES

1 A preheating time of 5 min is the standardized time for evenly spread material charges sufficient for sheets up to 2 mm thickness. For thicker mouldings, this time has to be adjusted accordingly.

2 At contact pressure, the press is just closed with a pressure low enough to avoid flow of the material. Full pressure means a pressure sufficient to shape the material and squeeze out the excess material.

## 5.3 Cooling

### 5.3.1 General

With some thermoplastics, the cooling rate affects the ultimate physical properties. For this reason, the cooling methods are specified in the table.

The method of cooling shall always be stated together with the final physical properties. The appropriate cooling method is normally given in the relevant International Standard for the material. If no method is indicated, method B shall be used.

### 5.3.2 Cooling methods

The appropriate cooling method shall be selected from the table.

**Table — Cooling methods**

| Cooling method | Average cooling rate (see 3.5) | Cooling rate (see 3.6) | Remarks |
|---|---|---|---|
| | $K \cdot min^{-1}$ | $K \cdot h^{-1}$ | |
| A | $10 \pm 5$ | | |
| B | $15 \pm 5$ | | |
| C | $60 \pm 30$ | | Quench cooling |
| D | | $5 \pm 0,5$ | Slow cooling |

In the case of quench cooling (see method C in the table), transfer the mould assembly from the heating press to the cooling press as quickly as possible by suitable means, for example, using a pair of tongs.

The demoulding temperature shall be $< 40$ °C if no other instructions are given.

NOTES

1 The use of two presses is required for method C (see 4.1).

2 Method D is recommended for producing test specimens free of any internal stress or for slow cooling after annealing of previously prepared sheets.

## 6 Inspection of the moulded specimens or sheets

After cooling, check the moulded specimens or sheets for appearance (i.e. for sink marks, shrink holes, discolorations) and for conformance to specified dimensions. If any moulding defects are found, the test specimens or sheets shall be discarded.

Make sure that there is no degradation or unwanted cross-linking, using the method specified in the relevant International Standard or as agreed between the interested parties.

## 7 Test report

The test report shall contain the following information:

a) reference to this International Standard;

b) dimension of the specimen and its intended use;

c) complete identification of moulding material (type, designation, etc.);

d) preparation of moulding material:

1) drying conditions for granulates and powder,

2) processing conditions used in the preparation of preforms and their average thickness;

e) type of mould and foil used;

f) moulding conditions:

1) preheating time,

2) moulding temperature, pressure and time,

3) cooling method used,

4) demoulding temperature;

g) state of specimen, if applicable;

h) any other observations.

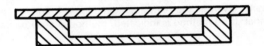 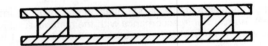

**Figure 1 — Types of flash ("picture frame") moulds**

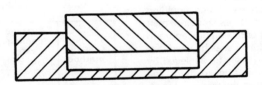

**Figure 2 — Positive-type mould**

**DRAFT INTERNATIONAL STANDARD** ISO/DIS 294-1

ISO/TC **61**/SC **9**

Secretariat: **ANSI**

Voting begins on
1995-03-23

Voting terminates on
1995-09-23

INTERNATIONAL ORGANIZATION FOR STANDARDIZATION • МЕЖДУНАРОДНАЯ ОРГАНИЗАЦИЯ ПО СТАНДАРТИЗАЦИИ • ORGANISATION INTERNATIONALE DE NORMALISATION

# Plastics — Injection moulding of test specimens of thermoplastic materials —

# Part 1:
General principles and multipurpose test specimens (ISO type A mould) and bars (ISO type B mould)

(Revision in part of ISO 294:1995)

*Plastiques — Moulage par injection des éprouvettes de matériaux thermoplastiques —*

*Partie 1: Principes généraux, éprouvettes à usages multiples (moule ISO de type A) et barreaux (moule ISO de type B)*

ICS 83.080.20

Descriptors: plastics, thermoplastic resins, moulding materials, test specimens, injection moulding, specimen preparation, reference data, generalities.

ISO/DIS 294-1

FOREWORD

This ~~third~~ edition cancels and replaces the *second* edition (ISO 294: 199**5**), which has been extended to improve the definitions of the injection moulding parameters and has been rearranged presenting the general principles suitable to define four ISO mould types for processing basic specimen types required for the determination of comparable test data.

Care has been taken to allow that all ISO mould types described in three parts of this International Standard can be fitted resulting in a common injection moulding equipment with a system of interchangeable cavity plates.

ISO 294 will consist of the following parts, under the general title

Plastics - Injection moulding of test specimens of thermoplastic materials

- Part 1: General principles, multipurpose-test specimens (ISO mould type A) and bars (ISO mould type B)

- Part 2: Small tensile bars (ISO mould type C)

- Part 3: Plates (ISO mould type D)

- Part 4: Determination of moulding shrinkage

Annexes A, B, C and D of this part of ISO 294 are for information only.

## Introduction

Many factors in the injection moulding process may influence the properties of moulded specimens and the values measured using various test procedures. The mechanical properties of test specimens are strongly dependent on the conditions of the moulding process used to prepare the specimens. An exact definition for each of the essential parameters of the moulding process is a basic requirement for the standardization of reproducible and comparable operating conditions.

It is important in defining moulding conditions to consider any influence of the conditions used in the moulding process on the material properties to be determined. Thermoplastics may show differences in molecular orientation (important mainly for amorphous polymers), in crystallisation morphology (for crystalline and semicrystaline polymers), in phase morphology (for heterogeneous thermoplastics) as well as in orientation of anisotropic fillers (for short fibers e. g.). Residual ("frozen-in") stresses in the moulded specimens and thermal degradation of the polymer during moulding may also influence properties. Each of these phenomena must be controlled to avoid fluctuation of numerical values of measured properties.

## 1    SCOPE

This part of this International Standard provides the general principles to be followed for injection moulding test specimens of thermoplastics and gives details of mould designs for preparing specimen types for reference data, i. e. for multi-purpose-test specimens according to ISO 3167 and for bars 80 mm x 10 mm x 4 mm. It provides a basis for establishing reproducible moulding conditions. Its purpose is to promote uniformity in describing the various essential parameters of the moulding operation and also to establish a uniform practice in reporting moulding conditions. The special conditions required to prepare specimens in a comparable and reproducible state will vary for each material used. These conditions are given in the International Standard for the relevant material or are to be agreed upon between interested parties.

NOTE 1 - ISO round-robin tests with acrylonitrile/buta-diene/styrene (ABS), styrene/butadiene (S/B) and polyme-thylmethacrylate (PMMA) showed that design of the mould is one of the important factors in preparation of test specimens.

## 2    NORMATIVE REFERENCES

The following standards contain provisions which, through reference in this text, constitute provisions of this International Standard. At the time of publication, the editions indicated were valid. All standards are subject to revision and parties to agreements based upon this International Standard are encouraged to use the most recent edition of the standards listed below. Members of IEC and ISO meintain registers of current valid International Standards.

ISO 294-2[2), Plastics - Injection moulding of test specimens of thermoplastic materials - Part 2: Small tensile bars (ISO mould type C)

ISO 294-3[2), Plastics - Injection moulding of test specimens of thermoplastic materials - Part 3: Plates (ISO mould type D)

ISO 294-4[2), Plastics - Injection moulding of test specimens of thermoplastic materials - Part 4: Determination of moulding shrinkage

ISO 3167: 1993, Plastics - Multipurpose test specimens

ISO 10350: 1993, Plastics - The acquisition and presentation of comparable single-point data

ISO 11403-1: 1994[1], Plastics - Acquisition and presentation of comparable multipoint data - Part 1: Mechanical properties

ISO/DIS 11403-2.2: 1994, Plastics - Acquisition and presentation of comparable multipoint data - Part 2: Thermal and processing properties

ISO 11403-3[2], Plastics - Acquisition and presentation of comparable multipoint data - Part 3: Environmental influences on properties

[1]    To be published
[2]    In preparation


## 3    DEFINITIONS

For the purposes of this International Standard the following definitions apply.

3.1  **mould temperature** $T_C$: The average temperature of the mould cavity surfaces measured after the system has attained stationary equilibrium under operating conditions and immediately after opening the mould (see 4.2.5 and 5.3).

It is expressed in degrees celcius, °C.

3.2  **melt temperature** $T_M$: The temperature of the molten plastic in a free shot (see 4.2.5 and 5.4).

It is expressed in degrees celsius, °C.

**3.3  melt pressure p:** The pressure of the plastic material in front of the screw at any time of the moulding process, (see figure 1).

It is expressed in megapascals, MPa.

The melt pressure is calculated based upon the force $F_s$ acting axially upon the screw, generated e. g. hydraulically, see equation (1).

$$ p = \frac{4 \cdot 10^3 \cdot F_s}{\pi \cdot D^2} \tag{1} $$

where

p     is the melt pressure, in MPa

$F_s$    is the axial force acting upon the screw, in kN

D     is the screw diameter, in mm

**3.4  hold pressure $p_H$:** The melt pressure, see 3.3, during the hold time interval (see figure 1).

It is expressed in megapascals, MPa.

**3.5  moulding cycle:** The complete sequence of operations in the moulding process required for the production of one set of test specimens (see figure 1). The time required for a complete moulding cycle is related to the following times as indicated on figure 1.

**3.6 cycle time $t_T$:** The total time used to carry out the complete sequence of operations making up the moulding cycle.

It is expressed in seconds, s.

The sum of cooling time $t_c$ and mould open time $t_o$ is the total time required to perform one moulding cycle (see 3.7 and 3.10).

3.7 **cooling time** $t_c$: The time from the beginning of screw forward movement until the mould starts to open.

It is expressed in seconds, s.

3.8 **injection time** $t_I$: The time from the beginning of screw forward movement until switching over from the injection period to the hold period.

It is expressed in seconds, s.

3.9 **hold time** $t_H$: The time interval of hold pressure, (see 3.4).

It is expressed in seconds, s.

3.10 **mould open time** $t_o$: The time interval from the instant the mould starts to open until the mould is closed and attains full clamping pressure.

It is expressed in seconds, s.

This includes the time for removing the moulded parts from the mould.

3.11 **cavity**: The part of hollow space of the mould that forms one specimen.

3.12 **single-cavity mould**: Mould with one cavity only (see figure 4).

**3.13 multi-cavity mould:** Mould that contains two or more identical cavities in a parallel flow arrangement (see figure 2 and 3).

The number of cavities of one multi-cavity mould is n.

Identical geometries of flow paths and symmetrical positioning of cavities in the mould ensure that all test specimens from one shot are equivalent in their properties. No series arrangement of cavities is permitted.

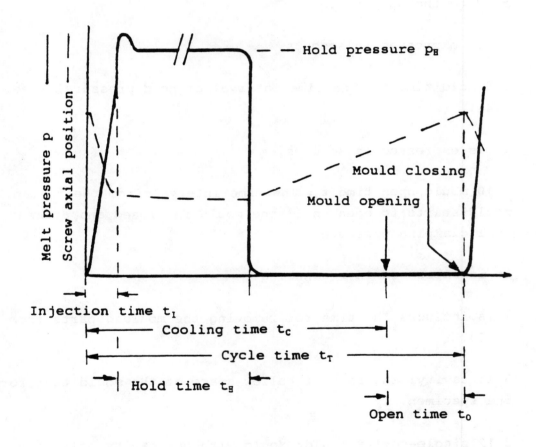

**Figure 1 – Schematic diagram of an injection moulding cycle. Melt pressure (full line) and screw axial position (dotted line) as function of time**

**3.14 family mould:** Mould that contains more than one cavity, not all of which are identical (see figure 5).

**3.15 ISO mould:** Fixed plate with central sprue, combined with a cavity plate forming the central part of a preferred multi-cavity mould, see 3.13.

It is recommended to install interchangeable cavity plates, ejector pins and pressure sensors, which, however, are not mandatory for an ISO mould, see 4.1.1.4. For a complete mould see e. g. annex C.

**3.16 cross section $A_c$:** Cross-sectional area of a cavity of a single- or multi-cavity mould at the position that forms the critical portion of the test specimen.

It is expressed in square millimetres, $mm^2$.

The critical portion of the test specimen for tensile bars e. g. is their narrow section, which bears the most stress during testing.

**3.17 shot volume $V_s$:** The total hollow volume of the mould including the cavity (ies), runner(s) and the sprue.

It is expressed in cubic millimetres, $mm^3$.

For the purpose of this standard it is estimated being the ratio between the mass of the moulding and the density of the solid plastic, i. e. neglecting the influence of moulding shrinkage.

**3.18 projected area $A_p$:** The projected area of the shot volume (see 3.17), projected to the parting surface.

It is expressed in square millimetres, $mm^2$.

**3.19 clamping force $F_M$:** The force clamping the plates of the mould.

It is expressed in kilonewton, kN.

The clamping force is recommended to be higher than the product of the projected area and the maximum melt pressure, see equation (2).

$$F_M \geq A_P \cdot p_{max} \cdot 10^{-3} \qquad\qquad (2)$$

where

$F_M$   is the clamping force (see 3.19), in kN

$A_P$   is the projected area (see 3.18), in mm²

$p_{max}$ is the maximum value of the melt pressure (see 3.3), in MPa

It is expressed in millimetres per second, mm/s.

**3.20 injection velocity $v_I$:** The average velocity of the melt as it passes through the cross section(s) $A_C$ (see 3.16).

It is expressed in millimetres per second, mm/s.

It can be used for single- and multi-cavity moulds only and is calculated according to equation (3).

$$v_I = \frac{V_S}{t_I \cdot A_C \cdot n} \qquad\qquad (3)$$

where

$v_I$   is the injection velocity, in  mm/s

$n$    is the number of cavities, see 3.13

$A_C$   is the cross section, see 3.16, in mm²

$V_S$   is the shot volume, see 3.17, in mm³

$t_I$   is the injection time, see 3.8, in s

# 4 APPARATUS

## 4.1 Moulds

### 4.1.1 ISO moulds (multi-cavity)

4.1.1.1 ISO moulds, see 3.15, are strongly recommended for the use of test specimens and are suitable for use in disputes involving International Standards.

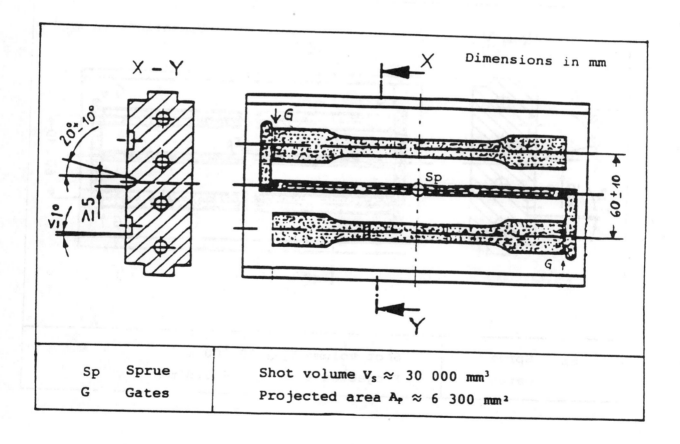

Figure 2 — Cavity plate of ISO mould type A

**4.1.1.2** Multipurpose test specimens conforming to ISO 3167 shall be moulded in the two-cavity ISO mould type A using a Z or a T runner (see annex A) as shown in figure 2 and conforming to 4.1.1.4. From these two types of runners the Z runner is preferred with respect to its better symmetry for the clamping force. The T type may be deleted on occasion of the next revision of this standard. The bars moulded shall have the dimensions of the specimen type A, ISO 3167.

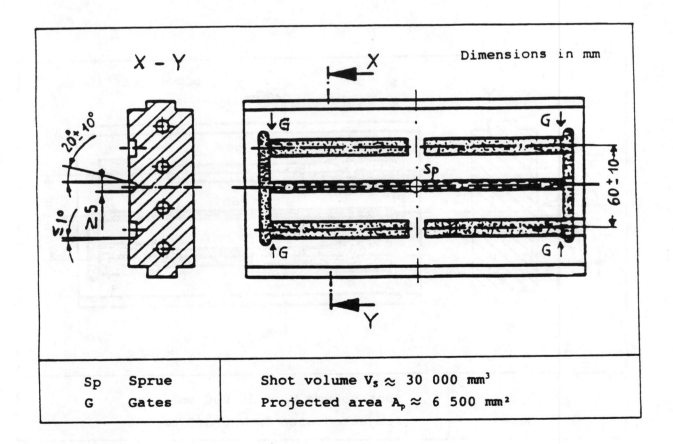

| | | |
|---|---|---|
| Sp | Sprue | Shot volume $V_s \approx 30\ 000$ mm³ |
| G | Gates | Projected area $A_p \approx 6\ 500$ mm² |

Figure 3 - Cavity plate of ISO mould type B

4.1.1.3 Rectangular bars (80 mm x 10 mm x 4 mm) shall be moulded in the four-cavity ISO mould type B with "double-T runner" as shown in figure 3 and conforming to 4.1.1.4. The bars moulded shall have the same cross-section dimensions as the tensile bars at their central part, see ISO 3167, and the length 80 mm ± 2 mm, see note 2.

4.1.1.4 The main construction details of the ISO moulds type A and B shall conform to figures 2 and 3 and to the following common requirements:

a)   The sprue diameter on the nozzle side shall be at least 4 mm.

b)   Both width and height (or diameter) of the runner system shall be at least 5 mm.

c)   The cavities shall be one-end gated as shown in the relevant figures 2 and 3.

d)   The height of the gate shall be at least 2/3 of the height of the cavity at the point where the gate enters the cavity, and the width of the gate shall equal that of the cavity.

e)   The gate shall be as short as possible, not exceeding 3 mm.

f)   The draft angle of runners shall be at least 10°, but not more than 30°. The cavity shall have a draft angle not greater than 1°, except in the area of tensile-specimen shoulders where the draft angle shall be not greater than 2°.

g)   The dimensions of the cavities shall result in test specimens, whose dimensions conform to the requirements given in the relevant testing standard. To allow for different levels of moulding shrinkage select the dimensions of the cavity between the nominal and the upper values of the relevant fixed dimensions for the test specimen. In case of the ISO moulds A and B the essential dimensions of the cavities in mm are as follows (see ISO 3167).

| | |
|---|---|
| Thickness | 4,0 to 4,2 |
| (Central) Width | 10,0 to 10,2 |
| Length (mould B) | 80 to 82 |

h) Ejector pins, if used, shall be located outside the test area of the test specimen, i. e. at the shoulders of tensile bars of the ISO mould type A and of type C, see ISO 294-2, outside the central 20 mm length of the bars of the ISO mould type B and outside the central area of 50 mm diameter of the plates of the ISO mould type D, see ISO 294-3.

i) The heat transfer system of the mould plates shall be designed so that differences in temperature at any point on a cavity surface and between both plates of the mould, under operating conditions are less than 5 °C.

k) Interchangeable cavity plates and gate inserts are recommended to permit rapid changes from production of one type of test specimen to production of another type. This is made easy by taking values of the shot volume $V_s$ as similar as possible. A perspective drawing of an example is given in Annex A.

l) It is recommended to install a pressure sensor within the central runner, which gives proper controlling of the injection period and which is mandatory for Part 4 of this International Standard. For a position suitable for common use with different ISO moulds, see 294-3, 4.1, l) and figure 2.

86

m) For a common layout of interchangeable cavity plates suitable for the different ISO moulds take notice of the figures 2 and 3, the parts 2 and 3 of this International Standard and additional details as follows.

- It is recommended to use the cavity length of 170 mm for the multipurpose-test specimens to be moulded in the ISO mould type A. This results in the maximum length of 180 mm that commonly can be used for hollow volume within the cavity plates.

- The value for a common width of the mould plates results from the minimum distance of couplings of the cooling channels. Additionally the optional installation of inserts may be respected for the ISO mould type B, suitable for the injection moulding of notched bars, see ISO 179.

- Common cutting lines, e. g. 170 mm apart, may be provided in order to separate the test specimens from the runners for the ISO moulds type A, B and C (see Part 2). A second pair of cutting lines 80 mm apart may be suitable for taking bars from multipurpose-test specimens (mould type A) as well as for separating injection moulded plates, see Part 3.

n) For critical checking of the symmetry of multi-cavity moulds it is recommended to mark the individual cavities but outside the test area of the test specimen, see h). This e. g. may simply be provided by suitably engraving symbols on the heads of the ejector pins, thus avoiding any damage of the surface of the cavity plate.

o) A mould consists of many parts described in other International Standards, a survey of which is given in annex B.

## 4.1.2 Single cavity moulds

The cavity of a single cavity mould (see figure 4 and 3.12) may be that of a dumbbell bar, a disc, or other desired shape. A single cavity mould generally gives values for some specific properties different from those based upon the recommended ISO moulds, see note 2.

> NOTE 2 - This may occur because the ratio of the cavity volume to the total volume of the mould may be different from those of the ISO moulds. Also, the smaller total volume of this type of mould makes conformance with the volume ratio requirements of 4.2.1 different and failure to conform to these requirements may contribute to erratic values.

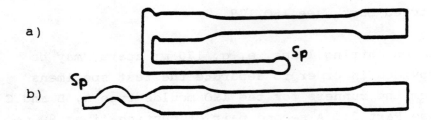

Figure 4 - Examples of single cavity mouldings, with sprue Sp
a)   normal to the moulding plate and
b)   parallel, at the parting surface. (The deflecting runner prevents from jetting)

## 4.1.3 Family moulds

A family mould (see figure 5 and 3.13) may contain, for example, flat bars combined with dumbbell bars and discs. A family mould may be used when the properties of test specimens obtained correspond to those obtained when using ISO moulds, see note 3.

NOTE 3 - In most cases, simultaneous and steady filling of the different types of cavities is not possible under different moulding conditions with a family mould. This is why this type of mould is not suitable in principle for reference test specimens preparation. Additionally the injection velocity $v_I$ (see 3.20) can not be defined using family moulds.

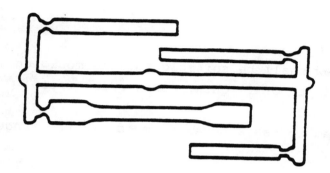

Figure 5 - Example of a family moulding

## 4.2  Injection moulding machine
### 4.2.1 Screw-stroke volume [1]

For preparation of reproducible and comparable test specimens, only reciprocating screw injection moulding machines equipped with the necessary devices for control and maintenance of conditions shall be used. The ratio of the shot volume $V_S$ of the mould, see 3.17, to the screw-stroke volume of the machine should be between 20 % and 80 %, unless a higher ratio may be required in the relevant International Standard for the material or in the manufacturers' recommendations.

### 4.2.2 Control system

The control system of the machine shall be capable of maintaining the operating conditions within the following ranges:

| | | | |
|---|---|---|---|
| Injection time | $t_i$ (see 3.8) | ± | 0,1 s |
| Hold pressure | $p_H$ (see 3.4) | ± | 5 % |
| Hold time | $t_H$ (see 3.9) | ± | 5 % |
| Melt temperature | $T_M$ (see 3.2) | ± | 3 °C |
| Mould temperature | $T_C$ (see 3.1) | ± | 3 °C up to 80 °C |
| | | ± | 5 °C above 80 °C |
| Mass of the moulding | | ≤ | 2 % |

### 4.2.3 Screw

The type of screw shall be suitable for the moulding material
(e. g. length, depth of flights, compression ratio)

It is recommended to use a screw diameter in the range between
18 mm and 40 mm.

### 4.2.4 Clamping force

The clamping force $F_M$ of the machine shall be great enough to
prevent flashing under all operating conditions.

According to clause 3.19 the minimum clamping force $F_M$ for the
ISO moulds type A and B is $F_M/kN \geq 6,5 \cdot p_{max}/MPa$, e. g. 520 kN
for 80 MPa.

For an injection moulding system with interchangeable moulds
also take into account Part 3 of this International Standard
i. e. the ISO mould type D with $A_P \approx 11\ 000\ mm^2$.

### 4.2.5 Calibrated thermometers

A calibrated, needle-type thermometer accurate to ± 1 °C shall
be used to measure the melt temperature $T_M$ (see 3.2). A calibrated surface thermometer accurate to ± 1 °C shall be used to
measure the temperatures of the surface of the mould, which
gives the mould temperature $T_C$ according to 3.1.

# 5 PROCEDURE

## 5.1 Conditioning of materials

Condition the pellets or granules of thermoplastics materials prior to moulding, as required in the relevant International Standard for the material, or according to the manufacturers' recommendations, if no International Standard covers this subject.

Avoid exposing materials to an atmosphere with temperatures significantly below the temperature of the workshop to prevent condensation of moisture onto the plastics material.

## 5.2 Injection moulding

5.2.1 Set the machine to the conditions specified in the relevant material International Standard or by agreement of the interested parties if no International Standard covers this subject.

5.2.2 For many thermoplastics, the suitable range of the injection velocity $v_I$ is 200 mm/s ± 100 mm/s using the ISO moulds type A and type B.

5.2.3 For adjustment of the commonly not specified parameter hold pressure $p_H$ modify the pressure until the mouldings are free from sink marks, voids and other visible faults and have minimal flash.

5.2.4 Ensure that the hold pressure is maintained constant during the hold time $t_H$ which is selected until the plastic material in the gate section has frozen, i. e. until the mass of the moulding has reached an upper limiting value under these conditions. Changes of the screw speed during the injection period are to be kept as low as possible.

5.2.5 Discard the mouldings until the machine has reached steady state conditions. Then record the operating conditions and begin specimen collection.

During the moulding process, maintain steady state conditions by suitable means, e. g. by control of the mass of the moulding.

5.2.6 After any change of material, clean the machine carefully and thoroughly.

## 5.3  Measurement of mould temperature

Determine the mould temperature $T_c$ after the system has attained thermal equilibrium and immediately after opening the mould. Carry out measurements of the temperature of the mould cavity surface at several cavity points on both the opposite sides of the mould using a calibrated surface pyrometer. Between each pairs of these multiple readings the mould shall be recycled for a minimum of ten cycles. Record each measurement individually and calculate the mould temperature as the average of these measurements.

## 5.4  Measurement of the melt temperature

Measure the melt temperature $T_M$ as follows. After thermal equilibrium or steady state temperature has been attained inject a free shot of 30 ccm minimum into a non-metallic container of suitable size and insert the needle of a rapid response, preheated, needle thermometer immediately into the centre of the molten plastic cake and move it gently until the reading of the thermometer has reached a maximum. Ensure that the preheating temperature is sufficiently close to the melt temperature to be measured. Use the same injection conditions for the free shot as those to be used to mould the specimens including allowance of identical cycle time between free shots.

The melt temperature may alternatively be measured by means of a suitable temperature sensor, provided the values obtained are the same as those by the free shot method. The sensor should allow only low heat loss and should respond rapidly to melt temperature changes. The sensor has to be mounted in a suitable place, such as in the nozzle of the injection moulding machine. In case of doubt, however, the method described in the first paragraph counts.

## 5.5  Post-moulding treatment of test specimens

Allow test specimens removed from the mould to cool gradually and uniformly to room temperature in order to avoid any differences in treatment of individual test specimens. Specimens made of thermoplastics sensitive to atmospheric exposure are to be protected by use of impermeable containers perhaps adding a desiccant.

## 6  PRECISION

The precision of this test method is not known, because interlaboratory data are not available. When interlaboratory data are obtained, a precision statement will be added with the next revision.

## 7  REPORT

The report shall include the following items:
a)   a reference to this part of this International Standard
b)   date, time, and place of moulding;
c)   full description of material (type, designation, manufacturer, lot number);
d)   conditioning of the material prior to moulding, if applicable;
e)   mould type A, B or details in case of another mould

(specimen type, relevant International Standard, number of cavities, gate size and location);

f)  injection moulding machine details (manufacturer, maximum stroke volume, clamping force, control systems);

g)  moulding conditions:

-   melt temperature $T_M$, see 3.2, in degrees celsius, °C
-   mould temperature $T_C$, see 3.1, in degrees celsius, °C
-   injection velocity $v_I$, see 3.20, in millimetres per second, mm/s
-   injection time $t_I$, see 3.8, in seconds, s
-   hold pressure $p_H$, see 3.4, in megapascals, MPa
-   hold time $t_H$, see 3.9, in seconds, s
-   cooling time $t_C$, see 3.7, in seconds, s
-   cycle time $t_T$, see 3.6, in seconds, s
-   mass of the moulding, in grams, g

h)  other relevant details (e. g. number of mouldings initially discarded, number retained, post-moulding treatment).

# Annex A
## (informative)
## Example for runner configurations

The layout of the mould may be changed by means of gate inserts as shown in figure A1

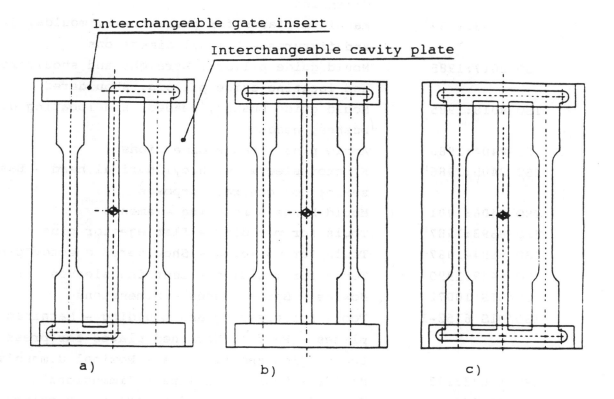

**Figure A 1 – Variations of the gate position**

a)   Injection mould as recommended in this International Standard (Z-runner)

b)   Variation with T-runner (not recommended)

c)   Variation with double-T-runner (e. g. for studying weld line strength .

## Annex B
### (informative)
### Standardized parts of injection moulds

The following list of International Standards for parts of mouldings has been selected from the document ISO/TC 29/SC 8, N 314 (1993-04-13) "Report of the work of ISO/TC 29/SC 18 - Tools for pressing and moulding"

| | |
|---|---|
| ISO 6751:1986[1] | Ejector pins with cylindrical head - Basic dimensions |
| ISO 6753:1982[2] | Machined plates for press tools, moulds, jiggs and fixtures - Nominal dimensions |
| ISO 8017:1985 | Mould guide pillars, straight and shouldered, and locating guide pillars, shouldered |
| ISO 8018:1985 | Mould guide bushes, headed, and locating guide bushes, headed |
| ISO 8404:1986 | Angle pins - Basis dimensions |
| ISO 8405:1986[1] | Ejector sleeves with cylindrical head - Basic series for general purposes |
| ISO 8406:1991 | Mould bases - Locating elements |
| ISO 8693:1987[1] | Tools for moulding - Flat ejector pins |
| ISO 8694-1987[1] | Tools for moulding - Shouldered ejector pins |
| ISO 9449:1990 | Tools for moulding - Centring sleeves |
| ISO/DIS 10072 | Moulds - Sprue bushes - Dimensions |
| ISO/DIS 6753-1 | Tools for pressing and moulding - Machined plates - Part 1: Machined plates for press tools, jigs and fixtures - Nominal dimensions |
| ISO 10073:1991 | Moulds - Support pillars - Dimensions |
| ISO/CD 6753-2 | Tools for pressing and moulding - Machined plates - Part 2: Machined plates for moulds - Type B, C, D, E and F - Nominal dimensions |
| ISO/CD 10907-1 | Tools for moulding - Locating rings - Part 1: Locating rings for mounting without thermal insulating sheets in small or medium moulds - Type A and B |

---

1) Proposed for revision

2) Under revision: ISO/DIS 6753-1 and ISO/CD 6753-2

## Annex C
## (informative)
## Example of a mould system

Figure C. 1 shows a spatial view of a mould with the interchangeable two-cavity plate: ISO mould type A.

## Annex D
## (informative)
## Bibliography

[1]  Johannaber, F., Kunststoffe, German Plastics 79 (1989), 1, 15 - 28

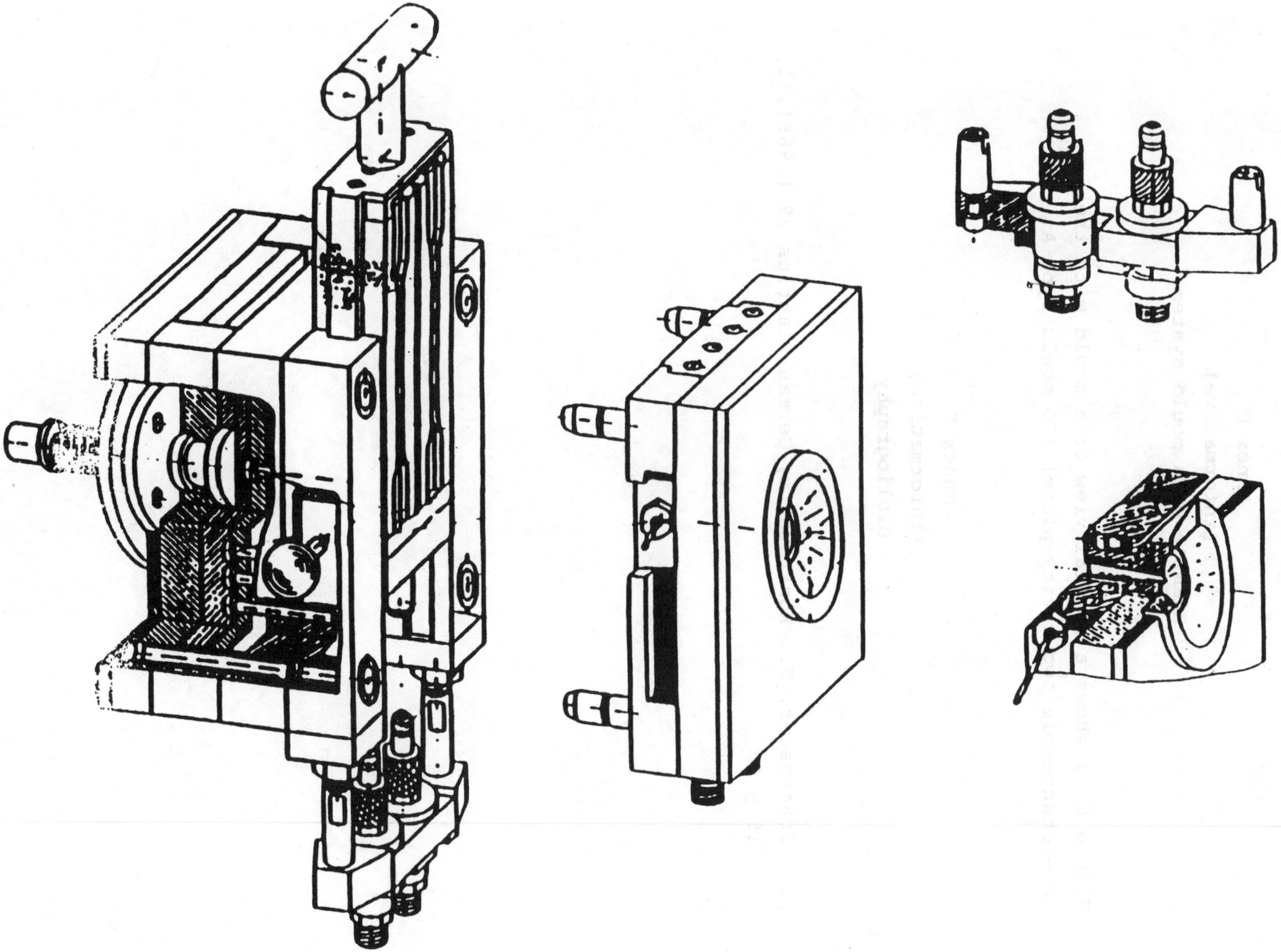

Figure C 1 – A spatial view of an example of the ISO mould type A

**EXPLANATORY REPORT**
**RAPPORT EXPLICATIF**    ISO/ DIS 294-1

will supersede:
remplacera:

ISO/TC 61 /SC 9     Secretariat

ANSI (USA)

| This form should be sent to the ISO Central Secretariat, together with the English and French versions of the committee draft, by the secretariat of the technical committee or sub-committee concerned (see 2.5.9 of part 1 of the ISO/IEC Directives) | Ce formulaire doit être envoyé au Secrétariat central de l'ISO en même temps que les versions anglaise et française du projet de comité, par le secrétariat du comité technique ou du sous-comité concerné (voir 2.5.9 de la partie 1 des Directives ISO/CEI) |

| The accompanying document is submitted for circulation to member body vote as a DIS, following consensus of the P-members of the committee obtained | Le document ci-joint est soumis, pour diffusion comme DIS, au vote comité membre, suite au consensus des membres (P) du comité obtenu |

on
le   1994-09-23

[X] at the meeting of TC.61. /SC..9...: see resolution No. ............... in document 61/9 N 740
à la réunion du    voir   n° ............... dans le   document 61/9 N 740

             61 N 4934

[ ] by postal ballot initiated on
par un vote par correspondance démarré le    19........

P-members in favour:    Canada, Czech Republic, Finland, France, Germany, India
Membres (P) approuvant le projet:   Italy, Japan, Netherlands, Switzerland, UK, USA
SC Plenary

P-members voting against:
Membres (P) désapprouvant:

P-members abstaining:
Membres (P) s'abstenant:

P-members who did not vote:
Membres (P) n'ayant pas voté:

**Remarks Remarques**

The results of voting and comments received on CD 294-1, CD 294-2, CD 294-3 and CD 294-4 were discussed at the meeting of SC 9/WG 18 in Tokyo, September 1994. All negatives were resolved and the WG recommended that modified texts be advanced to DIS level. Thus, we hereby submit these texts for your consideration.

I hereby confirm that this draft meets the requirements of part 3 of the ISO/IEC Directives
Je confirme que ce projet satisfait aux prescriptions de la partie 3 des Directives ISO/CEI

Date           Name and signature of the secretary
                Nom et signature du secrétaire
1995-02-10                    Julia Lindsay
                          Secretariat ISO/TC 61/SC 9

FORM 8A (ISO)                                       FORMULAIRE 8A (IS

ISO/TC **61**/SC **9**

Secretariat: **ANSI**

Voting begins on
1995-03-23

Voting terminates on
1995-09-23

INTERNATIONAL ORGANIZATION FOR STANDARDIZATION • МЕЖДУНАРОДНАЯ ОРГАНИЗАЦИЯ ПО СТАНДАРТИЗАЦИИ • ORGANISATION INTERNATIONALE DE NORMALISATION

# Plastics — Injection moulding of test specimens of thermoplastic materials —

## Part 3:
Plates (ISO type D moulds)

(Revision in part of ISO 294:1995)

*Plastiques — Moulage par injection des éprouvettes de matériaux thermoplastiques —*

*Partie 3: Plaques (moules ISO de type D)*

ICS 83.080.20

Descriptors: plastics, thermoplastic resins, moulding materials, test specimens, injection moulding, specimen preparation, reference data.

---

**To expedite distribution, this document is circulated as received from the committee secretariat. ISO Central Secretariat work of editing and text composition will be undertaken at publication stage.**

**Pour accélérer la distribution, le présent document est distribué tel qu'il est parvenu du secrétariat du comité. Le travail de rédaction et de composition de texte sera effectué au Secrétariat central de l'ISO au stade de publication.**

---

ISO/DIS 294-3

Introduction

See ISO 294-1

## 1 SCOPE

See ISO 294-1, clause 1

In this part of this International Standard the two-cavity ISO moulds type D1 and D2 are defined for injection moulding plates 60 mm x 60 mm with the preferred thicknesses of 1 mm (type D1) and 2 mm (type D2) which can be used for a variety of tests, see annex A. The moulds may additionally be equipped by inserts for studying the action of weld lines, see annex B.

## 2 NORMATIVE REFERENCES

The following standards contain provisions which, through reference in this text, constitute provisions of this International Standard. At the time of publication, the editions indicated were valid. All standards are subject to revision and parties to agreements based upon this International Standard are encouraged to use the most recent edition of the standards listed below. Members of IEC and ISO maintain registers of current valid International Standards.

ISO 294-1[1], Plastics - Injection moulding of test specimens of thermoplastic materials - Part 1: General principles, multipurpose-test specimens (ISO mould type A) and bars (ISO mould type B)

ISO 294-2[2], Plastics - Injection moulding of test specimens of thermoplastic materials - Part 2: Small tensile bars (ISO mould type C)

ISO 294-4[2], Plastics - Injection moulding of test specimens of thermoplastic materials - Part 4: Determination of moulding shrinkage

ISO 6603-1: 1985, Plastics - Determination of multiaxial impact behaviour of rigid plastics - Part 1: Falling dart method

ISO 6603-2: 1989, Plastics - Determination of multiaxial impact behaviour of rigid plastics - Part 2: Instrumented puncture test

[1]    Revision of ISO 294: 1994
[2]    In preparation


## 3    DEFINITIONS

See ISO 294-1, clause 3

# 4    APPARATUS

## 4.1   ISO moulds type D

Plates shall be moulded in the two-cavity ISO moulds type D, see figure 1. The plates moulded by type D2 shall have the dimensions as given in figure 2.

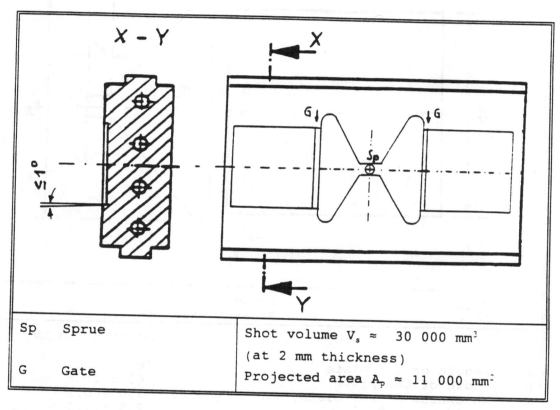

| Sp | Sprue | Shot volume $V_s \approx$ 30 000 mm³ |
| | | (at 2 mm thickness) |
| G | Gate | Projected area $A_p \approx$ 11 000 mm² |

Figure 1 - Cavity plates of ISO moulds type D

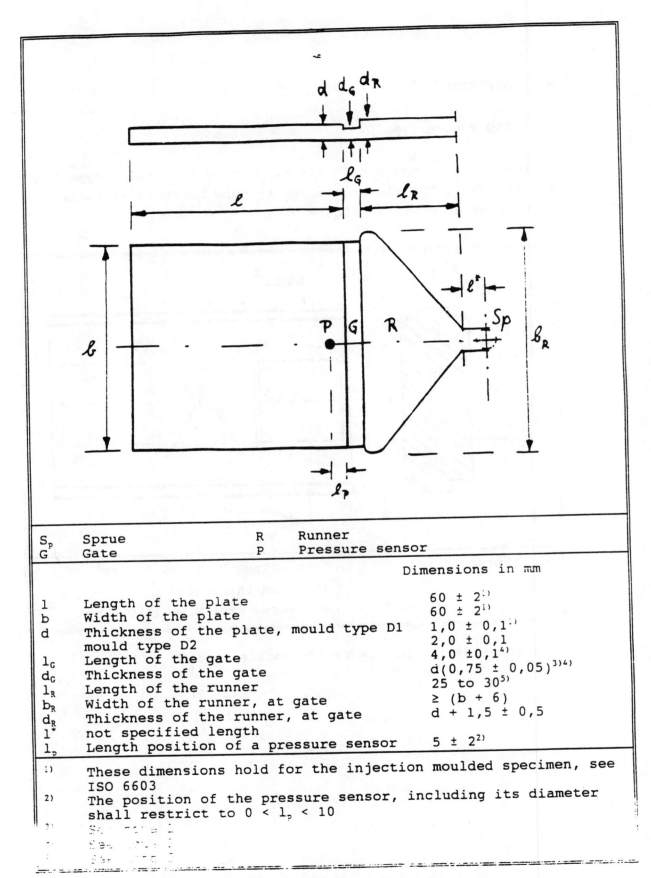

| | | Dimensions in mm |
|---|---|---|
| $S_p$ | Sprue | R Runner |
| G | Gate | P Pressure sensor |

| | | Dimensions in mm |
|---|---|---|
| l | Length of the plate | $60 \pm 2^{1)}$ |
| b | Width of the plate | $60 \pm 2^{1)}$ |
| d | Thickness of the plate, mould type D1 | $1,0 \pm 0,1^{1)}$ |
| | mould type D2 | $2,0 \pm 0,1$ |
| $l_G$ | Length of the gate | $4,0 \pm 0,1^{4)}$ |
| $d_G$ | Thickness of the gate | $d(0,75 \pm 0,05)^{3)4)}$ |
| $l_R$ | Length of the runner | $25$ to $30^{5)}$ |
| $b_R$ | Width of the runner, at gate | $\geq (b + 6)$ |
| $d_R$ | Thickness of the runner, at gate | $d + 1,5 \pm 0,5$ |
| $l^*$ | not specified length | |
| $l_p$ | Length position of a pressure sensor | $5 \pm 2^{2)}$ |

[1] These dimensions hold for the injection moulded specimen, see ISO 6603

[2] The position of the pressure sensor, including its diameter shall restrict to $0 < l_p < 10$

[3] See ...

[4] See ...

Figure 2 – Details of the ISO mouldings type D1 and D2

104

The main construction details of the ISO moulds type D shall conform to figures 1 and 2 and to the following requirements.

a)   See ISO 294-1, clause 4.1.1.4, a)
b)   Not relevant
c)   See ISO 294-1, clause 4.1.1.4, c)
d)   and e) Not relevant
f)   See ISO 294-1, clause 4.1.1.4, f)
g)   See ISO 294-1, clause 4.1.1.4, g) but relating to
     ISO 6603

   The essential dimensions in mm are as follows (see figure 2)

   Length                60 to 62

   Width                 60 to 62

   Thickness, type D2   2,0 to 2,1

   Thickness, type D1   1,0 to 1,1

h), i) and k) See ISO 294-1, clause 4.1.1.4, h), i) and k
l)   Figure 2 shows the position of a pressure sensor P within the
     cavity, which is mandatory for the measurement of moulding
     shrinkage only, see ISO 294-4. It may be suitable, however, for
     controlling the injection period using any ISO mould, see Part
     1, clause 4.1.1.4, l). The pressure sensor shall be coplanar
     with the mould surface in order to prevent disturbance from
     flow.
m) to o)  See Part 1, clause 4.1.1.4, m) to o).

NOTES

1 - Gates with strongly reduced thickness have a great influence on the orientation of the material within the plate up to large distances. The thickness at the gate, therefore, has been chosen at a minimum value to facilitate mechanically measuring the mould shrinkage, see ISO 294-4.

2 - The thickness and the length of the gate strongly influence the process of freezing of the melt flow into the cavity and thus the mould shrinkage, see Part 4 of this International Standard. The dimensions of the gate, therefore, are defined with low tolerances.

3 - The indicated value of the gate length $l_G$ allows separating the plates by stamping or sawing using fixed distances even for materials with different shrinkages.

4 - Taking the distance between cutting lines at 80 mm gives the advantage that the cutting machine can be used too for taking bars 80 mm x 10 mm x 4 mm from the central part of the multipurpose-test specimen, see ISO 294-1, clause 4.1.1.4, m).

## 4.2  Injection moulding machine

See ISO 294-1, clause 4.2 with the following exception in clause 4.2.4 Clamping force:

The recommended minimum clamping force $F_M$ for the ISO moulds type D is

$F_M/kN \geq 11 \cdot p_{max}/MPa$, e. g. $F_M \geq 880$ kN for $P_{max} = 80$ MPa.

## 5   PROCEDURE

### 5.1  Conditioning of materials

See ISO 294-1, clause 5.1

### 5.2  Injection moulding

See ISO 294-1, clause 5.2 but with the following change in clause 5.2.2.

For the ISO moulds type D1 and D2 the suitable ranges of injection velocity are recommended to be selected such that the injection times $t_I$ are comparable to that used for the ISO mould type A.

## 6   PRECISION

see ISO 294-1, clause 6

## 7   REPORT

see ISO 294-1, clause 7

## Annex A
### (informative)

Recommended applications for the plates or parts thereof. For references see clause 2 and annex C.

The ISO mould type D2 is recommended for preparation of test specimens to be used for determining multiaxial impact strength according to ISO 6603, see note 5, moulding shrinkage according to ISO 294-4, optical properties, see note 6, for preparing samples for coloured plastics, see note 7, for studying mechanical anisotropy, see note 8, and may be equipped by inserts for studying the effects of weld lines, see annex B.

The ISO mould type D1 is especially suitable for processing test specimens to be used for determining electrical properties, see note 9, for estimating the absorption of water, see note 10, and dynamic-mechanical properties, see note 11.

NOTES

5 - It is proposed to include the multiaxial impact strength into the mechanical, comparable multipoint data according to ISO 11403-1 [9]. The recommended specimen thickness is 2 mm.

6 - Plates produced from natural or essentially colourless materials are suitable for determining the refractive index [2] and luminous transmittance [10,11].

7 - Plates produced from coloured or natural materials are suitable for determining optically and mechanically indicative properties in order to study the influence of weathering according e. g. to ISO 4892 [5].

8 - Hour-glass specimens type 4 according to ISO 8256 [7], taken at different positions and directions from plates by machining according to ISO 2818 [4] are suitable for studying mechanical anisotropy by tensile and tensile-impact testing according to ISO 527-1 [3] and ISO 8256 [7] respectively.

9 - ISO 10350 [8] recommends the measurement of the electrical properties: relative permittivity and dissipation factor, volume resistivity, surface resistivity and electric strength using a plate of 1 mm thickness and ≥ 80 mm square shape. Almost all these properties, however, can be estimated too based upon 60 mm plates, for surface resistivity e. g. by fitting the geometry of the electrodes to the smaller plate and for electric strength e. g. by using spherical instead of cylindrical electrodes.

10 - ISO 62 [1] describes the determination of water absorption using a plate of 3 mm thickness immersed in water during a defined time. Plates of lower thickness, however, for many plastics allow the estimation of saturation values.

11 - ISO 6721-2 [6] describes the determination of the complex shear modulus using a torsion pendulum and specimens with the preferred thickness of 1 mm. These can be taken from plates processed by the ISO mould type D1.

# Annex B
## (informative)
## Weld lines

For references see clause 2 and annex C

Providing suitable inserts in the cavities, the effects of weld lines on mechanical properties can be studied, see figures B.1 and B.2.

Figure B.1 shows a single insert near gate, the weld line of which is generated by parallel melt flow. As shown by dotted lines, small tensile specimens type 4 according to ISO 8256 [7] ("hour-glass specimens") can be taken from the plate, which allow the effect of the weld line to be studied, using tensile or tensile impact testing according to ISO 527-1 [3] and ISO 8256 [7] and dependent on the distance from the insert.

Figure B.2 shows the use of multi-inserts,which generate weld lines with normal flow at contact and with different length of flow path prior to contact. The parallel melt flow of figure B.1 and the normal one of figure B.2 represent the basic types of weld lines. In each case only symmetric arrangements of the two-cavity moulds shall be used.

Figure B.1 - Moulding with single inserts hatched and machined hour glass specimens indicated by dotted lines.

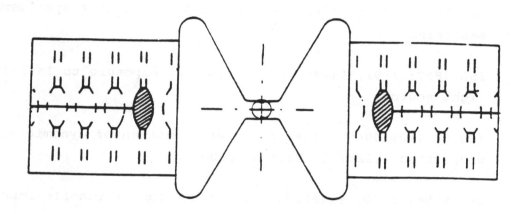

Figure B.2 - Moulding with multi-inserts hatched and machined hour glass specimens indicated by dotted lines.

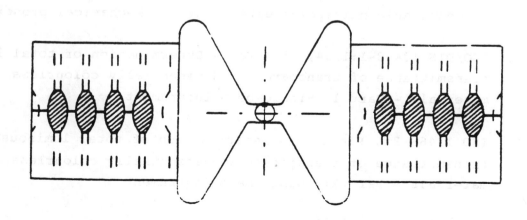

# Annex C
## (informative)
## Bibliography

[1]   ISO 62:1980, Plastics - Determination of water absorption

[2]   ISO 489:1983, Plastics - Determination of the refractive index
      of transparent plastics

[3]   ISO 527-1: 1993, Plastics - Determination of tensile
      properties - Part 1: General principles

[4]   ISO 2818: 1994[1], Plastics - Preparation of test specimens by
      machining

[5]   ISO 4892: 1991, Plastics - Methods of exposure to laboratory
      light sources

[6]   ISO 6721-2:1994[1], Plastics - Determination of dynamic-mechanical
      properties - Part 2: Torsion pendulum

[7]   ISO 8256: 1990, Plastics - Determination of tensile-impact
      strength

[8]   ISO 10350:1994[1], Plastics - The acquisition and presentation of
      camparable single-point data

[9]   ISO 11403-1: 1994[1], Plastics - The acquisition and presentation
      of comparable multipoint data - Part 1: Mechanical properties

[10]  ISO/DIS 13468-1:1994, Plastics - Determination of total luminous
      transmittance of transparent and essentially colourless
      materials - Part 1: Single-beam instrument

[11]  ISO 13468-2[2], Plastics - Determination of total luminous
      transmittance of transparent and essentially colourless
      materials - Part 2: Double-beam instrument

      [1]   To be published.
      [2]   In preparation

112

EXPLANATORY REPORT
RAPPORT EXPLICATIF

ISO/ DIS 294-3

will supersede :
remplacera :

ISO/TC 61 /SC 9

Secretariat
ANSI (USA)

This form should be sent to the ISO Central Secretariat, together with the English and French versions of the committee draft, by the secretariat of the technical committee or sub-committee concerned (see 2.5.9 of part 1 of the ISO/IEC Directives)

Ce formulaire doit être envoyé au Secrétariat central de l'ISO en même temps que les versions anglaise et française du projet de comité, par le secrétariat du comité technique ou du sous-comité concerné (voir 2.5.9 de la partie 1 des Directives ISO/CEI)

The accompanying document is submitted for circulation to member body vote as a DIS, following consensus of the P-members of the committee obtained

Le document ci-joint est soumis, pour diffusion comme DIS, au vote comité membre, suite au consensus des membres (P) du comité obtenu

on
le   1994=09=23

[X]  at the meeting of TC.61. /SC.9...:  see  resolution No. ............  in  document  61 N 4934
     à la réunion du                     voir            n°             dans le           61./9 N 740

[ ]  by postal ballot initiated on
     par un vote par correspondance démarré le       19 .. - .. ..

P-members in favour:
Membres (P) approuvant le projet:
SC Plenary

Canada, Czech Republic, Finland, France, Germany, India
Italy, Japan, Netherlands, Switzerland, UK, USA

P-members voting against:
Membres (P) désapprouvant:

P-members abstaining:
Membres (P) s'abstenant:

P-members who did not vote:
Membres (P) n'ayant pas voté:

Remarks/Remarques

The results of voting and comments received on CD 294-1, CD 294-2, CD 294-3 and CD 294-4 were discussed at the meeting of SC 9/WG 18 in Tokyo, September 1994. All negatives were resolved and the WG recommended that modified texts be advanced to DIS level.  Thus, we hereby submit these texts for your consideration.

I hereby confirm that this draft meets the requirements of part 3 of the ISO/IEC Directives
Je confirme que ce projet satisfait aux prescriptions de la partie 3 des Directives ISO/CEI

Date

1995-02-10

Name and signature of the secretary
Nom et signature du secrétaire

Julia Lindsay
Secretariat ISO/TC 61/SC 9

FORM 8A (ISO)

FORMULAIRE 8A (ISO)

F/F 92 01

ISO/TC **61**/SC **9**

Secretariat: **ANSI**

Voting begins on
1995-03-23

Voting terminates on
1995-09-23

INTERNATIONAL ORGANIZATION FOR STANDARDIZATION • МЕЖДУНАРОДНАЯ ОРГАНИЗАЦИЯ ПО СТАНДАРТИЗАЦИИ • ORGANISATION INTERNATIONALE DE NORMALISATION

# Plastics — Injection moulding of test specimens of thermoplastic materials —

## Part 4:
Determination of moulding shrinkage

(Revision in part of ISO 294:1995)

*Plastiques — Moulage par injection des éprouvettes de matériaux thermoplastiques —*

*Partie 4: Détermination du retrait au moulage*

ICS 83.080.20

Descriptors: plastics, thermoplastic resins, moulding materials, test specimens, injection moulding, specimen preparation, determination, shrinkage.

Project

ISO/DIS 294-4        Plastics - Injection moulding of test specimens
of thermoplastic materials - Part 4: Determination of moulding
shrinkage


Foreword

The measurement of mould and post shrinkage in case of thermoplastic
materials is described in this part 4 of the International Standard
ISO 294. The reason for this is that shrinkage is strongly dependant
on the geometry of the mould and the conditions of injection moulding. Thus the part 1 (general principles) and the part 3 (mould for
plates) of this International Standard form the necessary conditions
for defining data for shrinkage.

In contrast to the New Work Item Proposal NP 605 (ISO/TC 61, N 4849)
the present proposal does not include the measurement of filling
pressure. A German steering committee, which presently prepares the
documents for all parts of ISO 294 (editor: Breuer) proposes to delay
the standardisation of filling pressure with respect to the following
two reasons:

-    For comparable values of the filling pressure the injection
     moulding apparatus needs precise controlling of the screw speed,
     which is not yet available in many laboratories worldwide.

-    The reliability of commonly used pressure sensors is not yet
     sufficiently high for defining calibrated, comparable pressure
     measurements.

For the cavity pressure $p_c$, which must be used for shrinkage measurements only, the rather wide range of $\pm 5 \%$ has been defined.

## Introduction

See ISO 294-1.

In moulding of thermoplastic materials the difference between the dimensions of the cavity and of the moulded part produced therein from a given material may vary according to the design and operation of the mould. The difference may vary with the size of the moulding machine, the shape and dimensions of moulded sections, the degree and direction of flow or movement of material in the mould, the size of the nozzle, sprue, runner and gate, the cycle on which the machine is operated, the temperature of the melt and the mould, and the height and duration of hold pressure. Shrinkages will approach a minimum where design and operation are such, that a maximum amount of material is forced into the cavity and while still under pressure, as a result of the use of a runner, sprue and nozzle of proper size, along with proper hold conditions. Mould shrinkage and post shrinkage are generated by crystallisation, thermal contraction, volume relaxation and orientation relaxation of the material. The post shrinkage additionally may be influenced by humidity uptake.


## 1    SCOPE

See ISO 294-1, clause 1.

This Part of this International Standard defines a method of determining the moulding shrinkage and the post shrinkage during conditioning and/or treatment of injection moulded test specimens of thermoplastic materials, with regard to anisotropy.

In contrast to ISO 2577 that describes the determination of shrinkage for thermosetting materials, the present standard excludes the effects of humidity uptake on moulding shrinkage. As far as post shrinkage is generated by the uptake of humidity only, see also ISO 175.

Moulding shrinkage and post shrinkage are usefull for checking uniformity of manufacture of thermoplastic material. The method should not be used as a source of data for design calculations of components. Information on the typical behaviour of a material can be obtained, however, by testing at different melt and mould temperatures, injection velocities, hold pressures and values of other injection moulding parameters.

This experience about moulding shrinkage is important for the construction of moulds and knowledge of post shrinkage for establishing the suitability of the moulding material for the production of moulded parts with accurate dimensions.


## 2    NORMATIVE REVERENCES

The following standards contain provisions which, through reference in this text, constitute provisions of this International Standard. At the time of publication, the editions indicated were valid. All standards are subject to revision and parties to agreements based upon this International Standard are encouraged to use the most recent edition of the standards listed below. Members of IEC and ISO maintain registers of current valid International Standards.

ISO 175: 1981, Plastics - Determination of resistance to liquid chemicals

ISO 291: 1977, Plastics - Standard atmospheres for conditioning and testing

ISO 294-1[1], Plastics - Injection moulding of test specimens of thermoplastic materials - Part 1: General principles, multipurpose-test specimens (ISO mould type A) and bars (ISO mould type B).

ISO 294-3[2], Plastics - Injection moulding of test specimens of thermoplastic materials - Part 3: Plates (ISO mould type D).

ISO 2577: 1984, Plastics - Thermosetting moulding materials - Determination of shrinkage.

1)    Revision of ISO 294: 1994
2)    In preparation

## 3      DEFINITIONS

3.1 to 3.20: See ISO 294-1, clause 3.

3.21 moulding shrinkage $S_M$: The difference in dimensions between a dry moulding and the mould cavity in which it was moulded, both the mould and the moulding being at room temperature when measured, normalized to the dimension of the mould.

It is expressed in percent, %.

The moulding shrinkage shall be measured parallel to the melt flow, $S_{Mp}$, at half width of the test specimen and normal to the flow, $S_{Mn}$ at half length of the specimen, both taken at the same side of the test specimen.

**3.22 post shrinkage $S_P$:** The relative difference in dimensions of a moulded part according to 3.21, but taken between the beginning and the end of a post treatment, following the measurement of moulding shrinkage.

It is expressed in percent, %.

For the post shrinkage $S_{Pp}$ parallel and $S_{Pn}$ normal to the flow, see 3.21.

**3.23 total shrinkage $S_T$:** The relative difference in dimensions according to 3.21, but taken between the mould and the specimen after post treatment.

It is expressed in percent, %.

For the total shrinkage $S_{Tp}$ parallel and $S_{Tn}$ normal to the flow, see 3.21.

**3.24 cavity pressure $p_C$:** The pressure of the melt in the cavity, measured centrally near gate.

It is expressed in megapascals, MPa.

**3.25 cavity pressure at hold $p_{CH}$:** The cavity pressure, see 3.24, taken at one second after the injection time $t_I$.

It is expressed in megapascals, MPa.

For proper selecting the injection time $t_I$ see 5.2.4.

## 4 APPARATUS

### 4.1 Specimen type and injection mould

The measurement of shrinkage shall be carried out using the plate 60 mm x 60 mm x 2 mm and the appropriate ISO mould type D2 defined in Part 3 of this International Standard, clause 4.1.

For optically measuring the dimensions, the mould may be equipped with engraved marks, see Part 1, clause 4.1.1.4, n).

Installation of the pressure sensor P, recommended for the parts 1 to 3 of this International Standard (see Part 1, clause 4.1.1.4, l) and Part 3, figure 2), is mandatory for shrinkage measurements.

### 4.2 Injection moulding machine

See Part 1 and Part 3, relevant clauses 4.2, with the following addition in clause 4.2.2 Control system: Ranges of operating conditions.

Cavity pressure      $p_c$ (see 3.24)      ± 5 %

### 4.3 Equipment for measuring dimensions

Equipment suitable for measuring the lengths and the widths of the test specimens and of the corresponding cavity of the mould to within 0,02 mm, taken between the relevant centres of opposite sides.

Mechanical measuring systems shall contact the three moulded sides of the test specimen at the centres of the thickness. Pay attention to the thickness step of the gate that must be used for the length measurement and has a height of only 0,5 mm. Insure that the feelers of the measuring system do not show markable indentation.

It is recommended that a calibration plate be used for periodically checking the measuring equipment.

## 4.4  Standard atmosphere

See ISO 291

## 4.5  Oven

For post shrinkage only, according to the agreement between the interested parties.


## 5    PROCEDURE

## 5.1  Conditioning of materials

See Part 1, clause 5.1.

## 5.2  Injection moulding

5.2.1 For basic parameters of injection moulding see Part 1, clauses 5.2.1 and 5.2.2, and Part 3, clause 5.2.2.

5.2.2 Perform the measurement of mould shrinkage for one or more of the following preferred values of the cavity pressure at hold $p_{CH}$ (see 3.27) of 20 MPa, 40 MPa, 60 MPa, 80 MPa and 100 MPa. Interpolation may be used.

> NOTE 1 - For values of the pressure at hold $p_H$ higher than 80 MPa take into account the correspondingly high values of the clamping force $F_M$ (see Part 1, equation (2)), not given in commonly commercial equipment.

**5.2.3** Determine a hold pressure $p_H$ (see Part 1, clause 3.4), which results in the selected value of the cavity pressure at hold $p_{CH}$. Keep the hold pressure constant during the hold period.

**5.2.4** Carefully select the change-over point between injection and hold period carefully, i. e. avoiding a depression in the time elapse of the cavity pressure curve (see figure 1, curve c), as well as a maximum of more than 20 % beyond the smooth curve (see curve b and note 2).

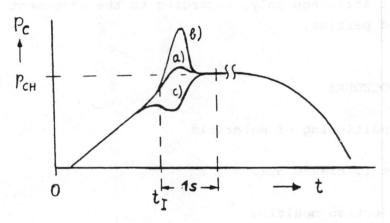

Figure 1 – Schematic diagram of the cavity pressure $p_c$ versus time t, for the injection time $t_I$ selected a) proper, b) too long and c) too short.

> NOTE 2 – Strong peaks in the cavity pressure generate overloading of the cavity followed by backflow of the melt: The orientation of the material near gate will be disturbed.

**5.2.5** Select the cooling time $t_c$ to be the minimum value sufficient for removing the mouldings from the mould without distortion.

**5.2.6** For the hold time $t_H$, see Part 1, clause 5.2.4 and for the thermal equilibrium of the processing equipment, see Part 1, clause 5.2.5.

## 5.3 Measuring of the mould temperature

See Part 1, clause 5.3.

## 5.4 Measuring of the melt temperature

See Part 1, clause 5.4.

## 5.5 Treatment of specimens after demoulding

5.5.1 In order to minimize warpage, separate the test specimens from the runners immediately after demoulding. Take care that the edges used for the measurement of dimensions are not damaged by cutting.

5.5.2 Allow the test specimens to cool to room temperature by placing them on a material with low thermal conductivity and under an appropriate load to avoid warping. Store them at a temperature of 23 °C ± 2 °C for between 16 h and 24 h. For materials which show markable differences of the moulding shrinkage if stored in humid or dry atmosphere, respectively, must be stored in dry atmosphere (e. g. in an impermeable box together with desiccant).

## 5.6 Measurement of moulding shrinkage

5.6.1 If not already known, measure the length $l_c$ and the width $b_c$ of the cavity taken between the relevant centres of opposite sides (or the distances between the engraved marks in the mould) to the nearest 0,02 mm at a temperature of 23 °C ± 2 °C (ISO 291 "atmosphere 23").

Record these measurements for use in the calculation of shrinkage.

> NOTE 3 - From time to time, the cavity should be checked for wear, etc.

5.6.2 Before measuring the dimensions of the specimen, place it on a flat surface or against a straight edge in order to determine any warpage or distortion. Any test specimen that has a warpage exceeding 3 % of its length shall be discarded.

5.6.3 For the determination of moulding shrinkage, measure, to the nearest 0,02 mm, the length $l_1$ and width $b_1$ of the plate parallel to its axis' between the centres of opposite end faces and gate respectively or the distances between the gauge marks, at a temperature of 23 °C ± 2 °C (ISO 291, "atmosphere 23").

5.6.4 Carry out the measurements using at least 5 test specimens for each set of moulding parameters.

5.7  Post treatment

The conditions of treatment (temperature, humidity or other environments) for the time period between the measurements of moulding and post shrinkage shall be taken from the relevant material International Standard or as agreed between the interested parties, see note 4.

NOTE 4 - The conditions of post treatment may result from storage or use.

5.8  Measurement of post shrinkage

After the post treatment, measure the test specimens again, at a temperature of 23 °C ± 2 °C (ISO 291, "atmosphere 23") to the nearest 0,02 mm, see 5.6.2 to 5.6.4, resulting in the length $l_2$ and the width $b_2$.

# 6    EXPRESSION OF RESULTS

## 6.1  Moulding shrinkage

The moulding shrinkage $S_{Mp}$ parallel and $S_{Mn}$ normal to the flow are given, as percentages, according to the equations (1) and (2).

$$S_{Mp} = 100 \ (l_c - l_1)/l_c \tag{1}$$
$$S_{Mp} = 100 \ (b_c - b_1)/b_c \tag{2}$$

where $l_c$ and $b_c$ are the central length and width, in millimetres, of the cavity, see 5.6.1, $l_1$ and $b_1$ are the corresponding sizes, in millimetres, of the test specimen, see 5.6.3.

## 6.2  Post shrinkage

The post shrinkage $S_{Pp}$ parallel and $S_{Pn}$ normal to the flow are given, as percentages, according to the equations (3) and (4)

$$S_{Pp} = 100 \ (l_1 - l_2)/l_1 \tag{3}$$
$$S_{Pn} = 100 \ (b_1 - b_2)/b_1 \tag{4}$$

where $l_2$ and $b_2$ are the sizes, in millimetres, of the test specimen after the post treatment, see 5.8.

## 6.3  Total shrinkage

the total shrinkage $S_{Tp}$ parallel and $S_{Tn}$ normal to the flow are given, as percentages, according to the equations (5) and (6)

$$S_{Tp} = 100 \ (l_c - l_2)/l_c \tag{5}$$
$$S_{Tn} = 100 \ (b_c - b_2)/b_c \tag{6}$$

where the relevant sizes are given in 5.6.1 and 5.8.

Mould, post and total shrinkage are interrelated according to equation (7)

$$S_T = S_M + S_P (1 - S_M/100)$$ (7)

## 7 PRECISION

See ISO 294-1, clause 6.

## 8 TEST REPORT

The test report shall include the following information:

a)    a reference to this part of this International Standard

a)    to h): See Part 1, clause 7, a) to h), but replacing in g) the hold pressure by the cavity pressure at hold, see 3.27.

i)    moulding, post and total shrinkage, parallel and normal to the flow, see 3.23 to 3.25, in percent, %, to the nearest 0,05 %.

EXPLANATORY REPORT
RAPPORT EXPLICATIF    ISO/ DIS 294-4

will supersede:
remplacera:

ISO/TC 61 /SC 9        Secretariat

ANSI (USA)

| This form should be sent to the ISO Central Secretariat, together with the English and French versions of the committee draft, by the secretariat of the technical committee or sub-committee concerned (see 2.5.9 of part 1 of the ISO/IEC Directives) | Ce formulaire doit être envoyé au Secrétariat central de l'ISO en même temps que les versions anglaise et française du projet de comité, par le secrétariat du comité technique ou du sous-comité concerné (voir 2.5.9 de la partie 1 des Directives ISO/CEI) |
|---|---|

The accompanying document is submitted for circulation to member body vote as a DIS, following consensus of the P-members of the committee obtained

Le document ci-joint est soumis, pour diffusion comme DIS, au vote comité membre, suite au consensus des membres (P) du comité obtenu

on
le    1994-09-23

[X]   at the meeting of   TC 61 /SC 9 :   see   resolution   No. ...............   in      61 N 4934
      à la réunion du               voir           n° ...............   dans le   document 61/9 N 740

[ ]   by postal ballot initiated on
     par un vote par correspondance démarré le     19 . . - . . - . .

P-members in favour:            Canada, Czech Republic, Finland, France, Germany, India
Membres (P) approuvant le projet:   Italy, Japan, Netherlands, Switzerland, UK, USA
SC Plenary

P-members voting against:
Membres (P) désapprouvant:

P-members abstaining:
Membres (P) s'abstenant:

P-members who did not vote:
Membres (P) n'ayant pas voté:

Remarks/Remarques

The results of voting and comments received on CD 294-1, CD 294-2, CD 294-3 and CD 294-4 were discussed at the meeting of SC 9/WG 18 in Tokyo, September 1994. All negatives were resolved and the WG recommended that modified texts be advanced to DIS level. Thus, we hereby submit these texts for your consideration.

I hereby confirm that this draft meets the requirements of part 3 of the ISO/IEC Directives
Je confirme que ce projet satisfait aux prescriptions de la partie 3 des Directives ISO/CEI

Date           Name and signature of the secretary
          Nom et signature du secrétaire
1995-02-10
                                    Julia Lindsay
                                    Secretariat ISO/TC 61/SC 9

# INTERNATIONAL STANDARD

**ISO
295**

Second edition
1991-11-01

# Plastics — Compression moulding of test specimens of thermosetting materials.

*Plastiques — Moulage par compression des éprouvettes en matières thermodurcissables*

Reference number
ISO 295:1991(E)

# Foreword

ISO (the International Organization for Standardization) is a worldwide federation of national standards bodies (ISO member bodies). The work of preparing International Standards is normally carried out through ISO technical committees. Each member body interested in a subject for which a technical committee has been established has the right to be represented on that committee. International organizations, governmental and non-governmental, in liaison with ISO, also take part in the work. ISO collaborates closely with the International Electrotechnical Commission (IEC) on all matters of electrotechnical standardization.

Draft International Standards adopted by the technical committees are circulated to the member bodies for voting. Publication as an International Standard requires approval by at least 75 % of the member bodies casting a vote.

International Standard ISO 295 was prepared by Technical Committee ISO/TC 61, *Plastics*, Sub-Committee SC 12, *Thermosetting materials*.

This second edition cancels and replaces the first edition (ISO 295:1974), of which it constitutes a technical revision.

# Plastics — Compression moulding of test specimens of thermosetting materials.

## 1 Scope

This International Standard specifies the general principles and the procedures to be followed for the preparation of test specimens from thermosetting compounds moulded under heat and pressure and for the establishment of comparable test reports from different testing organizations. It is applicable only to thermosetting materials based upon phenolics (ISO 800), aminoplastics (ISO 2112), melamine phenolics (ISO 4896), epoxides and un-saturated polyesters.

Because the properties of the specimens moulded from thermosetting materials depend on the conditions of preparation of the specimens, this International Standard also specifies the details of specimen preparation to be included with test reports of the properties of such specimens.

It may often be necessary to prepare specimens by special methods because of their composition, their flow properties or other variable factors. In this case, an agreement shall be made between the interested parties. The tables giving the specimen properties shall refer to these specific methods.

## 2 Normative references

The following standards contain provisions which, through reference in this text, constitute provisions of this International Standard. At the time of publication, the editions indicated were valid. All standards are subject to revision, and parties to agreements based on this International Standard are encouraged to investigate the possibility of applying the most recent editions of the standards indicated below. Members of IEC and ISO maintain registers of currently valid International Standards.

ISO 468:1982, *Surface roughness — Parameters, their values and general rules for specifying requirements.*

ISO 800:—[1], *Plastics — Phenolic moulding materials — Specification.*

ISO 1183:1987, *Plastics — Methods for determining the density and relative density of non-cellular plastics.*

ISO 2112:1990, *Plastics — Aminoplastic moulding materials — Specification.*

ISO 3167:1983, *Plastics — Preparation and use of multipurpose test specimens.*

ISO 4896:1990, *Plastics — Melamine/phenolic moulding materials — Specification.*

## 3 Definitions

For the purposes of this International Standard, the following definitions apply.

**3.1 deviations of temperature in position:** Deviations of temperature existing simultaneously between various points inside the mould after the temperature adjustment device has been set at a given temperature and after a permanent thermal equilibrium has been reached.

**3.2 deviations of temperature in time:** Deviations of temperature that may occur at a single given point on the inside of the mould at various times after the temperature adjustment device has been set at a given temperature and after a permanent thermal equilibrium has been reached.

---

1) To be published. (Revision of ISO 800:1977)

## 4 Apparatus

**4.1 Compression mould**, made of steel, able to withstand the specified temperatures and pressures. The mould shall be designed so that the compression force is transmitted to the moulding material with no appreciable loss. It may be of a single-cavity or a multi-cavity type. Figure 1 shows an example of a single-cavity positive mould. The cavity of the mould may have the shape of the multi-purpose test specimen described in ISO 3167. In some cases (aminoplastics for instance), a semi-positive mould is more suitable, even though the pressure on the moulding material is not as well defined. In this case, the specimen thickness shall be adjusted using spacers on the mould parting line.

The mould surface shall be free from superficial damage or contamination and have a shiny surface finish of $R_{aH}$ 0,4 µm to 0,8 µm (see ISO 468). Chrome plating is not always necessary, but it will prevent sticking.

The edge taper angle shall not be greater than 3° (see figure 1). Clearance between the vertical wall of the cavity and the punch shall be not greater than 0,1 mm (see figure 1).

The mould shall have a loading chamber (see figure 1) large enough to allow the whole charge to be fed in one operation. Moulding material in bulk form is from 2 to 10 times as voluminous as the moulded object.

The mould may be fitted with an ejector. If ejector pins are used [see the example in figure 2a)], they shall not deform the specimen in any way. If the parts are ejected by the movable bottom of the mould [see the example in figure 2b)], there shall be no significant leakage of material at the joint between the bottom and the cavity wall.

Because the face of the moulded part facing the lower die is heated for a longer time during the period between filling and compression, it may be useful to distinguish between the two faces by means of a fixed mark in the cavity.

Dimensions in millimetres

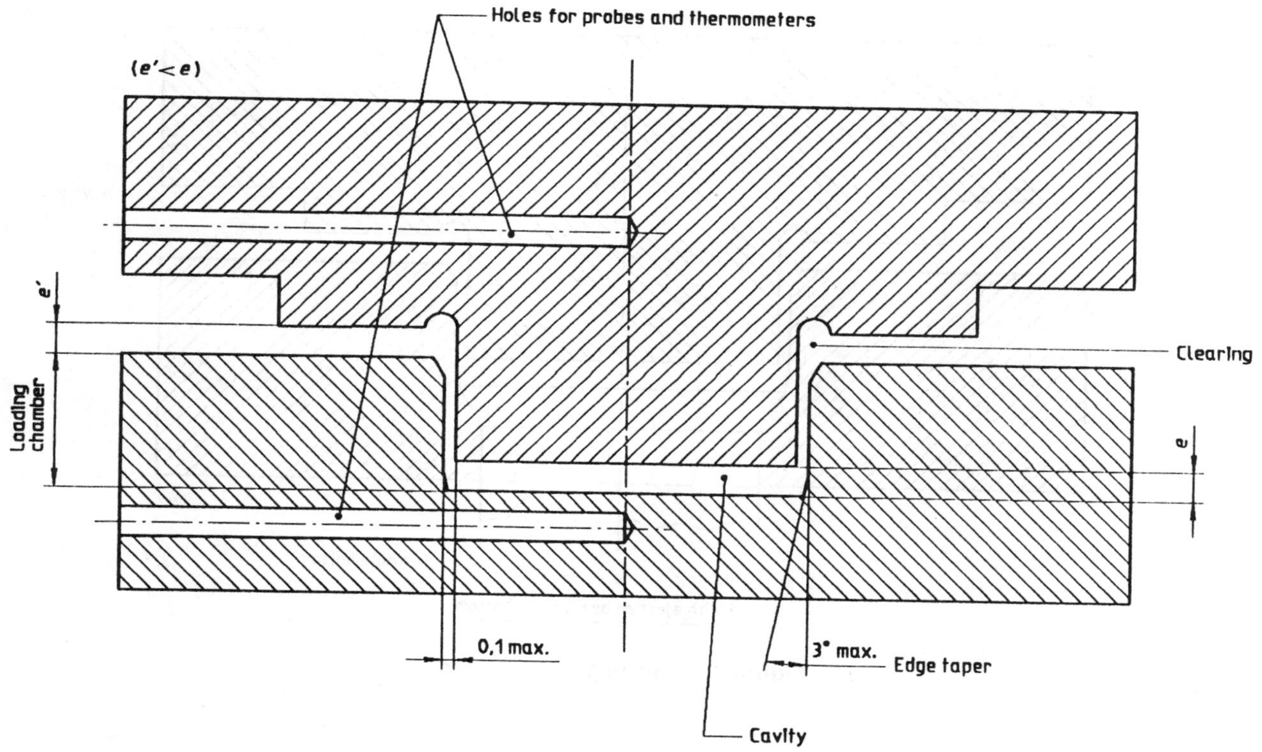

NOTE — Dimension $e'$ shall be calculated so that there is no risk of the piston damaging the die if there is no material present.

**Figure 1 — Example of single-cavity positive mould**

131

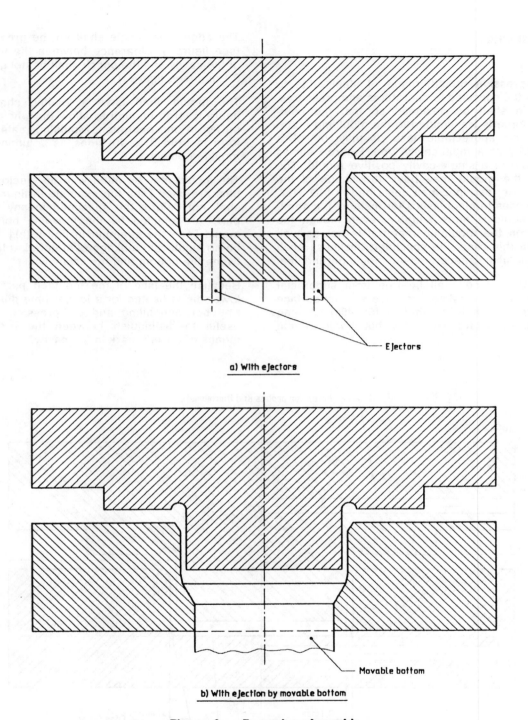

a) With ejectors

b) With ejection by movable bottom

Figure 2 — Examples of moulds

**4.2 Heating device**, capable of heating the mould so that the moulding temperature remains constant and uniform over all parts of the mould within the specified tolerances.

The mould may be heated either through the platen or by means of a built-in device (for example, circulating fluid or electric heating elements). In the latter case, the mould shall be insulated from the press platens with a sheet of insulating material. For practical reasons, it is generally preferable to heat the mould electrically.

**4.3 Mould temperature adjustment device**, capable of ensuring that the optimum required temperature is maintained constant over the whole mould with a permissible deviation of $\pm 3$ °C, i.e. the mould temperature shall not vary with time and position by more than $\pm 3$ °C (see 3.1 and 3.2).

**4.4 Compression-moulding press**, capable of ensuring that the specified pressure is applied and maintained during the whole of the curing time. The press may be hand-operated or programmed.

It is preferable to use a press having two closing speeds:

— a fast approach speed (for example 200 mm/s to 400 mm/s) to avoid precure of the material before closing;

— a slow closing speed (for example 5 mm/s) to prevent air or gases from being entrapped.

NOTE 1   The oil pressure $p_0$, in megapascals, to be applied, to obtain the specified pressure $p$, in megapascals, is given by the equation

$$p_0 = \frac{p \times A_1}{A}$$

where

$A$      is the area, in square metres, of the press piston head.

$A_1$      is the total area, in square metres, of the cavities.

**4.5 Stopwatch**, capable of being read to an accuracy of 1 s.

**4.6 Mould temperature measurement device**, such as a pyrometer or fusible salts.

**4.7 Balance**, having an accuracy of 0.1 g.

**4.8 Metal plate**, about 20 mm thick and having at least the same area as the specimen, for use as a cooling fixture after stripping (see clause 7).

# 5 Material conditioning prior to moulding

## 5.1 Storage

Moulding materials that require storage in a sealed container shall be so maintained at a temperature of 23 °C $\pm$ 3 °C or as prescribed by the supplier until immediately prior to preforming (see 5.2), drying (see 6.2), preheating (see 6.3) or moulding (see clause 7), as applicable. In those cases where materials need to be put back into storage, this shall be done in accordance with instructions given by the material supplier.

## 5.2 Preforming

If the volume of the moulding material is too great for the capacity of the loading chamber of a conventional mould, the material may be preformed; the conditions used for such preforming shall be stated in the moulding report.

# 6 Moulding conditions

## 6.1 General

Unless special conditions are specified, the moulding conditions given in table 1 shall be used.

133

## Table 1 — Moulding conditions

| Conditions | Type of moulding material | | | | | | |
|---|---|---|---|---|---|---|---|
| | Phenolics — Structure of filler | | Aminoplastics | | | Epoxides | Unsaturated Polyesters |
| | Fine | Coarse | Urea-formal-dehyde | Melamine-formaldehyde — General purpose | Melamine-formaldehyde — For food contact | | |
| **Pretreatment:** | | | | | | | |
| Drying | Permissible if specimens are to undergo electric tests | | | | | Not recommended | Not recommended |
| Preforming | Permissible | | Permissible | Permissible | Permissible | Permissible | Permissible |
| High-frequency preheating | Permissible to reduce curing time, but modifies material properties | | | | | | |
| Preplastification | Permissible | | Permissible | Permissible | Permissible | Not recommended | Not recommended |
| Breathing | Permissible | | Permissible | Permissible | Permissible | Not necessary | Not recommended |
| **Moulding:** | | | | | | | |
| Temperature (°C) | $165 \pm 3$ | | $150 \pm 3$ | $150 \pm 3$ | $150 \pm 3$ | 150 to 180 | 130 to 170 |
| Pressure (MPa) | 25 to 40 | 40 to 60 | 20 to 40 | 20 to 40 | 20 to 40 | 20 to 30 | 6 to 30 |
| Cure time (s) | 20 to 60 per millimetre of thickness | | | | | | |
| **Mould:** | | | | | | | |
| Surface finish | Surface finish $R_{aH}$ 0,4 µm to 0,8 µm | | | | | | |
| Chrome plating | Preferable | Preferable | Preferable | Preferable | Preferable | Required | Required |

## 6.2 Drying

Phenolics and aminoplastics may be dried prior to electrical tests. For drying, the material shall be spread out in a thin layer and heated in accordance with the following temperature and time schedules:

— phenolics: 30 min at 90 °C $\pm$ 3 °C, or 15 min at 105 °C $\pm$ 3 °C;

— aminoplastics: 60 min at 90 °C $\pm$ 3 °C.

The material shall be moulded immediately upon removal from the oven.

## 6.3 High-frequency preheating

High-frequency preheating is permissible in the case of phenolics and aminoplastics and pelletized or granular polyesters. It permits a reduction in curing time. The preheated material shall be moulded immediately after preheating.

## 6.4 Preplastification

Preplastification is permissible in the case of phenolics and aminoplastics. It ensures thermal and mechanical homogenization of the material. The preplasticized material shall be moulded imme-diately after preplastification. For the conditions for the preplastification, an agreement shall be made between the interested parties and the conditions shall be stated in the moulding report.

## 6.5 Release agents

Release agents, i.e. products designed to facilitate the release of the moulding from the mould, may be used only if it has been proved that they have no influence on the moulded-specimen properties. This requirement applies particularly when the speci-mens are to be tested for electrical properties, spectroscopic analysis or adverse taste and colour.

## 6.6 Breathing

If it is necessary to open the mould for the purpose of breathing, this shall be noted in the moulding re-port.

## 7 Procedure

Select the moulding conditions to be used (see clause 6). Allow the moulding temperature to reach equilibrium at $\pm$ 3 °C of the required value.

Check the temperature in the cavity (see 4.1) using the temperature measurement device (4.6).

Weigh out the required quantity of material to obtain the specified thickness of specimen. This quantity is the product of the moulded-part density and the volume of the test specimen, to which is added flash losses as determined by previous testing. Load the material, as powder or preform, in the cavity. Close the press (4.4). Allow to breathe if necessary (see note 2).

Start the stopwatch (4.5) when the pressure has reached the specified value. When the curing time is completed, open the press. Remove the specimen immediately and, unless otherwise specified in the test method (see note 3), allow it to cool on the metal plate used as the cooling fixture (4.8).

Check that the moulding is satisfactory as regards filling of the mould, appearance, absence of porosity, discoloration, flash and warpage. If necessary, check the density as determined in accordance with ISO 1183.

NOTES

2   In the case of a programme-controlled press, degasifying and opening may be performed automatically.

3   For some test methods, such as the determination of shrinkage, the requirement may be to place the hot moulding on a material of low thermal conductivity and under an appropriate load.

## 8   Moulding report

The moulding report shall include a reference to this International Standard and all information specified in table 2.

Table 2 — Information to be included in the moulding report

| | | | |
|---|---|---|---|
| **Physical form of material** | | Granules | |
| | | Powder | |
| | | Fine powder | |
| | | Other | |
| **Pretreatment** | Drying | Without | |
| | | Time | |
| | | Temperature | |
| | Preforming | Pressure | |
| | | Temperature | |
| | | Weight of preform | |
| | | Size of preform | |
| | High-frequency preheating | Preheater power | |
| | | Time | |
| | | Amperage | |
| | | Number of preforms | |
| | | Temperature of preforms | |
| | Preplastification | Cylinder temperature | |
| | | Dynamic pressure | |
| | | Screw speed | |
| | | Temperature of material | |
| **Compression moulding** | | Temperature | |
| | | Temperature measurement device | |
| | | Pressure | |
| | | Cure time | |
| | | Breathing | |
| **Mould** | | Type | |
| | | Number of cavities | |
| | | Chrome plated | |
| | | Heating device | |

# INTERNATIONAL STANDARD

**ISO
306**

Third edition
1994-08-01

## Plastics — Thermoplastic materials — Determination of Vicat softening temperature (VST)

*Plastiques — Matières thermoplastiques — Détermination de la
température de ramollissement Vicat (VST)*

Reference number
ISO 306:1994(E)

# Foreword

ISO (the International Organization for Standardization) is a worldwide federation of national standards bodies (ISO member bodies). The work of preparing International Standards is normally carried out through ISO technical committees. Each member body interested in a subject for which a technical committee has been established has the right to be represented on that committee. International organizations, governmental and non-governmental, in liaison with ISO, also take part in the work. ISO collaborates closely with the International Electrotechnical Commission (IEC) on all matters of electrotechnical standardization.

Draft International Standards adopted by the technical committees are circulated to the member bodies for voting. Publication as an International Standard requires approval by at least 75 % of the member bodies casting a vote.

International Standard ISO 306 was prepared by Technical Committee ISO/TC 61, *Plastics*, Subcommittee SC 2, *Mechanical properties*.

This third edition cancels and replaces the second edition (ISO 306:1987), which has been technically revised.

# Plastics — Thermoplastic materials — Determination of Vicat softening temperature (VST)

## 1 Scope

**1.1** This International Standard specifies four methods for the determination of the Vicat softening temperature (VST) of thermoplastic materials:

— Method A50 using a force of 10 N and a heating rate of 50 °C/h

— Method B50 using a force of 50 N and a heating rate of 50 °C/h

— Method A120 using a force of 10 N and a heating rate of 120 °C/h

— Method B120 using a force of 50 N and a heating rate of 120 °C/h

**1.2** The methods specified are applicable only to thermoplastics, for which they give a measure of the temperature at which the thermoplastics start to soften rapidly.

## 2 Normative references

The following standards contain provisions which, through reference in this text, constitute provisions of this International Standard. At the time of publication, the editions indicated were valid. All standards are subject to revision, and parties to agreements based on this International Standard are encouraged to investigate the possibility of applying the most recent editions of the standards indicated below. Members of IEC and ISO maintain registers of currently valid International Standards.

ISO 291:1977, *Plastics — Standard atmospheres for conditioning and testing.*

ISO 293:1986, *Plastics — Compression moulding test specimens of thermoplastic materials.*

ISO 294:—[1], *Plastics — Injection moulding of test specimens of thermoplastic materials.*

ISO 2818:—[2], *Plastics — Preparation of test specimens by machining.*

ISO 3167:1993, *Plastics — Multipurpose test specimens.*

## 3 Principle

Determination of the temperature at which a standard indenter penetrates 1 mm into the surface of a plastic test specimen under one of the loads given in 1.1 when the temperature is raised at a uniform rate.

The temperature at 1 mm penetration is quoted as the VST in degrees Celsius.

## 4 Apparatus

The apparatus consists essentially of:

**4.1 Rod**, provided with a **load-carrying plate** (4.4), held in a **rigid metal frame** so that it can move freely in the vertical direction, the base of the frame serving to support the test specimen under the indenting tip at the end of the rod (see figure 1).

Unless the rod and frame members have the same linear thermal expansion coefficient, the differential

---

1) To be published. (Revision of ISO 294:1975)

2) To be published. (Revision of ISO 2818:1980)

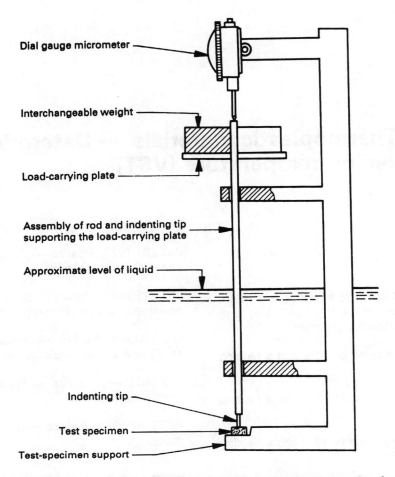

**Figure 1 — Example of apparatus with a heating bath for the determination of the VST**

change in the length of these parts introduces an error in the readings of the apparent deformation of the test specimen. A blank test shall be made on each apparatus using a test specimen made of rigid material having a low coefficient of expansion.[3] The temperature ranges to be used shall be covered and a correction term determined for each temperature. If the correction term is 0,02 mm or greater, its algebraic sign shall be noted and the term applied to each test by adding it algebraically to the reading of apparent penetration. It is recommended that the apparatus be constructed of low thermal expansion alloy.

**4.2 Indenting tip**, preferably of hardened steel, 3 mm long, of circular cross-section, and area 1,000 mm$^2$ ± 0,015 mm$^2$, fixed at the bottom of the rod (4.1). The lower surface of the indenting tip shall be plane and perpendicular to the axis of the rod and free from burrs.

**4.3 Calibrated micrometer dial gauge** (or other suitable measuring instrument), to measure to ± 0,01 mm the penetration of the indenting tip into the test specimen. The thrust of the dial gauge, which contributes to the thrust on the test specimen, shall be recorded (see 4.4).

NOTES

1 In certain types of apparatus, the force of the dial gauge spring is directed upwards and is subtracted from the load; in other forms, this force acts downwards and is added to the load.

2 Since the force exerted by the spring in certain dial gauges varies considerably over the stroke, this force is measured in that part of the stroke which is to be used.

**4.4 Load-carrying plate**, fitted to the rod (4.1), and **suitable weights** added centrally so that the total thrust applied to the test specimen can be made up to 10 N ± 0,2 N for methods A50 and A120 and

3) Invar and borosilicate glass have been found suitable for this purpose.

50 N ± 1 N for methods B50 and B120. The combined downward thrust due to the rod, indenting tip, load-carrying plate and the force of the dial gauge spring shall not exceed 1 N.

### 4.5 Heating equipment, consisting of a heating bath (4.5.1) containing a liquid or an oven (4.5.2) with forced circulation of air or nitrogen.

The heating equipment shall be provided with a means of control so that the temperature can be raised at a uniform rate of either 50 °C/h ± 5 °C/h or 120 °C/h ± 10 °C/h, as required. The requirement for the heating rate shall be considered satisfied if, over every 6 min interval during the test, the temperature change is 5 °C ± 0,5 °C or 12 °C ± 1 °C, respectively.

The apparatus may be arranged to shut off the heat automatically and sound an alarm when the specified indentation has been reached (see 7.5).

### 4.5.1 Heating bath, containing a liquid in which the test specimen can be immersed to a depth of at least 35 mm. An efficient stirrer shall be provided. It shall be established that the liquid chosen is stable at the temperature used and does not affect the material under test, for example by swelling or cracking.

When a heating bath is used, the temperature of the liquid, measured close to the test specimen, shall be taken as the VST (see 7.5).

NOTE 3    Liquid paraffin, transformer oil, glycerol and silicone oil are suitable liquid heat-transfer media, but other liquids may be used.

### 4.5.2 Oven, with forced air or nitrogen circulation of about 60 times per minute, with a volume of not less than 10 litres for each apparatus and in which the air or nitrogen flow is directed perpendicular to the upper surface of the test specimen at a speed of 1,5 m/s to 2 m/s.

The result of the test will depend on the rate of transfer of heat from the circulating air or nitrogen to the surface of the test specimen. Because of the relatively small test specimen and the fact that the lower surface is in contact with the test-specimen support, the air or nitrogen temperature shall not be taken as the VST.

Take the temperature indicated by a sensor in the rod close to the indenting tip, or in the test-specimen support, as the VST.

For an initial calibration, verify by experiment that the temperature indicated by the sensor is within ± 1 °C of the temperature that is indicated by an additional sensor embedded within a blank test specimen.

NOTE 4    Commercially available ovens are often furnished with means for suitable air or nitrogen circulation. If not, the necessary heat-transfer rate may be ensured by fitting stream plates which direct the circulating air or nitrogen perpendicular to the upper surface of the test specimen.

### 4.6 Temperature-measuring instrument.

#### 4.6.1 For a heating bath

Mercury-in-glass thermometer of the partial-immersion type or other suitable temperature-measuring instrument of appropriate range and accurate to within 0,5 °C. Mercury-in-glass thermometers shall be calibrated at the depth of immersion required by 7.2.

#### 4.6.2 For an oven with air or nitrogen circulation

Suitable temperature-measuring instrument of appropriate range and accurate to within 0,5 °C. The sensor (thermocouple or Pt 100) shall be positioned in the rod close to the indenting tip or in the test-specimen support (see 4.5.2).

## 5   Test specimens

### 5.1   At least two test specimens shall be used to test each sample. The test specimens shall be between 3 mm and 6,5 mm thick and at least 10 mm square or of 10 mm diameter. Their surfaces shall be flat and parallel and free from flash. They shall be made in accordance with the specifications, if any, for the material under test. In the absence of such specifications, any suitable procedure may be used for the preparation of test specimens.

### 5.2   If the samples submitted for test are in the form of moulding materials (for example, powder or granulated materials), these shall be moulded into specimens 3 mm to 6,5 mm thick, in accordance with the specifications relating to the material under test, or in accordance with ISO 293, ISO 294 or ISO 3167 if no material specification exists. If these are not applicable, any other reproducible procedure may be followed that modifies the properties of the material as little as possible.

**5.3** For sheet materials, the thickness of the test specimens shall be equal to the thickness of the sheet, except as follows:

a) If the thickness exceeds 6,5 mm, the test specimens shall be reduced in thickness to 3 mm to 6,5 mm by machining one surface (see ISO 2818), the other surface being left intact. The test surface shall be the intact one.

b) If the thickness of the sheet is less than 3 mm, not more than three pieces shall be stacked together in direct contact to give a total thickness between 3 mm and 6,5 mm and the thickness of the upper (measured) piece shall be at least 1,5 mm. Stacking of pieces of lesser thickness does not always give the same test result.

**5.4** The test results obtained may depend on the moulding conditions used in the preparation of the test specimens, although such a dependence is not common. When testing materials for which the results do depend on the moulding conditions, special annealing or preconditioning procedures may be used before testing provided they are agreed to by the interested parties.

## 6 Conditioning

Unless otherwise required by the specification for the material being tested, the specimens shall be conditioned in accordance with ISO 291.

## 7 Procedure

**7.1** Mount the test specimen horizontally under the indenting tip (4.2) of the unloaded rod (4.1). The indenting tip shall at no point be nearer than 3 mm to the edge of the test specimen. The surface of the test specimen in contact with the base of the apparatus shall be flat.

**7.2** Place the assembly in the heating equipment (4.5). The temperature of the heating equipment shall be 20 °C to 23 °C at the start of each test, unless previous tests have shown that, for the material under test, no error is caused by starting at another temperature. When a heating bath (4.5.1) is used, the bulb of the thermometer or the sensitive part of the temperature-measuring instrument (4.6.1) shall be at the same level as, and as close as possible to, the test specimen.

**7.3** After 5 min, with the indenting tip still in position, add a sufficient weight to the load-carrying plate (4.4) so that the total thrust on the test specimen is

10 N ± 0,2 N for methods A50 and A120 and 50 N ± 1 N for methods B50 and B120. Then note the reading of the micrometer dial gauge (or other indentation-measuring instrument) (4.3) or set the instrument to zero.

**7.4** Increase the temperature of the heating equipment at a uniform rate of 50 °C/h ± 5 °C/h or, alternatively, 120 °C/h ± 10 °C/h; when a heating bath is used stir the liquid well during the test. For referee tests, a rate of 50 °C/h shall be used.

NOTE 5 For some materials at the higher rate (120 °C/h), Vicat softening temperatures up to 10 °C higher can be observed.

**7.5** Note the temperature of the bath (see 4.6.1) or the built-in sensor (see 4.6.2) at which the indenting tip has penetrated into the test specimen by 1 mm ± 0,01 mm beyond its starting position defined in 7.3, and record it as the VST of the test specimen.

**7.6** Express the VST of the material under test as the arithmetic mean of the VSTs of the specimens tested. If the range of individual results exceeds 2 °C, record the individual results [see clause 9, h)] and repeat the test once using a further set of at least two specimens (see 5.1).

## 8 Precision

The precision of this test method is not known because interlaboratory data are not available. When interlaboratory data are obtained, a precision statement will be added at the following revision.

## 9 Test report

The test report shall include the following information:

a) a reference to this International Standard;

b) full identification of the material tested;

c) the method employed (A50, A120, B50 or B120);

d) the thickness and the number of layers of composite test specimens (i.e. specimens consisting of more than one layer) if these are used;

e) the method of preparation of the test specimens used;

f) the heat-transfer medium used;

g) the conditioning and annealing procedures used, if any;

h) the Vicat softening temperature (VST) of the material, in degrees Celsius (if the individual results after two measurements differ by more than the limit given in 7.6, all individual results shall be reported);

i) any unusual characteristics of the test specimen noted during the test or after removal from the apparatus.

# INTERNATIONAL STANDARD

**ISO**

**527-1**

First edition
1993-06-15

# Plastics — Determination of tensile properties —

## Part 1:
General principles

*Plastiques — Détermination des propriétés en traction —*

*Partie 1: Principes généraux*

Reference number
ISO 527-1:1993(E)

# Foreword

ISO (the International Organization for Standardization) is a worldwide federation of national standards bodies (ISO member bodies). The work of preparing International Standards is normally carried out through ISO technical committees. Each member body interested in a subject for which a technical committee has been established has the right to be represented on that committee. International organizations, governmental and non-governmental, in liaison with ISO, also take part in the work. ISO collaborates closely with the International Electrotechnical Commission (IEC) on all matters of electrotechnical standardization.

Draft International Standards adopted by the technical committees are circulated to the member bodies for voting. Publication as an International Standard requires approval by at least 75 % of the member bodies casting a vote.

International Standard ISO 527-1 was prepared by Technical Committee ISO/TC 61, *Plastics*, Sub-Committee SC 2, *Mechanical properties*.

Together with the other parts of ISO 527, it cancels and replaces ISO Recommendation R 527:1966, which has been technically revised.

ISO 527 consists of the following parts, under the general title *Plastics — Determination of tensile properties*:

— *Part 1: General principles*

— *Part 2: Test conditions for moulding and extrusion plastics*

— *Part 3: Test conditions for sheet and film*

— *Part 4: Test conditions for isotropic and orthotropic fibre-reinforced plastic composites*

— *Part 5: Test conditions for unidirectional fibre-reinforced plastic composites*

Annex A of this part of ISO 527 is for information only.

# Plastics — Determination of tensile properties —

# Part 1:
## General principles

## 1 Scope

**1.1** This part of ISO 527 specifies the general principles for determining the tensile properties of plastics and plastic composites under defined conditions.

Several different types of test specimen are defined to suit different types of material which are detailed in subsequent parts of ISO 527.

**1.2** The methods are used to investigate the tensile behaviour of the test specimens and for determining the tensile strength, tensile modulus and other aspects of the tensile stress/strain relationship under the conditions defined.

**1.3** The methods are selectively suitable for use with the following range of materials:

— rigid and semirigid thermoplastics moulding and extrusion materials, including filled and reinforced compounds in addition to unfilled types; rigid and semirigid thermoplastics sheets and films;

— rigid and semirigid thermosetting moulding materials, including filled and reinforced compounds; rigid and semirigid thermosetting sheets, including laminates;

— fibre-reinforced thermoset and thermoplastics composites incorporating unidirectional or non-unidirectional reinforcements such as mat, woven fabrics, woven rovings, chopped strands, combination and hybrid reinforcements, rovings and milled fibres; sheets made from pre-impregnated materials (prepregs);

— thermotropic liquid crystal polymers.

The methods are not normally suitable for use with rigid cellular materials or sandwich structures containing cellular material.

**1.4** The methods are applied using specimens which may be either moulded to the chosen dimensions or machined, cut or punched from finished and semifinished products such as mouldings, laminates, films and extruded or cast sheet. In some cases a multipurpose test specimen (see ISO 3167:1993, *Plastics — Preparation and use of multipurpose test specimens*), may be used.

**1.5** The methods specify preferred dimensions for the test specimens. Tests which are carried out on specimens of different dimensions, or on specimens which are prepared under different conditions, may produce results which are not comparable. Other factors, such as the speed of testing and the conditioning of the specimens, can also influence the results. Consequently, when comparative data are required, these factors must be carefully controlled and recorded.

## 2 Normative references

The following standards contain provisions which, through reference in this text, constitute provisions of this part of ISO 527. At the time of publication, the editions indicated were valid. All standards are subject to revision, and parties to agreements based on this part of ISO 527 are encouraged to investigate the possibility of applying the most recent editions of the standards indicated below. Members of IEC and ISO maintain registers of currently valid International Standards.

ISO 291:1977, *Plastics — Standard atmospheres for conditioning and testing.*

ISO 2602:1980, *Statistical interpretation of test re-*

sults — *Estimation of the mean — Confidence interval.*

ISO 5893:1985, *Rubber and plastics test equipment — Tensile, flexural and compression types (constant rate of traverse) — Description.*

# 3  Principle

The test specimen is extended along its major longitudinal axis at constant speed until the specimen fractures or until the stress (load) or the strain (elongation) reaches some predetermined value. During this procedure the load sustained by the specimen and the elongation are measured.

# 4  Definitions

For the purposes of this part of ISO 527, the following definitions apply.

**4.1  gauge length, $L_0$:** Initial distance between the gauge marks on the central part of the test specimen; see figures of the test specimens in the relevant part of ISO 527.

It is expressed in millimetres (mm).

**4.2  speed of testing, $v$:** Rate of separation of the grips of the testing machine during the test.

It is expressed in millimetres per minute (mm/min).

**4.3  tensile stress, $\sigma$ (engineering):** Tensile force per unit area of the original cross-section within the gauge length, carried by the test specimen at any given moment.

It is expressed in megapascals (MPa) [see 10.1, equation (3)].

**4.3.1  tensile stress at yield; yield stress, $\sigma_y$:** First stress at which an increase in strain occurs without an increase in stress.

It is expressed in megapascals (MPa).

It may be less than the maximum attainable stress (see figure 1, curves b and c).

**4.3.2  tensile stress at break, $\sigma_B$:** The tensile stress at which the test specimen ruptures (see figure 1).

It is expressed in megapascals (MPa).

**4.3.3  tensile strength, $\sigma_M$:** Maximum tensile stress sustained by the test specimen during a tensile test (see figure 1).

It is expressed in megapascals (MPa).

**4.3.4  tensile stress at $x$ % strain** (see 4.4), $\sigma_x$: Stress at which the strain reaches the specified value $x$ expressed in percentage.

It is expressed in megapascals (MPa).

It may be measured for example if the stress/strain curve does not exhibit a yield point (see figure 1, curve d). In this case, $x$ shall be taken from the relevant product standard or agreed upon by the interested parties. However, $x$ must be lower than the strain corresponding to the tensile strength, in any case.

**4.4  tensile strain, $\varepsilon$:** Increase in length per unit original length of the gauge.

It is expressed as a dimensionless ratio, or in percentage (%) [see 10.2, equations (4) and (5)].

It is used for strains up to yield point (see 4.3.1); for strains beyond yield point see 4.5.

**4.4.1  tensile strain at yield, $\varepsilon_y$:** Tensile strain at the yield stress (see 4.3.1 and figure 1, curves b and c).

It is expressed as a dimensionless ratio, or in percentage (%).

**4.4.2  tensile strain at break, $\varepsilon_B$:** Tensile strain at the tensile stress at break (see 4.3.2), if it breaks without yielding (see figure 1, curves a and d).

It is expressed as a dimensionless ratio, or in percentage (%).

For breaking after yielding, see 4.5.1.

**4.4.3  tensile strain at tensile strength, $\varepsilon_M$:** Tensile strain at the point corresponding to tensile strength (see 4.3.3), if this occurs without or at yielding (see figure 1, curves a and d).

It is expressed as a dimensionless ratio or in percentage (%).

For strength values higher than the yield stress, see 4.5.2.

**4.5  nominal tensile strain, $\varepsilon_t$:** Increase in length per unit original length of the distance between grips (grip separation).

It is expressed as a dimensionless ratio, or in percentage (%) [see 10.2, equations (6) and (7)].

It is used for strains beyond yield point (see 4.3.1). For strains up to yield point, see 4.4. It represents the total relative elongation which takes place along the free length of the test specimen.

**4.5.1  nominal tensile strain at break, $\varepsilon_{tB}$:** Nominal tensile strain at the tensile stress at break (see 4.3.2), if the specimen breaks after yielding (see figure 1, curves b and c).

It is expressed as a dimensionless ratio, or in percentage (%).

For breaking without yielding, see 4.4.2.

**4.5.2 nominal tensile strain at tensile strength,** $\varepsilon_{tM}$: Nominal tensile strain at tensile strength (see 4.3.3), if this occurs after yielding (see figure 1, curve b).

It is expressed as a dimensionless ratio, or in percentage (%).

For strength values without or at yielding, see 4.4.3.

**4.6 modulus of elasticity in tension; Young's modulus,** $E_t$: Ratio of the stress difference $\sigma_2$ minus $\sigma_1$ to the corresponding strain difference values $\varepsilon_2 = 0{,}002\ 5$ minus $\varepsilon_1 = 0{,}000\ 5$ (see figure 1, curve d and 10.3, equation (8)].

It is expressed in megapascals, (MPa).

This definition does not apply to films and rubber.

NOTE 1   With computer-aided equipment, the determination of the modulus $E_t$ using two distinct stress/strain points can be replaced by a linear regression procedure applied on the part of the curve between these mentioned points.

**4.7 Poisson's ratio,** $\mu$: Negative ratio of the tensile strain $\varepsilon_n$, in one of the two axes normal to the direction of pull, to the corresponding strain $\varepsilon$ in the direction of pull, within the initial linear portion of the longitudinal versus normal strain curve.

It is expressed as a dimensionless ratio.

Poisson's ratio is indicated as $\mu_b$ (width direction) or $\mu_h$ (thickness direction) according to the relevant axis. Poisson's ratio is preferentially used for long-fibre-reinforced materials.

# 5 Apparatus

## 5.1 Testing machine

### 5.1.1 General

The machine shall comply with ISO 5893, and meet the specifications given in 5.1.2 to 5.1.5, as follows.

### 5.1.2 Speeds of testing

The tensile-testing machine shall be capable of maintaining the speeds of testing (see 4.2) as specified in table 1.

**Table 1 — Recommended testing speeds**

| Speed | Tolerance |
|---|---|
| mm/min | % |
| 1 | ± 20 [1] |
| 2 | ± 20 [1] |
| 5 | ± 20 |
| 10 | ± 20 |
| 20 | ± 10 |
| 50 | ± 10 |
| 100 | ± 10 |
| 200 | ± 10 |
| 500 | ± 10 |

1) These tolerances are smaller than those indicated in ISO 5893.

### 5.1.3 Grips

Grips for holding the test specimen shall be attached to the machine so that the major axis of the test specimen coincides with the direction of pull through the centreline of the grip assembly. This can be achieved, for example, by using centring pins in the grips. The test specimen shall be held such that slip relative to the grips is prevented as far as possible and this shall preferably be effected with the type of grip that maintains or increases pressure on the test specimen as the force applied to the test specimen increases. The clamping system shall not cause premature fracture at the grips.

### 5.1.4 Load indicator

The load indicator shall incorporate a mechanism capable of showing the total tensile load carried by the test specimen when held by the grips. The mechanism shall be essentially free from inertia lag at the specified rate of testing, and shall indicate the load with an accuracy of at least 1 % of the actual value. Attention is drawn to ISO 5893.

### 5.1.5 Extensometer

The extensometer shall comply with ISO 5893. It shall be capable of determining the relative change in the gauge length on the test specimen at any time during the test. It is desirable, but not essential, that this instrument should automatically record this change. The instrument shall be essentially free from inertia lag at the specified speed of testing, and shall be capable of measuring the change of gauge length with an accuracy of 1 % of the relevant value or better. This corresponds to ± 1 µm for the measurement of the modulus, based on a gauge length of 50 mm.

When an extensometer is attached to the test specimen, care shall be taken to ensure that any distortion of or damage to the test specimen is minimal. It is essential that there is no slippage between the extensometer and the test specimen.

The specimens may also be instrumented with longitudinal strain gauges, the accuracy of which shall be 1 % of the relevant value or better. This corresponds to a strain accuracy of $20 \times 10^{-6}$ (20 microstrain) for the measurement of the modulus. The gauges, surface preparation and bonding agents should be chosen to exhibit adequate performance on the subject material.

## 5.2 Devices for measuring width and thickness of the test specimens

### 5.2.1 Rigid materials

A micrometer or its equivalent, capable of reading to 0,02 mm or less and provided with means for measuring the thickness and width of the test specimens, shall be used. The dimensions and shape of the anvils shall be suitable for the specimens being measured and shall not exert a force on the specimen such as to detectably alter the dimension being measured.

### 5.2.2 Flexible materials

A dial-gauge, capable of reading to 0,02 mm or less and provided with a flat circular foot which applies a pressure of 20 kPa ± 3 kPa, shall be used for measuring the thickness.

## 6 Test specimens

### 6.1 Shape and dimensions

See that part of ISO 527 relevant to the material being tested.

### 6.2 Preparation of specimens

See that part of ISO 527 relevant to the material being tested.

### 6.3 Gauge marks

If optical extensometers are used, especially for thin sheet and film, gauge marks on the specimen are necessary to define the gauge length. These shall be approximately equidistant from the midpoint, and the distance between the marks shall be measured to an accuracy of 1 % or better.

Gauge marks shall not be scratched, punched or impressed upon the test specimen in any way that may damage the material being tested. It must be ensured that the marking medium has no detrimental effect

on the material being tested and that, in the case of parallel lines, they are as narrow as possible.

## 6.4 Checking the test specimens

The specimens shall be free of twist and shall have mutually perpendicular pairs of parallel surfaces. The surfaces and edges must be free from scratches, pits, sink marks and flash. The specimens shall be checked for conformity with these requirements by visual observation against straightedges, squares and flat plates, and with micrometer calipers. Specimens showing observed or measured departure from one or more of these requirements shall be rejected or machined to proper size and shape before testing.

## 6.5 Anisotropy

See that part ISO 527 relevant to the material being tested.

## 7 Number of test specimens

7.1 A minimum of five test specimens shall be tested for each of the required directions of testing and for the properties considered (modulus of elasticity, tensile strength etc.). The number of measurements may be more than five if greater precision of the mean value is required. It is possible to evaluate this by means of the confidence interval (95 % probability, see ISO 2602).

7.2 Dumb-bell specimens that break within the shoulders or the yielding of which spreads to the width of the shoulders shall be discarded and further specimens shall be tested.

7.3 Data from parallel-sided specimens where jaw slippage occurs, or where failure occurs within 10 mm of either jaw, or where an obvious fault has resulted in premature failure, shall not be included in the analysis. Repeat tests shall be carried out on new test specimens.

Data, however variable, shall not be excluded from the analysis for any other reason, as the variability in such data is a function of the variable nature of the material being tested.

NOTE 2   When the majority of failures falls outside the criteria for an acceptable failure, the data may be analysed statistically, but it should be recognized that the final result is likely to be conservative. In such instances, it is preferable for the tests to be repeated with the dumb-bell specimens to reduce the possibility of unacceptable results.

## 8 Conditioning

The test specimen shall be conditioned as specified in the appropriate standard for the material concerned. In the absence of this information, the most appropriate condition from ISO 291 shall be selected, unless otherwise agreed upon by the interested parties.

# 9 Procedure

## 9.1 Test atmosphere

Conduct the test in the same atmosphere used for conditioning the test specimen, unless otherwise agreed upon by the interested parties, e.g. for testing at elevated or low temperatures.

## 9.2 Dimensions of test specimen

Measure the width $b$ to the nearest 0,1 mm and the thickness $h$ to the nearest 0,02 mm at the centre of each specimen and within 5 mm of each end of the gauge length.

Record the minimum and maximum values for width and thickness of each specimen and make sure that they are within the tolerances indicated in the standard applicable for the given material.

Calculate the arithmetic means for the width and thickness of each specimen, which shall be used for calculation purposes.

NOTES

3 In the case of injection-moulded specimens, it is not necessary to measure the dimensions of each specimen. It is sufficient to measure one specimen from each lot to make sure that the dimensions correspond to the specimen type selected (see the relevant part of ISO 527). With multiple-cavity moulds, ensure that the dimensions of the specimens are the same for each cavity.

4 For test specimens stamped from sheet or film material, it is permissible to assume that the mean width of the central parallel portion of the die is equivalent to the corresponding width of the specimen. The adoption of such a procedure should be based on comparative measurements taken at periodic intervals.

## 9.3 Clamping

Place the test specimen in the grips, taking care to align the longitudinal axis of the test specimen with the axis of the testing machine. To obtain correct alignment when centring pins are used in the grips, it is necessary to tension the specimen only slightly before tightening the grips (see 9.4). Tighten the grips evenly and firmly to avoid slippage of the test specimen.

## 9.4 Prestresses

The specimen shall not be stressed substantially prior to test. Such stresses can be generated during centring of a film specimen, or can be caused by the clamping pressure, especially with less rigid materials.

The residual stress $\sigma_0$ at the start of a test shall not exceed the following value, for modulus measurement:

$$|\sigma_0| \leqslant 5 \times 10^{-4} E_t \qquad \ldots (1)$$

which corresponds to a prestrain of $\varepsilon_0 \leqslant 0,05$ %, and for measuring relevant stresses $\sigma$, e.g. $\sigma = \sigma_Y$, $\sigma_M$ or $\sigma_B$:

$$\sigma_0 \leqslant 10^{-2}\sigma \qquad \ldots (2)$$

## 9.5 Setting of extensometers

After balancing the prestresses, set and adjust a calibrated extensometer to the gauge length of the test specimen, or provide longitudinal strain gauges, in accordance with 5.1.5. Measure the initial distance (gauge length) if necessary. For the measurement of Poisson's ratio, two elongation- or strain-measuring devices shall be provided to act in the longitudinal and normal axes simultaneously.

For optical measurements of elongation, place gauge marks on the specimen in accordance with 6.3.

The elongation of the free length of the test specimen, measured from the movement of the grips, is used for the values of the nominal tensile strain $\varepsilon_t$ (see 4.5).

## 9.6 Testing speed

Set the speed of testing in accordance with the appropriate standard for the material concerned. In the absence of this information, the speed of testing should be agreed between the interested parties in accordance with table 1.

It may be necessary or desirable to adopt different speeds for the determination of the elastic modulus, of the stress/strain properties up to the yield point, and for the measurement of tensile strength and maximum elongation. For each testing speed, seperate specimens shall be used.

For the measurement of the modulus of elasticity, the selected speed of testing shall provide a strain rate as near as possible to 1 % of the gauge length per minute. The resulting testing speed for different types of specimens is given in that part of ISO 527 relevant to the material being tested.

## 9.7 Recording of data

Record the force and the corresponding values of the increase of the gauge length and of the distance between grips during the test. It is preferable to use an automatic recording system which yields complete stress/strain curves for this operation [see clause 10, equations (3), (4) and (5)].

Determine all relevant stresses and strains defined in clause 4 from the stress/strain curve (see figure 1), or using other suitable means.

For failures outside the criteria for an acceptable failure, see 7.2 and 7.3.

# 10 Calculation and expression of results

## 10.1 Stress calculations

Calculate all stress values defined in 4.3 on the basis of the initial cross-sectional area of the test specimen:

$$\sigma = \frac{F}{A} \qquad \ldots (3)$$

where

$\sigma$     is the tensile stress value in question, expressed in megapascals;

$F$     is the measured force concerned, in newtons;

$A$     is the initial cross-sectional area of the specimen, expressed in square millimetres.

## 10.2 Strain calculations

Calculate all strain values defined in 4.4 on the basis of the gauge length:

$$\varepsilon = \frac{\Delta L_0}{L_0} \qquad \ldots (4)$$

$$\varepsilon\,(\%) = 100 \times \frac{\Delta L_0}{L_0} \qquad \ldots (5)$$

where

$\varepsilon$     is the strain value in question, expressed as a dimensionless ratio, or in percentage;

$L_0$     is the gauge length of the test specimen, expressed in millimetres;

$\Delta L_0$     is the increase in the specimen length between the gauge marks, expressed in millimetres.

The values of the nominal tensile strain, defined in 4.5, shall be calculated on the basis of the initial distance between the grips:

$$\varepsilon_t = \frac{\Delta L}{L} \qquad \ldots (6)$$

$$\varepsilon_t\,(\%) = 100 \times \frac{\Delta L}{L} \qquad \ldots (7)$$

where

$\varepsilon_t$     nominal tensile strain, expressed as a dimensionless ratio or percentage, %;

$L$     initial distance between grips, expressed in millimetres;

$\Delta L$     increase of the distance between grips, expressed in millimetres.

## 10.3 Modulus calculation

Calculate the modulus of elasticity (Young's modulus), defined in 4.6 on the basis of two specified strain values:

$$E_t = \frac{\sigma_2 - \sigma_1}{\varepsilon_2 - \varepsilon_1} \qquad \ldots (8)$$

where

$E_t$     is Young's modulus of elasticity, expressed in megapascals;

$\sigma_1$     is the stress, in megapascals, measured at the strain value $\varepsilon_1 = 0{,}000\,5$;

$\sigma_2$     is the stress, in megapascals, measured at the strain value $\varepsilon_2 = 0{,}002\,5$;

For computer-aided equipment, see 4.6, note 1.

## 10.4 Poisson's ratio

If required, calculate Poisson's ratio defined in 4.7 on the basis of two corresponding strain values perpendicular to each other:

$$\mu_n = -\frac{\varepsilon_n}{\varepsilon} \qquad \ldots (9)$$

where

$\mu_n$     is Poisson's ratio, expressed as a dimensionless ratio with $n = b$ (width) or $h$ (thickness) indicating the normal direction chosen;

$\varepsilon$     is the strain in the longitudinal direction;

$\varepsilon_n$     is the strain in the normal direction, with $n = b$ (width) or $h$ (thickness).

## 10.5 Statistical parameters

Calculate the arithmetic means of the test results and, if required, the standard deviations and the 95 % confidence intervals of the mean values according to the procedure given in ISO 2602.

## 10.6 Significant figures

Calculate the stresses and the modulus to three significant figures. Calculate the strains and Poisson's ratio to two significant figures.

## 11  Precision

See that part of ISO 527 relevant to the material being tested.

## 12  Test report

The test report shall include the following information:

a) a reference to the relevant part of ISO 527;

b) all the data necessary for identification of the material tested, including type, source, manufacturer's code number and history, where these are known;

c) description of the nature and form of the material in terms of whether it is a product, semifinished product, test panel or specimen. It should include the principal dimensions, shape, method of manufacture, succession of layers and any pretreatment;

d) type of test specimen, the width and thickness of the parallel section, including mean, minimum and maximum values;

e) method of preparing the test specimens, and any details of the manufacturing method used;

f) if the material is in product or semifinished product form, the orientation of the specimen in relation to the product or semifinished product from which it is cut;

g) number of test specimens tested;

h) standard atmosphere for conditioning and testing, plus any special conditioning treatment, if required by the relevant standard for the material or product concerned;

i) accuracy grading of the test machine (see ISO 5893);

j) type of elongation or strain indicator;

k) type of clamping device and clamping pressure, if known;

l) testing speeds;

m) individual test results;

n) mean value(s) of the measured property(ies), quoted as the indicative value(s) for the material tested;

o) standard deviation, and/or coefficient of variation, and/or confidence limits of the mean, if required;

p) statement as to whether any test specimens have been rejected and replaced, and, if so, the reasons;

q) date of measurement.

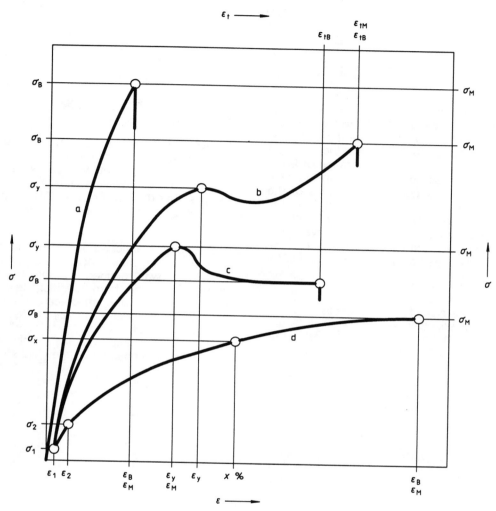

| Curve a | Brittle materials |
| Curves b and c | Tough materials with yield point |
| Curve d | Tough materials without yield point |

The points for the calculation of Young's modulus $E_t$ according to 10.3 are indicated by $(\sigma_1, \varepsilon_1)$ and $(\sigma_2, \varepsilon_2)$, shown only for curve d ($\varepsilon_1 = 0{,}000\ 5$; $\varepsilon_2 = 0{,}002\ 5$).

**Figure 1 — Typical stress/strain curves**

# Annex A
(informative)

## Young's modulus and related values

Due to their viscoelastic behaviour many properties of polymer materials depend not only on temperature but also on time. With regard to the tensile test, this causes nonlinear stress/strain curves (bending towards the strain axis) even within the range of linear viscoelasticity. This effect is pronounced in the case of tough polymers. Consequently, the values of the tangent modulus of tough materials taken from the initial part of the stress/strain curves often depend strongly on the scales used. Thus the conventional method (tangent at the initial point of the stress/strain curve) does not give reliable moduli for these materials.

The method for the measurement of Young's modulus prescribed in this part of ISO 527 is based, therefore, on two specified strain values, i.e. 0,25 % and 0,05 %. (The lower strain value has been set at not zero in order to avoid errors in the measured modulus caused by possible onset effects at the beginning of the stress/strain curve.)

In the case of brittle polymers, both the new and the conventional methods give the same values for the modulus. The new method, however, allows accurate and reproducible measurement of the moduli of tough plastics. The definition of the initial tangent modulus, therefore, has been deleted in the present part of ISO 527.

The aspects mentioned above for the modulus similarly relate to the "offset yield point", which in ISO/R 527 was defined by the deviation of the stress/strain curve from its initial linearity. The offset yield point, therefore, is replaced by a point of specified strain (stress at $x$ % strain, $\sigma_x$, see 4.3.4). Since the definition of such a "substitute" yield point is significant for tough materials only, the specified strain shall be chosen near the yield strain commonly found.

**INTERNATIONAL STANDARD ISO 527-1:1993**
TECHNICAL CORRIGENDUM 1

Published 1994-05-01

INTERNATIONAL ORGANIZATION FOR STANDARDIZATION· МЕЖДУНАРОДНАЯ ОРГАНИЗАЦИЯ ПО СТАНДАРТИЗАЦИИ· ORGANISATION INTERNATIONALE DE NORMALISATION

# Plastics — Determination of tensile properties —

# Part 1:
## General principles

TECHNICAL CORRIGENDUM 1

*Plastiques — Détermination des propriétés en traction —*

*Partie 1: Principes généraux*

*RECTIFICATIF TECHNIQUE 1*

Technical corrigendum 1 to International Standard ISO 527-1:1993 was prepared by Technical Committee ISO/TC 61, *Plastics*, Subcommittee SC 2, *Mechanical properties*.

———————

Throughout the text, delete the expression "Young's modulus" or replace it by "tensile modulus of elasticity", depending on the context. This concerns definition 4.6, subclause 10.3 (twice), figure 1 and annex A (title plus second paragraph).

UDC  [678.5/.8].017:620.172

Ref. No. ISO 527-1:1993/Cor.1:1994(E)

Descriptors: plastics, tests, tension tests, determination, tensile properties, generalities.

# INTERNATIONAL STANDARD

**ISO**

**527-2**

First edition
1993-06-15

# Plastics — Determination of tensile properties —

## Part 2:
Test conditions for moulding and extrusion plastics

*Plastiques — Détermination des propriétés en traction —*

*Partie 2: Conditions d'essai des plastiques pour moulage et extrusion*

Reference number
ISO 527-2:1993(E)

# Foreword

ISO (the International Organization for Standardization) is a worldwide federation of national standards bodies (ISO member bodies). The work of preparing International Standards is normally carried out through ISO technical committees. Each member body interested in a subject for which a technical committee has been established has the right to be represented on that committee. International organizations, governmental and non-governmental, in liaison with ISO, also take part in the work. ISO collaborates closely with the International Electrotechnical Commission (IEC) on all matters of electrotechnical standardization.

Draft International Standards adopted by the technical committees are circulated to the member bodies for voting. Publication as an International Standard requires approval by at least 75 % of the member bodies casting a vote.

International Standard ISO 527-2 was prepared by Technical Committee ISO/TC 61, *Plastics*, Sub-Committee SC 2, *Mechanical properties*.

Together with the other parts of ISO 527, it cancels and replaces ISO Recommendation R 527:1966, which has been technically revised.

Annex A of this part of ISO 527 cancels and replaces ISO 6239:1986, *Plastics — Determination of tensile properties by use of small specimens.*

ISO 527 consists of the following parts, under the general title *Plastics — Determination of tensile properties*:

— *Part 1: General principles*

— *Part 2: Test conditions for moulding and extrusion plastics*

— *Part 3: Test conditions for sheet and film*

— *Part 4: Test conditions for isotropic and orthotropic fibre-reinforced plastic composites*

— *Part 5: Test conditions for unidirectional fibre-reinforced plastic composites*

Annex A forms an integral part of this part of ISO 527.

International Organization for Standardization
Case Postale 56 • CH-1211 Genève 20 • Switzerland
Printed in Switzerland

# Plastics — Determination of tensile properties —

# Part 2:
Test conditions for moulding and extrusion plastics

## 1 Scope

**1.1** This part of ISO 527 specifies the test conditions for determining the tensile properties of moulding and extrusion plastics, based upon the general principles given in ISO 527-1.

**1.2** The methods are selectively suitable for use with the following range of materials:

— rigid and semirigid thermoplastics moulding, extrusion and cast materials, including compounds filled and reinforced by e.g. short fibres, small rods, plates or granules but excluding textile fibres (see ISO 527-4 and ISO 527-5) in addition to unfilled types;

— rigid and semirigid thermosetting moulding and cast materials, including filled and reinforced compounds but excluding textile fibres as reinforcement (see ISO 527-4 and ISO 527-5);

— thermotropic liquid crystal polymers.

The methods are not suitable for use with materials reinforced by textile fibres (see ISO 527-4 and ISO 527-5), with rigid cellular materials or sandwich structures containing cellular material.

**1.3** The methods are applied using specimens which may be either moulded to the chosen dimensions or machined, cut or punched from injection- or compression-moulded plates. The multipurpose test specimen is preferred (see ISO 3167:1993, *Plastics — Multipurpose test specimens*).

## 2 Normative references

The following standards contain provisions which, through reference in this text, constitute provisions of this part of ISO 527. At the time of publication, the editions indicated were valid. All standards are subject to revision, and parties to agreements based on this part of ISO 527 are encouraged to investigate the possibility of applying the most recent editions of the standards indicated below. Members of IEC and ISO maintain registers of currently valid International Standards.

ISO 37:1977, *Rubber, vulcanized — Determination of tensile stress-strain properties.*

ISO 293:1986, *Plastics — Compression moulding test specimens of thermoplastic materials.*

ISO 294:—[1], *Plastics — Injection moulding of test specimens of thermoplastic materials.*

ISO 295:1991, *Plastics — Compression moulding of test specimens of thermosetting materials.*

ISO 527-1:1993, *Plastics — Determination of tensile properties — Part 1: General principles.*

ISO 1926:1979, *Cellular plastics — Determination of tensile properties of rigid materials.*

ISO 2818:—[2], *Plastics — Preparation of test specimens by machining.*

---

1) To be published. (Revision of ISO 294:1975)

2) To be published. (Revision of ISO 2818:1980)

## 3 Principle

See ISO 527-1:1993, clause 3.

## 4 Definitions

For the purposes of this part of ISO 527, the definitions given in ISO 527-1 apply.

## 5 Apparatus

See ISO 527-1:1993, clause 5.

## 6 Test specimens

### 6.1 Shape and dimensions

Wherever possible, the test specimens shall be dumb-bell-shaped types 1A and 1B as shown in figure 1. Type 1A is preferred for directly-moulded multipurpose test specimens, type 1B for machined specimens.

NOTE 1   Types 1A and 1B test specimens having 4 mm thickness are identical to the multipurpose test specimens according to ISO 3167, types A and B, respectively.

For the use of small specimens, see annex A.

### 6.2 Preparation of test specimens

Test specimens shall be prepared in accordance with the relevant material specification. When none exists, or unless otherwise specified, specimens shall be either directly compression- or injection moulded from the material in accordance with ISO 293, ISO 294 or ISO 295, as appropriate, or machined in accordance with ISO 2818 from plates that have been compression- or injection-moulded from the compound.

All surfaces of the test specimens shall be free from visible flaws, scratches or other imperfections. From moulded specimens all flash, if present, shall be removed, taking care not to damage the moulded surface.

Test specimens from finished goods shall be taken from flat areas or zones having minimum curvature. For reinforced plastics, test specimens should not be machined to reduce their thickness unless absolutely necessary. Test specimens with machined surfaces will not give results comparable to specimens having non-machined surfaces.

### 6.3 Gauge marks

See ISO 527-1:1993, subclause 6.3.

### 6.4 Checking the test specimens

See ISO 527-1:1993, subclause 6.4.

## 7 Number of test specimens

See ISO 527-1:1993, clause 7.

## 8 Conditioning

See ISO 527-1:1993, clause 8.

## 9 Procedure

See ISO 527-1:1993, clause 9.

For the measurement of the modulus of elasticity, the speed of testing shall be 1 mm/min for specimen types 1A and 1B (see figure 1). For small specimens see annex A.

## 10 Calculation and expression of results

See ISO 527-1:1993, clause 10.

## 11 Precision

The precision of this test method is not known, because interlaboratory data are not available. When interlaboratory data are obtained, a precision statement will be added with the next revision.

## 12  Test report

The test report shall include the following information:

a) a reference to this part of ISO 527, including the type of specimen and the testing speed according to:

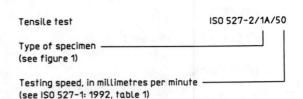

Tensile test                                    ISO 527–2/1A/50

Type of specimen ────────────────────────┐
(see figure 1)

Testing speed, in millimetres per minute ──────────┘
(see ISO 527–1: 1992, table 1)

For items b) to q) in the test report, see ISO 527-1:1993, 12 b) to q).

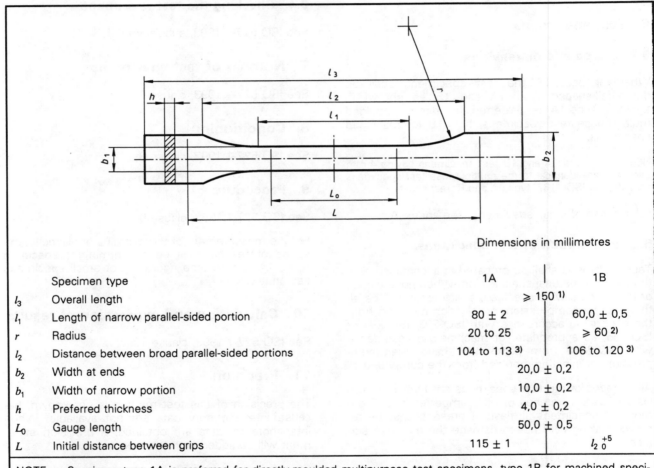

Dimensions in millimetres

| | Specimen type | 1A | 1B |
|---|---|---|---|
| $l_3$ | Overall length | $\geq 150$ [1] | |
| $l_1$ | Length of narrow parallel-sided portion | $80 \pm 2$ | $60{,}0 \pm 0{,}5$ |
| $r$ | Radius | 20 to 25 | $\geq 60$ [2] |
| $l_2$ | Distance between broad parallel-sided portions | 104 to 113 [3] | 106 to 120 [3] |
| $b_2$ | Width at ends | $20{,}0 \pm 0{,}2$ | |
| $b_1$ | Width of narrow portion | $10{,}0 \pm 0{,}2$ | |
| $h$ | Preferred thickness | $4{,}0 \pm 0{,}2$ | |
| $L_0$ | Gauge length | $50{,}0 \pm 0{,}5$ | |
| $L$ | Initial distance between grips | $115 \pm 1$ | $l_2 \,{}^{+5}_{\ 0}$ |

NOTE — Specimen type 1A is preferred for directly-moulded multipurpose test specimens, type 1B for machined specimens.

1) For some materials, the length of the tabs may need to be extended (e. g. $l_3 = 200$ mm) to prevent breakage or slippage in the testing jaws.

2) $r = [(l_2 - l_1)^2 + (b_2 - b_1)^2]/4\,(b_2 - b_1)$

3) Resulting from $l_1$, $r$, $b_1$ and $b_2$, but within the indicated tolerance.

**Figure 1 — Test specimen types 1A and 1B**

# Annex A
## (normative)

## Small specimens

If for any reason it is not possible to use a standard type 1 test specimen, specimens of the types 1BA, 1BB (see figure A.1), 5A or 5B (see figure A.2) may be used, provided that the speed of testing is adjusted to the value given in 5.1.2, table 1 of ISO 527-1:1993, which gives the nominal strain rate for the small test specimen closest to that used for the standard-sized specimen. The rate of nominal strain is the quotient of the speed of testing (see 4.2 in ISO 527-1:1993) and the initial distance between grips. Where modulus measurements are required, the test speed shall be 1 mm/min. It may be technically difficult to measure modulus on small specimens because of small gauge lengths and short testing times. Results obtained from small specimens are not comparable with those obtained from type 1 specimens.

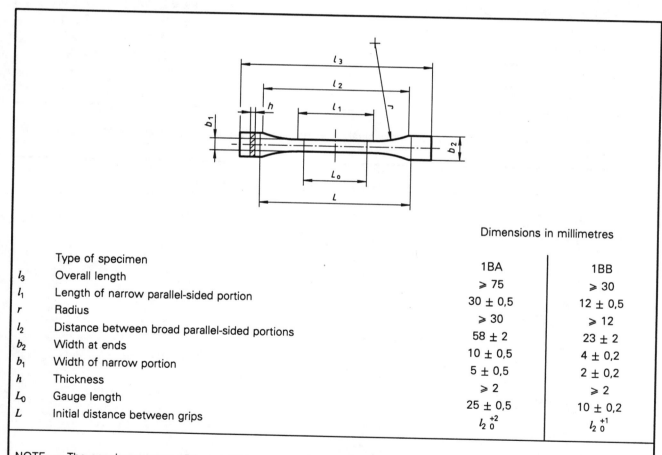

Dimensions in millimetres

| | Type of specimen | 1BA | 1BB |
|---|---|---|---|
| $l_3$ | Overall length | $\geqslant 75$ | $\geqslant 30$ |
| $l_1$ | Length of narrow parallel-sided portion | $30 \pm 0,5$ | $12 \pm 0,5$ |
| $r$ | Radius | $\geqslant 30$ | $\geqslant 12$ |
| $l_2$ | Distance between broad parallel-sided portions | $58 \pm 2$ | $23 \pm 2$ |
| $b_2$ | Width at ends | $10 \pm 0,5$ | $4 \pm 0,2$ |
| $b_1$ | Width of narrow portion | $5 \pm 0,5$ | $2 \pm 0,2$ |
| $h$ | Thickness | $\geqslant 2$ | $\geqslant 2$ |
| $L_0$ | Gauge length | $25 \pm 0,5$ | $10 \pm 0,2$ |
| $L$ | Initial distance between grips | $l_2 {}^{+2}_{\ 0}$ | $l_2 {}^{+1}_{\ 0}$ |

NOTE — The specimen types 1BA and 1BB are proportionally scaled to type 1B with a reduction factor of 1:2 and 1:5 respectively with the exception of thickness.

**Figure A.1 — Test specimen types 1BA and 1BB**

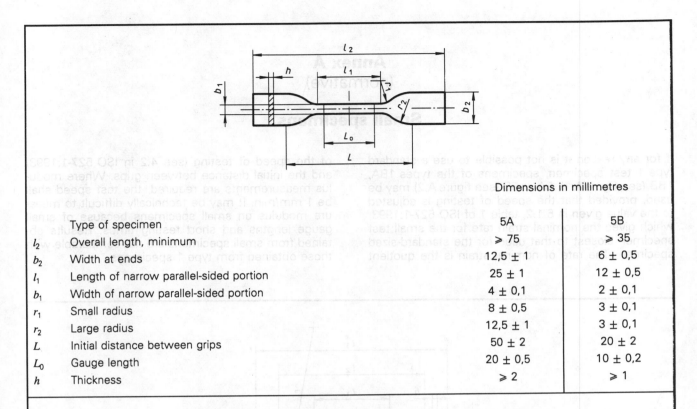

Dimensions in millimetres

| Type of specimen | | 5A | 5B |
|---|---|---|---|
| $l_2$ | Overall length, minimum | ⩾ 75 | ⩾ 35 |
| $b_2$ | Width at ends | 12,5 ± 1 | 6 ± 0,5 |
| $l_1$ | Length of narrow parallel-sided portion | 25 ± 1 | 12 ± 0,5 |
| $b_1$ | Width of narrow parallel-sided portion | 4 ± 0,1 | 2 ± 0,1 |
| $r_1$ | Small radius | 8 ± 0,5 | 3 ± 0,1 |
| $r_2$ | Large radius | 12,5 ± 1 | 3 ± 0,1 |
| $L$ | Initial distance between grips | 50 ± 2 | 20 ± 2 |
| $L_0$ | Gauge length | 20 ± 0,5 | 10 ± 0,2 |
| $h$ | Thickness | ⩾ 2 | ⩾ 1 |

NOTE — Test specimen types 5A and 5B are approximately proportional to type 5 of ISO 527-3 and represent respectively types 2 and 3 of ISO 37.

**Figure A.2 — Test specimen types 5A and 5B**

# INTERNATIONAL STANDARD

ISO
**899-1**

First edition
1993-12-15

# Plastics — Determination of creep behaviour —

## Part 1:
Tensile creep

*Plastiques — Détermination du comportement au fluage —*
*Partie 1: Fluage en traction*

Reference number
ISO 899-1:1993(E)

# Foreword

ISO (the International Organization for Standardization) is a worldwide federation of national standards bodies (ISO member bodies). The work of preparing International Standards is normally carried out through ISO technical committees. Each member body interested in a subject for which a technical committee has been established has the right to be represented on that committee. International organizations, governmental and non-governmental, in liaison with ISO, also take part in the work. ISO collaborates closely with the International Electrotechnical Commission (IEC) on all matters of electrotechnical standardization.

Draft International Standards adopted by the technical committees are circulated to the member bodies for voting. Publication as an International Standard requires approval by at least 75 % of the member bodies casting a vote.

International Standard ISO 899-1 was prepared by Technical Committee ISO/TC 61, *Plastics*, Sub-Committee SC 2, *Mechanical properties*.

Together with ISO 899-2, it cancels and replaces ISO 899:1981 and ISO 6602:1985, which have been technically revised.

ISO 899 consists of the following parts, under the general title *Plastics — Determination of creep behaviour*:

— *Part 1: Tensile creep*

— *Part 2: Flexural creep by three-point loading*

Annexes A and B of this part of ISO 899 are for information only.

International Organization for Standardization
Case Postale 56 • CH-1211 Genève 20 • Switzerland

Printed in Switzerland

# Plastics — Determination of creep behaviour —

# Part 1:
Tensile creep

## 1 Scope

**1.1** This part of ISO 899 specifies a method for determining the tensile creep of plastics in the form of standard test specimens under specified conditions such as those of pretreatment, temperature and humidity.

**1.2** The method is suitable for use with rigid and semi-rigid non-reinforced, filled and fibre-reinforced plastics materials (see ISO 472 for definitions) in the form of dumb-bell-shaped test specimens moulded directly or machined from sheets or moulded articles.

**1.3** The method is intended to provide data for engineering-design and research and development purposes.

**1.4** Tensile creep may vary significantly with differences in specimen preparation and dimensions and in the test environment. The thermal history of the test specimen can also have profound effects on its creep behaviour (see annex A). Consequently, when precise comparative results are required, these factors must be carefully controlled.

**1.5** If tensile-creep properties are to be used for engineering-design purposes, the plastics materials should be tested over a broad range of stresses, times and environmental conditions.

## 2 Normative references

The following standards contain provisions which, through reference in this text, constitute provisions of this part of ISO 899. At the time of publication, the editions indicated were valid. All standards are subject to revision, and parties to agreements based on this part of ISO 899 are encouraged to investigate the possibility of applying the most recent editions of the standards indicated below. Members of IEC and ISO maintain registers of currently valid International Standards.

ISO 291:1977, *Plastics — Standard atmospheres for conditioning and testing.*

ISO 472:1988, *Plastics — Vocabulary.*

ISO 527-1:1993, *Plastics — Determination of tensile properties — Part 1: General principles.*

ISO 527-2:1993, *Plastics — Determination of tensile properties — Part 2: Test conditions for moulding and extrusion plastics.*

## 3 Definitions

For the purposes of this part of ISO 899, the definitions given in ISO 472 and the following definitions apply.

**3.1 creep:** The increase in strain with time when a constant force is applied.

**3.2 initial stress, $\sigma$:** The tensile force per unit area of the initial cross-section within the gauge length.

It is given by the equation

$$\sigma = \frac{F}{A}$$

where

$F$    is the force, in newtons;

$A$    is the initial cross-sectional area of the specimen, in square millimetres.

The stress is expressed in megapascals.

**3.3 extension, $(\Delta L)_t$:** The increase in the distance between the gauge marks, expressed in millimetres, at time $t$.

It is given by the equation

$$(\Delta L)_t = L_t - L_0$$

where

$L_t$ is the gauge length, in millimetres, at any given time $t$ during the test;

$L_0$ is the original gauge length, in millimetres, of the specimen after application of a preload but prior to application of the test load.

**3.4 tensile-creep strain, $\varepsilon_t$:** The change in length per unit original length of the gauge length produced by the applied load at any given time during a creep test. It is expressed as a dimensionless ratio or as a percentage.

It is given by the equation

$$\varepsilon_t = \frac{(\Delta L)_t}{L_0}$$

or

$$\varepsilon_t = \frac{(\Delta L)_t}{L_0} \times 100 \ (\%)$$

**3.5 tensile-creep modulus, $E_t$:** The ratio of initial stress to creep strain, calculated as in 7.1.

**3.6 isochronous stress-strain curve:** A Cartesian plot of stress versus creep strain, at a specific time after application of the test load.

**3.7 time to rupture:** The period of time which elapses between the point in time at which the specimen is fully loaded and the rupture point.

**3.8 creep-strength limit:** That initial stress which will just cause rupture ($\sigma_{B,t}$) or will produce a specified strain ($\sigma_{\varepsilon,t}$) at a specified time $t$, at a given temperature and relative humidity.

**3.9 recovery from creep:** The decrease in strain at any given time after completely unloading the specimen, expressed as a percentage of the strain just prior to the removal of the load.

## 4 Apparatus

**4.1 Gripping device,** capable of ensuring that the direction of the load applied to the test specimen coincides as closely as possible with the longitudinal axis of the specimen. This ensures that the test specimen is subjected to simple stress and that the stresses in the loaded section of the specimen may be assumed to be uniformly distributed over cross-sections perpendicular to the direction of the applied load.

NOTE 1 It is recommended that grips be used that will allow the specimen to be fixed in place, correctly aligned, prior to applying the load. Self-locking grips which allow the specimen to move as the load increases are not suitable for this test.

**4.2 Loading system,** capable of ensuring that the load is applied smoothly, without causing transient overloading, and that the load is maintained to within ± 1 % of the desired load. In creep-to-rupture tests, provision shall be made to prevent any shocks which occur at the moment of rupture being transmitted to adjacent loading systems. The loading mechanism shall allow rapid, smooth and reproducible loading.

**4.3 Extension-measuring device,** comprising any contactless or contact device capable of measuring the extension of the specimen gauge length under load without influencing the specimen behaviour by mechanical effects (e.g. undesirable deformations, notches), other physical effects (e.g. heating of the specimen) or chemical effects. In the case of contactless (optical) measurement of the strain, the longitudinal axis of the specimen shall be perpendicular to the optical axis of the measuring device. The accuracy of the extension-measuring device shall be within ± 0,01 mm.

For creep-to-rupture tests, it is recommended that the extension be measured by means of a contactless optical system operating on the cathetometer principle. Automatic indication of time to rupture is highly desirable. The gauge length shall be marked on the specimen, either by attaching (metal) clips with scratched-on gauge marks, or by gauge marks ruled with an inert, thermally stable paint.

Electrical-resistance strain gauges are suitable only if the material tested is of such a nature as to permit such strain gauges to be attached to the specimen by means of adhesive and only if the adhesion quality is constant during the duration of the test.

**4.4 Time-measurement device,** accurate to 0,1 %.

**4.5 Micrometer,** reading to 0,01 mm or closer, for measuring the thickness and width of the test specimen.

## 5 Test specimens

Use test specimens of the same shape and dimensions as specified for the determination of tensile properties (see ISO 527-2).

# 6 Procedure

## 6.1 Conditioning and test atmosphere

Condition the test specimens as specified in the International Standard for the material under test. In the absence of any information on conditioning, use the most appropriate set of conditions specified in ISO 291, unless otherwise agreed upon by the interested parties.

NOTE 2  The creep behaviour will be affected not only by the thermal history of the specimen under test, but also by the temperature and (where applicable) humidity used in conditioning.

Conduct the test in the same atmosphere as used for conditioning, unless otherwise agreed upon by the interested parties, e.g. for testing at elevated or low temperatures. Ensure that the variation in temperature during the duration of the test remains within ± 2 °C.

## 6.2 Measurement of test-specimen dimensions

Measure the dimensions of the conditioned test specimens in accordance with ISO 527-1:1993, subclause 9.2.

## 6.3 Mounting the test specimens

Mount a conditioned and measured specimen in the grips and set up the extension-measuring device as required.

## 6.4 Selection of stress value

Select a stress value appropriate to the application envisaged for the material under test, and calculate, using the equation given in 3.2, the test load to be applied to the test specimen.

## 6.5 Loading procedure

### 6.5.1 Preloading

When it is necessary to preload the test specimen prior to increasing the load to the test load, for example in order to eliminate backlash by the test gear, take care to ensure that the preload does not influence the test results. Do not apply the preload until the temperature and humidity of the test specimen (gripped in the test apparatus) correspond to the test conditions. Measure the gauge length after application of the preload. Maintain the preload during the whole duration of the test.

### 6.5.2 Loading

Load the test specimen progressively so that full loading of the test specimen is reached between 1 s and 5 s after the beginning of the application of the load. Use the same rate of loading for each of a series of tests on one material.

Take the total load (including the preload) to be the test load.

## 6.6 Extension-measurement schedule

Record the point in time at which the specimen is fully loaded as $t = 0$. Unless the extension is automatically and/or continuously recorded, choose the times for making individual measurements as a function of the creep curve obtained from the particular material under test. It is preferable to use the following measurement schedule:

1 min, 3 min, 6 min, 12 min and 30 min;

1 h, 2 h, 5 h, 10 h, 20 h, 50 h, 100 h, 200 h, 500 h, 1 000 h, etc.

If discontinuities are suspected or encountered in the creep-strain versus time plot, take readings more frequently than recommended above.

## 6.7 Time measurement

Measure, to within ± 0,1 % or ± 2 s (whichever is the less severe tolerance), the total time which has elapsed up to each creep measurement.

## 6.8 Temperature and humidity control

Unless temperature and relative humidity (where applicable) are recorded automatically, record them at the beginning of the test and then at least three times a day initially. When it has become evident that the conditions are stable within the specified limits, they may be checked less frequently.

## 6.9 Measurement of recovery rate (optional)

Upon completion of non-rupture tests, remove the load rapidly and smoothly and measure the recovery rate using, for instance, the same schedule as was used for creep measurement.

# 7 Expression of results

## 7.1 Method of calculation

Calculate the tensile-creep modulus $E_t$ by dividing the initial stress $\sigma$ by the strain $\varepsilon_t$ at each of the selected measurement times.

It is given, in megapascals, by the equation

$$E_t = \frac{\sigma}{\varepsilon_t} = \frac{F \cdot L_0}{A \cdot (\Delta L)_t}$$

where

$F$    is the applied force, in newtons;

$L_0$   is the initial gauge length, in millimetres;

$A$    is the initial cross-sectional area, in square millimetres, of the specimen;

$(\Delta L)_t$ is the extension, in millimetres, at time $t$.

## 7.2    Presentation of results

### 7.2.1    Creep curves

If testing is carried out at different temperatures, the raw data should preferably be presented, for each temperature, as a series of creep curves showing the tensile strain plotted against the logarithm of time, one curve being plotted for each initial stress used (see figure 1).

The data may also be presented in other ways, e.g. as described in 7.2.2 and 7.2.3, to provide information required for particular applications.

### 7.2.2    Creep-modulus/time curves

For each initial stress used, the tensile-creep modulus, calculated in accordance with 7.1, may be plotted against the logarithm of the time under load (see figure 2).

If testing is carried out at different temperatures, plot a series of curves for each temperature.

### 7.2.3    Isochronous stress-strain curves

An isochronous stress-strain curve is a Cartesian plot showing how the strain depends on the applied load, at a specific point in time after application of the load. Several curves are normally plotted, corresponding to times under load of 1 h, 10 h, 100 h, 1 000 h, and 10 000 h. Since each creep test gives only one point on each curve, it is necessary to carry out the test at at least three different stresses, and preferably more, to obtain an isochronous curve.

To obtain an isochronous stress-strain curve for a particular time under load (say 10 h) from a series of creep curves as shown in figure 1, read off, from each creep curve, the strain at 10 h, and plot these strain values ($x$-axis) against the corresponding stress values ($y$-axis). Repeat the process for other times to obtain a series of isochronous curves (see figure 3).

If testing is carried out at different temperatures, plot a series of curves for each temperature.

### 7.2.4    Three-dimensional representation

A relationship of the form $\varepsilon = f(t, \sigma)$ exists between the different types of curve (see figures 1 to 3) that can be derived from the raw creep-test data. This relationship can be represented as a surface in a three-dimensional space (see reference [1], annex B).

All the curves that can be derived from the raw creep-test data form part of this surface. Because of the experimental error inherent in each measurement, the points corresponding to the actual measurements normally do not lie on the curves but just off them.

The surface $\varepsilon = f(t, \sigma)$ can therefore be generated by deriving a number of the curves which form it, but a number of sophisticated smoothing operations are usually necessary. Computer techniques permit this to be done rapidly and reliably.

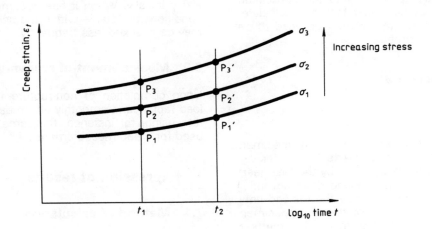

**Figure 1 — Creep curves**

### 7.2.5 Creep-to-rupture curves

Creep-to-rupture curves allow the prediction of the time to failure at any stress. They may be plotted as stress against log time (see figure 4) or log stress against log time.

### 7.3 Precision

The precision of this test method is not known because interlaboratory data are not available. When interlaboratory data are obtained, a precision statement will be added at the next revision.

## 8 Test report

The test report shall include the following particulars:

a) a reference to this part of ISO 899;

b) a complete description of the material tested, including all pertinent information on composition, preparation, manufacturer, tradename, code number, date of manufacture, type of moulding and any annealing;

c) the dimensions of each test specimen;

d) the method of preparation of the test specimens;

e) the directions of the principal axes of the test specimens with respect to the dimensions of the product or some known or inferred orientation in the material;

f) details of the atmosphere used for conditioning and testing;

g) the creep-test data for each temperature at which testing was carried out, presented in one or more of the graphical forms described in 7.2, or in tabular form;

h) if recovery-rate measurements are made, the time-dependent strain after unloading the test specimen (see 6.9).

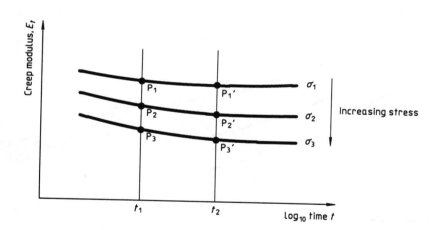

Figure 2 — Creep-modulus/time curves

169

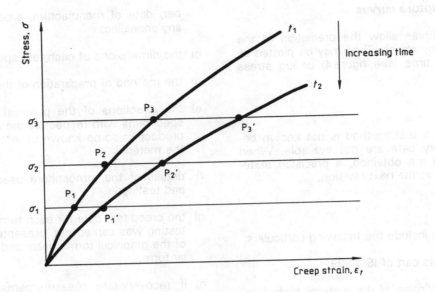

**Figure 3 — Isochronous stress-strain curves**

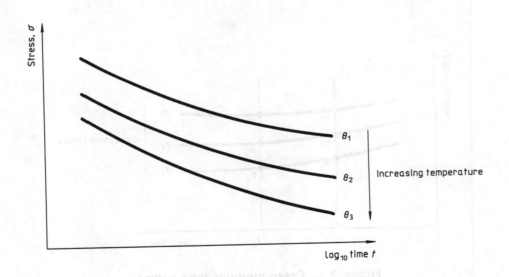

NOTE — The stress $\sigma$ may also be plotted on a logarithmic scale.

**Figure 4 — Creep-to-rupture curves**

# Annex A

(informative)

# Physical-ageing effects on the creep of polymers

## A.1  General

Physical ageing takes place when a polymer is cooled from an elevated temperature at which the molecular mobility is high to a lower temperature at which relaxation times for molecular motions are long in comparison with the storage time at that temperature. Under these circumstances, changes in the structure will take place over a long period of time, involving rearrangement in the shape and packing of molecules as the polymer approaches the equilibrium structural state for the lower temperature. Associated with this ageing process, there is a progressive decrease in the molecular mobility of the polymer, even when the temperature remains constant. As a direct consequence of this, the creep deformation produced by an applied stress will depend upon the age of the polymer, creep rates being lower in more highly aged material.

This is illustrated in figure A.1 which shows creep compliance curves for PVC specimens of different ages. Each of these specimens has been rapidly cooled from a temperature of 85 °C (close to $T_g$) and stored at the test temperature of 23 °C for different times $t_e$ prior to load application. The physical age of a specimen is then defined by the time $t_e$ and it can be seen that the older the specimen the further its creep curve is shifted on the time axis.

## A.2  Creep at elevated temperatures

The influence of physical ageing on creep behaviour is more complicated when measurements are made at elevated temperatures following a storage period at a lower (ambient) temperature. Under these cir-cumstances, the physical ageing that takes place during storage at the lower temperature is temporarily reversed when the specimen is heated to the test temperature. The rate at which this takes place depends on the size of the temperature change and the age of the specimen when the temperature is raised. Following the reduction in the apparent (or effective) age of the specimen, physical ageing is reactivated at the higher temperature. Again, the timescale over which this happens depends on the test conditions. One consequence of these changes in age state caused by the temperature increase is thus a dependence of the creep behaviour at the elevated temperature on the dwell time at this temperature prior to load application.

Typical ways in which this type of thermal history influences creep compliance are illustrated in figures A.2 and A.3. In figure A.2, specimens were stored for a period $t_{e1}$ of 200 h at a temperature of 23 °C prior to heating to the test temperature of 44 °C. Creep curves were then obtained after different periods $t_{e2}$ at 44 °C prior to load application. Despite the relatively long storage period $t_{e1}$ at ambient temperature, the creep behaviour shows a strong dependence on the dwell time $t_{e2}$.

In figure A.3, creep tests were carried out under the same conditions but following a storage period $t_{e1}$ of greater than 1 year at 23 °C prior to heating to the test temperature. The progressive reduction in the effective age of the specimens is actually observed here as a shift in the curves to shorter creep times, and results from the more extensive structural changes that have taken place in the specimens through physical ageing before heating.

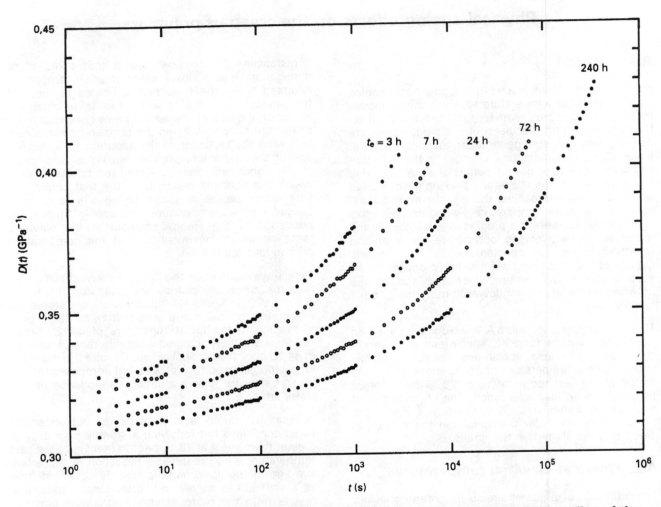

**Figure A.1 — Creep curves for PVC at 23 °C obtained at different times $t_e$ after rapid cooling of the specimen from 85 °C to 23 °C**

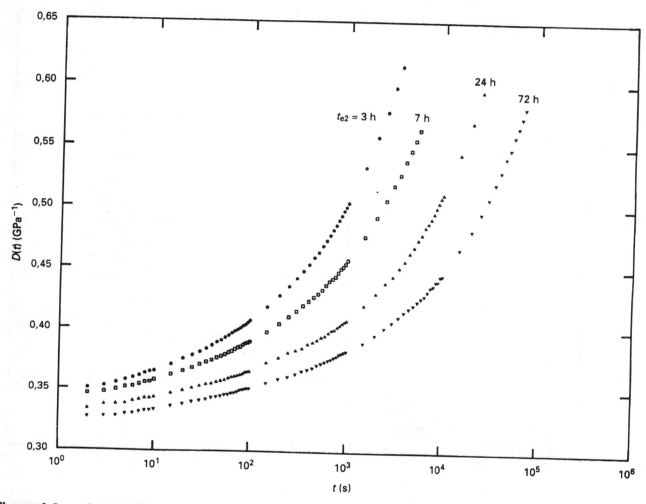

**Figure A.2 — Creep curves for PVC at 44 °C obtained by application of the load at different times $t_{e2}$ after heating from 23 °C** (the specimen had been stored for 200 h at 23 °C prior to heating)

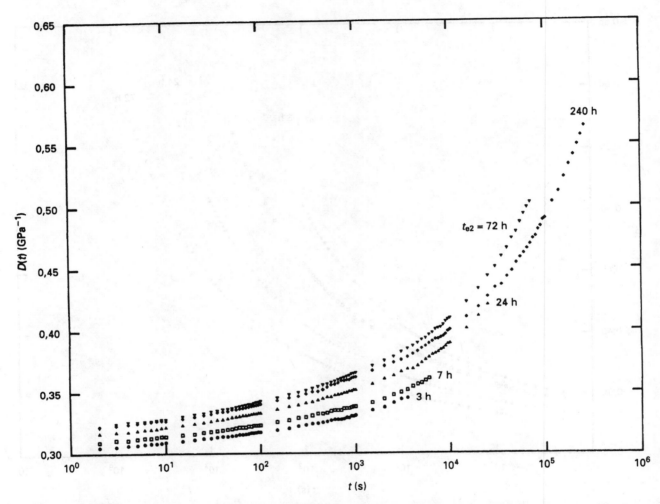

**Figure A.3 — As for figure A.2 but following storage for more than 1 year at 23 °C prior to heating**

# Annex B
(informative)

# Bibliography

[1]  TURNER, S. Creep in thermoplastics — Preliminary concepts and definitions, *British Plastics,* June (1964), pp. 322-324.

# INTERNATIONAL STANDARD

**ISO 1133**

Second edition
1991-03-15

## Plastics — Determination of the melt mass-flow rate (MFR) and the melt volume-flow rate (MVR) of thermoplastics

*Plastiques — Détermination de l'indice de fluidité à chaud des thermoplastiques, en masse (MFR) et en volume (MVR)*

Reference number
ISO 1133:1991(E)

# Foreword

ISO (the International Organization for Standardization) is a worldwide federation of national standards bodies (ISO member bodies). The work of preparing International Standards is normally carried out through ISO technical committees. Each member body interested in a subject for which a technical committee has been established has the right to be represented on that committee. International organizations, governmental and non-governmental, in liaison with ISO, also take part in the work. ISO collaborates closely with the International Electrotechnical Commission (IEC) on all matters of electrotechnical standardization.

Draft International Standards adopted by the technical committees are circulated to the member bodies for voting. Publication as an International Standard requires approval by at least 75 % of the member bodies casting a vote.

International Standard ISO 1133 was prepared by Technical Committee ISO/TC 61, *Plastics*.

This second edition cancels and replaces the first edition (ISO 1133:1981). This new edition includes, in addition to the previously described method for the determination of the melt mass-flow rate, a new procedure for the automatic measurement of both melt mass-flow rate and melt volume-flow rate.

Annex A forms an integral part of this International Standard.

# Plastics — Determination of the melt mass-flow rate (MFR) and the melt volume-flow rate (MVR) of thermoplastics

## 1 Scope

**1.1** This International Standard specifies a method for the determination of the melt mass-flow rate (MFR) and the melt volume-flow rate (MVR) of thermoplastic materials under specified conditions of temperature and load. Normally, the test conditions for measurement of melt flow rate are specified in the material standard with a reference to this International Standard. The test conditions normally used for thermoplastics are listed in annex A. The melt volume-flow rate will normally be found useful when comparing filled and unfilled thermoplastics. The melt mass-flow rate can now be determined by automatic measurement provided the melt density at the test temperature is known.

**1.2** The melt mass-flow rate and melt volume-flow rate of thermoplastics are dependent on the rate of shear. The rates of shear in this test are much smaller than those used under normal conditions of fabrication, and therefore data obtained by this method for various thermoplastics may not always correlate with their behaviour in actual use.

Both methods are useful in quality control.

## 2 Normative references

The following standards contain provisions which, through reference in this text, constitute provisions of this International Standard. At the time of publication, the editions indicated were valid. All standards are subject to revision, and parties to agreements based on this International Standard are encouraged to investigate the possibility of applying the most recent editions of the standards indicated below. Members of IEC and ISO maintain registers of currently valid International Standards.

ISO 1622-1:1985, *Plastics — Polystyrene (PS) moulding and extrusion materials — Part 1: Designation.*

ISO 1872-1:1986, *Plastics — Polyethylene (PE) and ethylene copolymer thermoplastics — Part 1: Designation.*

ISO 1873-1:1986, *Plastics — Polypropylene (PP) and propylene-copolymer thermoplastics — Part 1: Designation.*

ISO 2580-1:1990, *Plastics — Acrylonitrile/butadiene/styrene (ABS) moulding and extrusion materials — Part 1: Designation.*

ISO 2897-1:1990, *Plastics — Impact-resistant polystyrene (SB) moulding and extrusion materials — Part 1: Designation.*

ISO 4613-1:1985, *Plastics — Ethylene/vinyl acetate copolymer thermoplastics (E/VAC) — Part 1: Designation.*

ISO 4894-1:1990, *Plastics — Styrene/acrylonitrile (SAN) copolymer moulding and extrusion materials — Part 1: Designation.*

ISO 6402-1:1990, *Plastics — Impact-resistant acrylonitrile/styrene moulding and extrusion materials (ASA, AES, ACS), excluding butadiene-modified materials — Part 1: Designation.*

ISO 6507-1:1982, *Metallic materials — Hardness test — Vickers test — Part 1: HV 5 to HV 100.*

ISO 7391-1:1987, *Plastics — Polycarbonate moulding and extrusion materials — Part 1: Designation.*

ISO 7792-2:1988, *Plastics — Polyalkylene terephthalates — Part 2: Preparation of test specimens and determination of properties.*

ISO 8257-1:1987, *Plastics — Poly(methyl methacrylate) (PMMA) moulding and extrusion materials — Part 1: Designation.*

ISO 9988-1:1991, *Plastics — Polyoxymethylene (POM) moulding and extrusion materials — Part 1: Designation.*

# 3 Apparatus

## 3.1 Basic apparatus

The apparatus is basically an extrusion plastometer (capillary rheometer) operating at a fixed temperature. The general design is as shown in figure 1. The thermoplastic material, which is contained in a vertical metal cylinder, is extruded through a die by a loaded piston. The apparatus consists of the following essential parts:

**3.1.1 Steel cylinder**, fixed in a vertical position and suitably insulated for operation up to 400 °C. The cylinder length shall be between 115 mm and 180 mm and the internal diameter 9,55 mm ± 0,025 mm. The base of the cylinder shall be thermally insulated in such a way that the area of the exposed metal is less than 4 cm², and it is recommended that an insulating material such as $Al_2O_3$ ceramic fibre or another suitable material be used in order to avoid sticking of the extrudate.

The bore shall be suitably hardened to a Vickers hardness of no less than 500 (HV 5 to HV 100) (see ISO 6507-1). A piston guide shall be provided to prevent additional friction caused by misalignment of the piston.

**3.1.2 Steel piston**, having a working length at least as long as the cylinder. The piston shall have a head 6,35 mm ± 0,1 mm in length. The diameter of the head shall be less than the internal diameter of the cylinder by 0,075 mm ± 0,01 mm. The lower edge of the head shall have a radius of 0,4 mm and the upper edge shall have its sharp edge removed. Above the head, the piston shall be relieved to about 9 mm diameter. A stud may be added at the top of the piston to support the removable load, but the piston shall be thermally insulated from the load. Along the piston stem, two thin annular reference marks shall be scribed 30 mm apart and so positioned that the upper one is aligned with the top of the cylinder when the distance between the lower edge of the piston head and the top of the die is 20 mm. These annular marks on the piston are used as reference points during the determination (see 6.3 and 7.4.3).

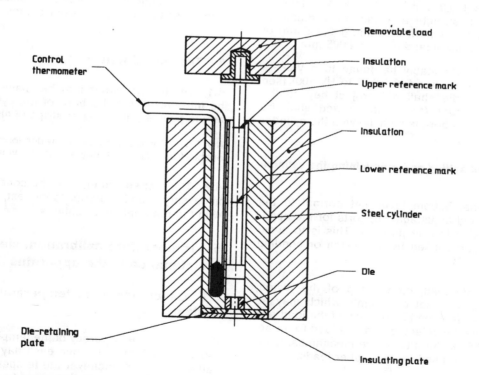

Control thermometer

Removable load

Insulation

Upper reference mark

Insulation

Lower reference mark

Steel cylinder

Die

Die-retaining plate

Insulating plate

**Figure 1 — Typical apparatus for determining melt flow rate** (showing one of the possible methods of retaining the die and one type of piston)

To ensure satisfactory operation of the apparatus, the cylinder and the piston shall be made of steel of different hardness. It is convenient for ease of maintenance and renewal to make the cylinder of the harder steel.

The piston may be either hollow or solid. In tests with lower loads, the piston shall be hollow, otherwise it may not be possible to obtain the lowest prescribed load. When the test is performed with the higher loads, the hollow piston is not desirable, as the higher load may distort such a piston. In such tests, a solid piston or a hollow piston with suitable guides shall be used. When using this latter modification, it is essential that the heat loss along the piston, which is generally longer than usual, does not alter the test temperature of the material.

**3.1.3  Temperature-control system**, such that the selected temperature of the material in the cylinder can be maintained to within ± 0,5 °C. Automatic temperature control is strongly recommended.

**3.1.4  Dies**, made of tungsten carbide or hardened steel, 8,000 mm ± 0,025 mm in length. The interior shall be circular, straight and uniform in diameter such that in all positions it is within ± 0,005 mm of a true cylinder of nominal diameter 2,095 mm.

The bore shall be suitably hardened to a Vickers hardness of no less than 500 (HV 5 to HV 100) (see ISO 6507-1). The die shall not project beyond the base of the cylinder (see figure 1) and shall be mounted so that its bore is co-axial with the cylinder bore.

**3.1.5  Means of setting and maintaining the cylinder truly vertical**.

A two-directional bubble level, set normal to the cylinder axis, and adjustable supports for the apparatus are suitable for the purpose. This is to avoid excessive friction caused by the piston or bending under heavy loads.

**3.1.6  Removable load**, on the top of the piston, which consists of a set of weights which may be adjusted so that the combined mass of the load and the piston gives the selected nominal load to an accuracy of ± 0,5 %. An alternative mechanical loading device may be used for the higher loads.

## 3.2  Accessory equipment

### 3.2.1  General

**3.2.1.1  Equipment for introducing samples into the cylinder**, consisting of a packing rod made of non-abrasive material.

**3.2.1.2  Cleaning equipment.**

**3.2.1.3  Mercury-in-glass thermometer** (calibration thermometer) or another temperature-measuring device. This measuring device shall be calibrated to permit temperature measurement to ± 0,1 °C at the temperature and immersion conditions to be used when calibrating the temperature-control system in accordance with 5.1

### 3.2.2  For procedure A

**3.2.2.1  Cutting tool**, for cutting off the extruded sample. A sharp-edged spatula has been found suitable.

**3.2.2.2  Stop-watch**, accurate to ± 0,1 s.

**3.2.2.3  Balance**, accurate to ± 0,5 mg.

### 3.2.3  For procedure B

**Measurement equipment**, for the automatic measurement of distance and time for the piston movement.

The equipment shall have the capacity to obtain three measurements for each sample in the cylinder.

# 4  Test specimen

**4.1**  The test specimen may be in any form that can be introduced into the bore of the cylinder, for example powder, granules or strips of films.

NOTE 1   Some materials in powder form do not give a bubble-free filament if they are not previously pressed.

**4.2**  The test specimen shall be conditioned and, if necessary, stabilized prior to the test, in accordance with the material specifications.

# 5  Temperature calibration, cleaning and maintenance of the apparatus

## 5.1  Calibration of the temperature-control system

**5.1.1**  Verify the accuracy of the temperature-control system (3.1.3) at least once each day that the apparatus is used or whenever the temperature of test is changed, whichever is the more frequent. For this purpose, adjust the cylinder temperature-control system until the cylinder will remain at the required temperature as indicated by the control thermometer. Preheat a calibration thermometer (3.2.1.3) to the same temperature. Then charge the cylinder with a small quantity (3 to 4 pellets) of the material to be tested, or a material representative thereof (see 5.1.2), using the same technique as for a test (see 6.2). Four minutes after completing the charging of the material, introduce the calibration thermometer into the sample chamber and immerse it in the material therein until the tip of the bulb is 10 mm from the upper face of the die. After a further interval of at least 4 min, correct the temperature indicated by the control thermometer by algebraic addition of the difference between the temperatures read on the two thermometers.

**5.1.2** It is essential that the material used during calibration be sufficiently fluid to permit, for instance, a mercury-filled thermometer bulb to be introduced without excessive force and risk of damage. A material with an MFR of greater than 45 g/10 min (2,16 kg charge) at the temperature of calibration has been found suitable.

If such a material is used for calibration purposes in place of a more viscous material which is to be tested, the dummy material shall have a thermal diffusivity similar to that of the material to be tested, so that warm-up behaviour is similar. It is necessary that the quantity charged for calibration be such that, when the calibration thermometer is subsequently introduced, an appropriate portion of the thermometer is immersed for accurate temperature measurement. This can be checked by inspecting the level of material coating the end of the calibration thermometer, removing the thermometer from the cylinder if necessary.

## 5.2 Cleaning the apparatus

The apparatus shall be cleaned thoroughly after each determination. The cylinder may be cleaned with cloth patches. The piston shall be cleaned while hot with a cloth. The die may be cleaned with a closely fitting brass reamer or wooden peg. Pyrolytic cleaning in a nitrogen atmosphere at about 550 °C may also be used. On no account shall abrasives or materials likely to damage the surface of the piston, cylinder or die be used.

## 5.3 Maintenance of apparatus

It is recommended that, at fairly frequent intervals, for example once a week for instruments in constant use, the insulating plate and the die-retaining plate, if fitted as in figure 1, be removed, and the cylinder cleaned throughout.

## 6 Procedure A

**6.1** Clean the apparatus (see 5.2). Before beginning a series of tests, ensure that the cylinder (3.1.1) and piston (3.1.2) have been at the selected temperature for not less than 15 min.

**6.2** Then charge the cylinder with 3 g to 8 g of the sample according to the anticipated melt flow rate (see, for example, table 1). During the charging, compress the material with the packing rod (3.2.1.1), using hand pressure. To ensure a charge as free from air as possible for material susceptible to oxidative degradation, complete the charging process in 1 min. Put the piston, loaded or unloaded according to the flow rate of the material, in the cylinder.

If the melt flow rate of the material is high, that is, more than 10 g/10 min, the loss of sample during preheating will be appreciable. In this case, use an unloaded piston or one carrying a smaller weight during the preheating period, and then change to the desired weight at the end of the 4 min preheating time.

### Table 1

| Melt flow rate<br><br>g/10 min | Mass of test portion in cylinder[1]<br><br>g | Extrudate cut-off time-interval<br><br>s |
|---|---|---|
| 0,1 to 0,5 | 3 to 5 | 240[2] |
| > 0,5 to 1 | 4 to 5 | 120 |
| > 1 to 3,5 | 4 to 5 | 60 |
| > 3,5 to 10 | 6 to 8 | 30 |
| > 10 | 6 to 8 | 5 to 15[3] |

1) When the density of the material is greater than 1,0 g/cm³, it may be necessary to increase the mass of the test portion.

2) It is recommended that melt flow rate should not be measured if the value obtained in this test is less than 0,1 g/10 min or greater then 100 g/10 min.

3) To achieve adequate repeatability when testing materials having an MFR greater than 25 g/10 min, it may be necessary either to control and measure cut-off intervals automatically to less than 0,1 s or to use procedure B.

**6.3** Four minutes after completing the introduction of the test portion, during which time the temperature shall have returned to that selected, place the selected load on the piston, if it was unloaded or under-loaded. Depending on the actual viscosity of the material, allow the piston to descend under gravity or push it down faster using hand pressure, until a bubble-free filament is extruded. The time for this operation shall not exceed 1 min. Cut off the extrudate with the cutting tool (3.2.2.1), and discard. Then allow the loaded piston to descend under gravity. When the lower reference mark has reached the top edge of the cylinder, start the stopwatch (3.2.2.2), and simultaneously cut off the extruded portion with the cutting tool and again discard.

Then collect successive cut-offs in order to measure the extrusion rate, at time-intervals, depending on the melt flow rate, so chosen that the length of a single cut-off is not less than 10 mm and preferably between 10 mm and 20 mm (see cut-off time-intervals in table 1 as a guide).

For low values of MFR (and MVR), it may not be possible to take a cut-off with a length of 10 mm or more within the maximum time-interval of 240 s. In this case, procedure B shall be used.

Stop cutting when the upper mark on the piston stem reaches the top edge of the cylinder. Discard any cut-off containing visible air bubbles. After cooling, weigh individually, to the nearest 1 mg, the remaining cut-offs, which shall number at least three, and calculate their average mass. If the difference between the maximum and the minimum value of the individual weighings exceeds 15 % of the average, discard the result and repeat the test on a fresh portion of the sample.

The time between charging the cylinder and the last measurement shall not exceed 25 min.

**6.4** The melt mass-flow rate (MFR), expressed in grams per 10 min, is given by the equation

$$MFR(\theta, m_{nom}) = \frac{t_{ref} \cdot m}{t}$$

where

$\theta$     is the test temperature, in degrees Celsius;

$m_{nom}$    is the nominal load, in kilograms;

$m$     is the average mass, in grams, of the cut-offs;

$t_{ref}$    is the reference time (10 min), in seconds (600 s);

$t$      is the cut-off time-interval, in seconds.

Express the result to two significant figures.

# 7 Procedure B

## 7.1 Principle

The melt mass-flow rate (MFR) and the melt volume-flow rate (MVR) are determined by using either of the following two principles:

a) measurement of the distance the piston moves in a specified time;

or

b) measurement of the time in which the piston moves a specified distance.

## 7.2 Optimum measurement accuracy

For repeatable determination of MFR between 0,1 g/10 min and 50 g/10 min or MVR between 0,1 cm$^3$/10 min and 50 cm$^3$/10 min, the movement of the piston has to be measured to the nearest $\pm$ 0,1 mm and the time to an accuracy of 0,1 s.

## 7.3 Pretreatment

Follow procedure A specified in 6.1 to 6.3 (to end of first paragraph).

## 7.4 Determination

**7.4.1** When the lower reference mark has reached the top edge of the cylinder, start the automatic measurement. Proceed as specified in 7.4.2 a) if using the principle given in 7.1 a) or as specified in 7.4.2 b) if using the principle given in 7.1 b).

**7.4.2** Measure

a) the distance moved by the piston at predetermined times (three or more)

or

b) the times taken by the reference mark to cover a specified distance (three or more).

Stop the measurement when the upper mark on the piston stem reaches the top edge of the cylinder.

**7.4.3** The time between charging the cylinder and the last measurement shall not exceed 25 min.

## 7.5 Expression of results

**7.5.1** The melt volume-flow rate (MVR), expressed in cubic centimetres per 10 min, is given by the equation

$$MVR(\theta, m_{nom}) = \frac{A \cdot t_{ref} \cdot l}{t} = \frac{427 l}{t}$$

where

$\theta$     is the test temperature, in degrees Celsius;

$m_{nom}$     is the nominal load, in kilograms;

$A$     is the mean cross-sectional area, in square centimetres, of the piston and the cylinder (= 0,711 cm2);

$t_{ref}$     is the reference time (10 min), in seconds (600 s);

$t$     is the predetermined time of measurement [see 7.4.2 a)] or the mean value of individual time measurements [see 7.4.2 b)], in seconds;

$l$     is the predetermined distance moved by the piston [see 7.4.2 b)] or the mean value of individual distance measurements [see 7.4.2 a)], in centimetres.

**7.5.2** The melt mass-flow rate (MFR), expressed in grams per 10 min, is given by the equation

$$MFR(\theta, m_{nom}) = \frac{A \cdot l \cdot \rho \cdot t_{ref}}{t} = \frac{427 l \cdot \rho}{t}$$

where

$\theta$, $m_{nom}$, $A$, $l$, $t_{ref}$ and $t$ are as defined in 7.5.1;

$\rho$     is the density, in grams per cubic centimetre, of the melt at the test temperature and is given by the equation

$$\rho = \frac{m}{0,711 l}$$

        $m$     being the mass, determined by weighing, of a known extruded volume of length $l$.

**7.5.3** Express the result to two significant figures.

## 8 Precision

When the method is used with certain materials, consideration shall be given to the factors leading to a decrease in repeatability. Such factors include

a) thermal degradation or crosslinking of the material, causing the melt flow rate to change during the preheating or test period; powdered materials requiring long preheating times are sensitive to this effect and, in certain cases, the inclusion of stabilizers is necessary to reduce the variability;

b) filled or reinforced materials, where the distribution or orientation of the filler may affect the melt flow rate.

The precision of the method is not known because inter-laboratory data are not available. A single precision statement would not be suitable because of the number of materials covered. However, a coefficient of variation of about $\pm$ 10 % could be expected.

## 9 Test report

The test report shall include the following particulars:

a) a reference to this International Standard;

b) all details necessary for the complete identification of the test sample, including the physical form of the material with which the cylinder was charged;

c) the details of conditioning;

d) the details of any stabilization (see 4.2);

e) the temperature and load used in the test;

f) the melt mass-flow rate, in grams per 10 min, or the melt volume-flow rate, in cubic centimetres per 10 min, expressed to two significant figures;

g) a report of any unusual behaviour of the test portion, such as discoloration, sticking, extrudate distortion or unexpected variation in melt flow rate.

# Annex A
## (normative)

## Test conditions for melt flow rate determination

The conditions used shall be as indicated in the appropriate material designation or specification. Table A.1 indicates test conditions that have been found useful. Table A.2 indicates test conditions that are presently specified in relevant International Standards. Other test conditions not listed here may be used, if necessary, for a particular material.

### Table A.1

| Conditions | | Test temperature, $\theta$ | Nominal load (combined), $m_{nom}$ |
|---|---|---|---|
| No. | Code-letter | °C | kg |
| 1 | A | 250 | 2,160 |
| 2 | B | 150 | 2,160 |
| 4 | D | 190 | 2,160 |
| 6 | F | 190 | 10,000 |
| 7 | G | 190 | 21,600 |
| 8 | H | 200 | 5,000 |
| 12 | M | 230 | 2,160 |
| 13 | N | 230 | 3,800 |
| 17 | S | 280 | 2,160 |
| 18 | T | 190 | 5,000 |
| 19 | U | 220 | 10,000 |
| 21 | W | 300 | 1,200 |
| 22 | Z | 125 | 0,325 |

NOTE — If, in the future, conditions other than those listed in this table are necessary, e.g. for new thermoplastics, only the loads already in use shall be chosen. Temperatures shall also be selected from those already in the table. If absolutely necessary, new temperatures might have to be taken because of the nature of the new thermoplastic. In this case, application to ISO/TC 61/SC 5 shall be made to include the new conditions. If approved, a suitable code-letter will provisionally be issued and the standard amended at the 5-year revision.

**Table A.2**

| International Standard (see clause 2) | Materials | Conditions | | Test temperature, $\theta$ °C | Nominal load (combined), $m_{nom}$ kg |
|---|---|---|---|---|---|
| | | No. | Code-letter | | |
| ISO 1622-1 | PS | 8 | H | 200 | 5,000 |
| ISO 1872-1 | PE | 4 | D | 190 | 2,160 |
| ISO 1872-1 | PE | 7 | G | 190 | 21,600 |
| ISO 1872-1 | PE | 18 | T | 190 | 5,000 |
| ISO 1873-1 | PP | 12 | M | 230 | 2,160 |
| ISO 2580-1 | ABS | 19 | U | 220 | 10,000 |
| ISO 2897-1 | Impact-resistant PS | 8 | H | 200 | 5,000 |
| ISO 4613-1 | E/VAC | 2 | B | 150 | 2,160 |
| ISO 4613-1 | E/VAC | 4 | D | 190 | 2,160 |
| ISO 4613-1 | E/VAC | 22 | Z | 125 | 0,325 |
| ISO 4894-1 | SAN | 19 | U | 220 | 10,000 |
| ISO 6402-1 | ASA | 19 | U | 220 | 10,000 |
| ISO 7391-1 | PC | 21 | W | 300 | 1,200 |
| ISO 7792-2 | PET | 17 | S | 280 | 2,160 |
| ISO 7792-2 | PBT | 1 | A | 250 | 2,160 |
| ISO 8257-1 | PMMA | 13 | N | 230 | 3,800 |
| [1] | PB | 4 | D | 190 | 2,160 |
| [1] | PB | 6 | F | 190 | 10,000 |
| ISO 9988-1 | POM | 4 | D | 190 | 2,160 |
| [1] | MM/ABS | 19 | U | 220 | 10,000 |

1) No designation standard available yet.

185

# ISO
# 1183

First edition
1987-07-15

# INTERNATIONAL STANDARD

INTERNATIONAL ORGANIZATION FOR STANDARDIZATION
ORGANISATION INTERNATIONALE DE NORMALISATION
МЕЖДУНАРОДНАЯ ОРГАНИЗАЦИЯ ПО СТАНДАРТИЗАЦИИ

## Plastics — Methods for determining the density and relative density of non-cellular plastics

*Plastiques — Méthodes pour déterminer la masse volumique et la densité relative des plastiques non alvéolaires*

Reference number
ISO 1183 : 1987 (E)

# Foreword

ISO (the International Organization for Standardization) is a worldwide federation of national standards bodies (ISO member bodies). The work of preparing International Standards is normally carried out through ISO technical committees. Each member body interested in a subject for which a technical committee has been established has the right to be represented on that committee. International organizations, governmental and non-governmental, in liaison with ISO, also take part in the work.

Draft International Standards adopted by the technical committees are circulated to the member bodies for approval before their acceptance as International Standards by the ISO Council. They are approved in accordance with ISO procedures requiring at least 75 % approval by the member bodies voting.

International Standard ISO 1183 was prepared by Technical Committee ISO/TC 61, *Plastics*.

It cancels and replaces ISO Recommendation R 1183 : 1970, of which it constitutes a technical revision.

Users should note that all International Standards undergo revision from time to time and that any reference made herein to any other International Standard implies its latest edition, unless otherwise stated.

# Plastics — Methods for determining the density and relative density of non-cellular plastics

## 1 Scope and field of application

**1.1** This International Standard specifies four methods for the determination of the density and relative density of non-cellular plastics in the form of sheet, film, tube, moulded objects, and moulding powders, granules and pellets.

— Method A

Immersion method for plastics in a finished condition, whether machined or otherwise formed (see 5.1.3), but not powders.

— Method B

Pyknometer method for plastics in the form of powder, granules, pellets, flake or moulded articles reduced to small particles.

— Method C

Titration method for plastics in forms similar to those required for method A, including pellets.

— Method D

Density gradient column method for plastics in forms similar to those required for method A, and including pellets. Density gradient columns are columns of liquid, the densities of which increase uniformly from top to bottom. They are particularly suited to measurement of small samples of products and to comparison of densities.

**1.2** Density and relative density are used frequently, both to follow the variations in the physical structures of specimens and in calculation of the amount of material necessary to fill a given volume. Density is the preferred property relating the mass and volume of an object, specimen or material. These properties may also be useful in assessing uniformity among samples or specimens. These methods are designed to yield results accurate to at least 0,2 % without applying corrections for weighings in air, and to 0,05 % with such corrections.

**1.3** Often the density of plastics will depend upon the methods employed in the preparation of test specimens. When this is the case, precise details of the methods of preparation shall be given; these are ordinarily included in the specifications for the material.

## 2 References

ISO 31-3, *Quantities and units of mechanics.*

ISO 291, *Plastics — Standard atmospheres for conditioning and testing.*

## 3 Definitions

**3.1** **density**, $\varrho_t$: The ratio of the mass of the sample to its volume $V_t$ (at the temperature $t$), expressed in kilograms per cubic metre, grams per cubic centimetre or grams per millilitre.

Ordinarily, $t$ will be one of the standard laboratory temperatures specified in ISO 291 (23 or 27 °C).

**3.2** **relative density**: The ratio of the mass of a given volume of material at the temperature $t_1$ to that of an equal volume of a reference material at the temperature $t_2$; it is expressed as relative density, at $t_1$ and $t_2$ (symbol: $d_{t_2}^{t_1}$) where $t$ is the temperature, in degrees Celsius.

Ordinarily, $t$ will be one of the standard laboratory temperatures specified in ISO 291 (23 or 27 °C).

Relative density may also be defined as the ratio of the density of a substance to the density of a reference substance under conditions that are specified for both substances.

NOTES

1 When the reference substance is water, the English term "specific gravity" is often used instead of "relative density", and the French term "densité" instead of "densité relative".

Density at $t_1$ °C may be converted to specific gravity, using the equation

$$d_{t_2}^{t_1} = \frac{\varrho_{S,t_1}}{\varrho_{W,t_2}}$$

where

$d_{t_2}^{t_1}$ is the specific gravity of the sample;

$\varrho_{S,t_1}$ is the density, in grams per cubic centimetre, of the specimen, at temperature $t_1$;

$\varrho_{W,t_2}$ is the density, in grams per cubic centimetre, of water, at temperature $t_2$; values at some laboratory temperatures are as follows:

| $t$ | $\varrho_{W,t}$ |
|---|---|
| °C | g/cm³ |
| 20 | 0,998 2 |
| 23[1] | 0,997 6 |
| 27 | 0,996 5 |

The International Bureau of Weights and Measures should be consulted for exact relationships between density and specific gravity at other temperatures.

2 The following equivalent terms, based upon ISO 31-3, are given here for clarification.

| English term | French term | Symbol | Formul-ation | Units |
|---|---|---|---|---|
| density (mass density) | masse volumique | $\varrho$ | $\dfrac{m}{V}$ | kg/m³ g/cm³ g/ml |
| relative density | densité relative | $d$ | $\dfrac{\varrho_1}{\varrho_2}$ | dimen-sionless |
| specific volume | volume massique | $v$ | $\dfrac{V}{m} = \dfrac{1}{\varrho}$ | m³/kg cm³/g ml/g |

## 4  Conditioning

In general, conditioning specimens to constant temperature will not be required because the determination itself brings the specimen to the constant temperature of the test. Conditioning specimens to constant moisture content, on the other hand, may be required.

Specimens which change in density during the test to such an extent that it may exceed the accuracy required of the density determination shall be conditioned before measurement in accordance with the applicable material specifications. When changes in density with time or atmospheric conditions are the primary purpose of the measurements, the speciments shall be conditioned as agreed upon by the interested parties.

## 5  Methods

### 5.1  Method A

#### 5.1.1  Apparatus

##### 5.1.1.1  Balance, accurate to 0,1 mg.

##### 5.1.1.2  Pan straddle or other stationary support.

##### 5.1.1.3  Pyknometer, of, for example, 50 ml capacity, with side-arm overflow capillary, for determining the density of the immersion liquid when this liquid is not water. The pyknometer shall be equipped with a thermometer graduated in 0,1 °C from 0 to 30 °C.

##### 5.1.1.4  Liquid bath, capable of being thermostatically controlled to within 0,1 °C.

#### 5.1.2  Immersion liquids: Freshly distilled water or other suitable liquid containing not more than 0,1 % of a wetting agent to help in removing air bubbles. The liquid or solution with which the specimen comes into contact during the measurement shall have no effect on the specimen and shall not be absorbed by the specimen in any significant quantity.

#### 5.1.3  Specimens

Specimens may consist of films, sheets, tubes, moulded objects, granules other than powder, or specimens taken from such forms with surfaces made smooth in an appropriate way to minimize the entrapment of air bubbles upon immersion in the liquid. The specimen shall be of any convenient size to give adequate clearance between the speciment and beaker (a mass of approximately 1 to 5 g is often convenient).

#### 5.1.4  Procedure

##### 5.1.4.1  Weigh the specimen suspended with a wire of diameter 0,125 mm or less[2].

##### 5.1.4.2  Immerse the specimen, still suspended by the wire, in the immersion liquid (5.1.2), contained in a beaker on the pan straddle or other stationary support (5.1.1.2). The temperature of the immersion liquid shall be $t \pm 0,1$ °C, where $t$ is 23 or 27 °C. Remove adhering air bubbles with a fine wire. It is convenient to mark the level of immersion, for example, by a shallow notch in the wire. Weigh the immersed specimen.

##### 5.1.4.3  Determine the density of immersion liquids other than water. Weigh the pyknometer (5.1.1.3) empty and then containing freshly distilled water at temperature $t$. Weigh the same pyknometer, after cleaning and drying, filled with the immersion liquid (also at temperature $t$). Calculate the density $\varrho_{IL}$ of the immersion liquid, using the equation

$$\varrho_{IL} = \frac{m_{IL}}{m_W} \times \varrho_{W,t}$$

where

$m_{IL}$  is the mass of the immersion liquid;

$m_W$  is the mass of water;

$\varrho_{W,t}$  is the density of water at temperature $t$.

##### 5.1.4.4  Calculate the density $\varrho_{S,t}$ of the specimen, using the equation

$$\varrho_{S,t} = \frac{m_{S,A}\,\varrho_{IL}}{m_{S,A} - m_{S,IL}}$$

where

$m_{S,A}$  is the mass, in grams, of the specimens in air;

---

1) Reference temperature : see ISO 291.
2) Such a fine wire renders corrections for the apparent loss in mass of the wire unnecessary.

$m_{S,IL}$ is the uncorrected mass, in grams, of the specimens in the immersion liquids;

$\varrho_{IL}$ is the density of the immersion liquid determined as in 5.1.4.3, expressed in grams per cubic centimetre or grams per millilitre.

NOTE — For specimens having a density less than that of the immersion liquid, the test may be performed in exactly the same way as described above, with the following exception : a sinker of lead or other dense material is attached to the wire, such that the sinker rests below the fluid level as does the specimen during immersion. The apparent loss in mass $\Delta m$ of the sinker on immersion may be considered as a part of the suspending wire; it should be subtracted from $m_{S,IL}$ in the above equation.

Thus,

$$\varrho_{S,t} = \frac{m_{S,A}\,\varrho_{IL}}{m_{S,A} - (m_{S,IL} - \Delta m)}$$

(if no sinker is used, $\Delta m = 0$)

where

$m_{S,A}$ and $m_{S,IL}$ have the same meaning as in the preceding equation;

$\Delta m$ is the apparent loss in mass of the sinker on immersion in the liquid.

For correction of buoyancy, see clause 6.

## 5.2 Method B

### 5.2.1 Apparatus

**5.2.1.1 Balance**, accurate to 0,1 mg.

**5.2.1.2 Pan straddle** or other stationary support.

**5.2.1.3 Pyknometer** (see 5.1.1.3).

**5.2.1.4 Liquid bath** (see 5.1.1.4).

### 5.2.2 Immersion liquids

See 5.1.2.

### 5.2.3 Specimens

Specimens of powders, granules, pellets or flake shall be measured in the form in which they are received. Specimen mass shall be in the range of 1 to 5 g.

### 5.2.4 Procedure

**5.2.4.1** Weigh the pyknometer (5.2.1.3) empty and dry. Weigh a suitable quantity of the plastic material in the pyknometer. Cover the test specimen with immersion liquid (5.2.2) and remove all air from the specimen by placing the pyknometer in a desiccator and applying a vacuum. Break the vacuum and fill the pyknometer with the immersion liquid. Bring it to constant temperature in the liquid bath (5.2.1.4) and

then complete filling exactly to the limits of the capacity of the pyknometer with its contents. Wipe dry and weigh the pyknometer with specimen and immersion liquid.

**5.2.4.2** Empty and clean the pyknometer. Fill it with boiled distilled water, removing air as above, and determine the mass of the pyknometer and contents at the temperature of test.

**5.2.4.3** Repeat the process with the immersion liquid if an immersion liquid other than water was used, and calculate its density as specified in 5.1.4.3.

**5.2.4.4** Calculate the density of the specimen, $\varrho_{S,t}$, using the equation

$$\varrho_{S,t} = \frac{m_S\,\varrho_{IL}}{m_1 - m_2}$$

where

$m_S$ is the mass, in grams, of the specimen;

$m_1$ is the mass, in grams, of liquid required to fill the pyknometer;

$m_2$ is the mass, in grams, of liquid required to fill the pyknometer containing the specimen;

$\varrho_{IL}$ is the density of the immersion liquid determined as specified in 5.1.4.3, in grams per cubic centimetre.

For correction of buoyancy, see clause 6.

## 5.3 Method C

### 5.3.1 Apparatus

**5.3.1.1 Thermostatically controlled bath** (see 5.1.1.4).

**5.3.1.2 Glass cylinder**, of capacity 250 ml.

**5.3.1.3 Thermometer**, graduated in 0,1 °C divisions, with a range suitable for the test temperature used.

**5.3.1.4 Measuring flask**, of capacity 100 ml.

**5.3.1.5 Draw glass stirrers**.

**5.3.1.6 Automatic burette**, of capacity 25 ml, graduated in 0,1 ml and kept in the thermostatically controlled bath (5.3.1.1).

**5.3.2 Immersion liquids**: Two miscible freshly distilled liquids of different densities. The densities given for various liquids in the table in the annex may serve as an appropriate guide.

The liquid or solution with which the specimen comes into contact during the measurement shall have no effect on the specimen and shall not be absorbed by the specimen in any significant quantity.

### 5.3.3 Specimens

Specimens shall be in any suitable solid form.

### 5.3.4 Procedure

**5.3.4.1** Choose a liquid whose density is next below the one of the material to be tested. If necessary, make a preliminary rapid test in a few millilitres of the liquid.

**5.3.4.2** By means of the measuring flask (5.3.1.4), accurately measure 100 ml of one of the immersion liquids and transfer it to the clean, dry 250 ml test glass cylinder (5.3.1.2). Secure the cylinder in the thermostatically controlled bath (5.3.1.1) at the temperature $t$. Ordinarily, $t$ will be one of the standard laboratory temperatures (23 or 27 °C).

**5.3.4.3** Place the pieces of the test specimen in the cylinder. They shall fall to the bottom and be free of air bubbles. Allow about 5 min for the cylinder and its contents to stabilize at the bath temperature, stirring several times.

**5.3.4.4** When the temperature of the liquid is $t$ °C, add the second immersion liquid (5.3.2) millilitre by millilitre supplied from the automatic burette (5.3.1.6). Stir the liquid vertically after each addition by means of a flat-tipped glass rod (5.3.1.5). and avoid producing air bubbles.

After each addition of the second liquid and mixing, observe the behavior of the test pieces. At first, they fall rapidly to the bottom, then their rate of fall becomes slower. At this point, add the second liquid in 0,1 ml fractions. Note the total amount of second liquid added when the first pieces keep afloat within the liquid, at the level to which they were brought by stirring, without moving up or down for at least 1 min.

Add more of the second liquid until the heaviest pieces reach a state of neutral equilibrium. For each sample, note the amount of the second liquid required. For each pair of liquids, the functional relationship between the amount of the second liquid added and the density shall be established and plotted in graphical form.

NOTES

1  It is recommended that the thermometer (5.3.1.3) be kept permanently in the solution. This makes it possible to check that thermal equilibrium is attained at the time of measurement and, in particular, that the heat of dilution has been dissipated.

2  The density of the liquid mixture at each point of the graph of functional relationship can be evaluated by the pyknometer method.

## 5.4  Method D

### 5.4.1  Apparatus

**5.4.1.1  Density gradient column,** consisting of a suitable tube, which may be graduated, not less than 40 mm in diameter, with cover.

**5.4.1.2  Thermostatically controlled bath** (see 5.1.1.4).

**5.4.1.3  Calibrated glass floats,** covering the density range to be studied and approximately evenly distributed throughout this range.

**5.4.1.4  Set of suitable hydrometers,** covering the range of densities to be studied, having density graduations of 0,001 g/cm³, or other suitable means for measuring densities of liquids.

**5.4.1.5  Balance** (see 5.1.1.1).

**5.4.1.6  Siphon** or **pipette assembly,** for filling the gradient tube (5.4.1.1), such as shown in the figure or any other which will yield an equivalent result.

**5.4.1.7  Cathetometer** (optional).

### 5.4.2  Immersion liquids

See 5.3.2.

The mixtures of the two liquids selected shall be prepared as specified in 5.4.4.2.

### 5.4.3  Specimens

Specimens shall consist of pieces of the material cut to any convenient shape for ease of identification. Dimensions shall be chosen to permit accurate measurement of the position of the centre of the volume. Surfaces shall be smooth and free from cavities to preclude the entrapment of air bubbles upon immersion.

### 5.4.4  Procedure

**5.4.4.1**  Preparation of glass floats

**5.4.4.1.1**  Glass floats (see 5.4.1.3), prepared by any convenient method, shall be fully annealed, approximately spherical and of diameter not greater than 5 mm.

**5.4.4.1.2**  Prepare a solution of about 500 ml of the immersion liquids (5.4.2) to be used in the gradient tube (5.4.1.1), such that the density of the solution as measured with hydrometers (5.4.1.4) is approximately equal to the lowest desired density. When the floats are at ambient temperature, drop them gently into the solution. Retain the floats that sink very slowly and discard those that sink rapidly or retain them for another tube. If necessary, in order to obtain a suitable range of floats,

a)  adjust selected floats to the desired density by rubbing the bead part of the float on a glass plate on which is spread a thin slurry of silicon carbide with less than 38 µm (400 mesh) particle size, or other appropriate abrasive, or

b)  etch floats with hydrofluoric acid.

Progress may be followed by dropping the float into the test solution at intervals and noting the changes in its rate of sinking.

**5.4.4.1.3** Determine the density of each standard glass float prepared as above by placing the float in a solution of two suitable liquids (5.4.2), the density of which can be varied over the desired range by the addition of either liquid to the mixture. If the float sinks, add the denser liquid, stirring it well. Allow the solution to rest and add no further liquid until the float shows signs of moving. If it does move, repeat the above procedure until the float remains stationary for at least 30 min. It is convenient to make these measurements in the bath (5.4.1.2) at the same temperature as is used for the density gradient tube. In any case, the solution for calibration of the floats shall be maintained at $t \pm 0,1$ °C where $t$ is 23 or 27 °C.

**5.4.4.1.4** Determine the density of the solution, to the nearest 0,000 1 g/ml, in which the float remains in equilibrium, using the pyknometer method (see 5.1.4.3) or other convenient method, for example, with hydrometers. Apply the buoyancy correction (see clause 6) if necessary. Record this density as the density of the float and repeat the procedure for each float. If it is convenient to place all floats in the liquid together, then calibrate them in turn, starting with the least dense.

Alternatively, calculate the density of the liquid mixture in which the float remains stationary from the volumes of liquids used, taking care to apply corrections for fluid contractions.

**5.4.4.2** Preparation of density gradient column

**5.4.4.2.1** Place the graduated tube in the thermostatically controlled bath (5.4.1.2), maintained at $t \pm 0,1$ °C. Select a suitable combination of liquids (5.4.2) from the table in the annex such that the resulting sensitivity of the column will preferably be no poorer than 0,001 g/ml per centimetre of tube length. Satisfactory density ranges for a column are, for example, 0,001 to 0,1 g/ml. The extreme upper and lower portions of the tube shall not be used, and readings shall not be taken outside the calibrated part.

**5.4.4.2.2** Any of several methods for preparing the gradient may be used, including the following:

Assemble the apparatus as shown in the figure, using two vessels of the same size. Then select an appropriate amount of two suitable liquids which previously have been carefully de-aerated by gentle heating or application of a vacuum. The volume of liquid used in the mixer (vessel 2 in the figure) shall be equal to at least one-half of the total volume desired in the gradient tube (see note 1).

Place an appropriate mass of the less dense liquid into vessel 2 of suitable size and begin to stir, using a magnetic stirrer. Adjust the speed of stirring so that the surface of the liquid does not fluctuate greatly. Place an equal mass of the denser liquid into vessel 1. Take care that no air is dispersed in the liquid.

Use the less dense liquid (starting liquid in vessel 2) to prime the siphon (5.4.1.6), which should be equipped with a capillary tip at the delivery end for flow control; then start the delivery of

the liquid to the gradient tube, Fill the tube to the top graduation desired. (See note 2.)

Allow the density gradient column so prepared to settle for at least 24 h.

NOTES

1  Calculate the density $\varrho_2$, of the liquid in vessel 2 (see the figure), used to prepare a desired gradient, using the equation

$$\varrho_2 = \varrho_{max} - \frac{2(\varrho_{max} - \varrho_{min})V_2}{V}$$

where

$\varrho_{min}$  is the lower limit of required density, chosen to be 0,01 g/ml lighter than the density of the least dense glass float calibrated for the indivisual gradient tube;

$\varrho_{max}$  is the upper limit of required density and starting density for the liquid in vessel 1, chosen to be 0,005 g/ml heavier than the density of the densest glass float calibrated for the individual gradient tube;

$V$  is the total volume required in the gradient tube;

$V_2$  is the volume of the starting liquid in vessel 2.

2  Preparation of a suitable gradient column may require 1 to 1½ h or longer, depending upon the volume required in the gradient tube.

**5.4.4.2.3** For every 25 cm of tube length, dip a minimum of five clean calibrated floats spanning the effective range of the column into the less dense liquid used in the preparation of the gradient column, and add them to the tube. If it is observed that the floats group together and do not spread out evenly in the tube, discard the solution and repeat the procedures.

Alternatively, place the floats in the tube immediately upon preparation of the column. If the floats appear to group together and do not spread evenly in the tube, discard the solution and repeat the preparation.

Cap the tube, and retain it in the constant temperature bath for 24 to 48 h. At the end of this time, measure the distances of floats from the bottom of the tube to the nearest millimetre and plot a curve of the density of the floats as a function of their distances. The curve so produced shall

a)  be monotonic;

b)  have no discontinuity;

c)  have not more than one point of inflection.

The solution shall otherwise be discarded.

NOTE — Density gradient columns normally remain stable for several months. A daily check of the original calibration will reveal when instability has been reached.

**5.4.4.3**  Measurement of density

Wet three representative test specimens with the less dense of the two liquids used in the tube and gently place them in the tube. Allow the tube and specimens to reach equilibrium, which will require 10 min or more. Films of thickness less than 0,05 mm require at least 1½ h to settle. Rechecking specimens of thin films after several hours is advisable.

192

NOTES

1 A fine wire carefully manipulated is suitable for removing air bubbles from specimens.

2 Old samples can be removed without destroying the gradient by slowly withdrawing a wire screen basket attached to a long wire. This can be done conveniently by means of a clock motor. Withdraw the basket from the bottom of the tube and, after cleaning, return it to the bottom of the tube. It is essential that this procedure be performed at a slow enough rate (approximately 10 mm length of column per minute) in order not to disturb the density gradient.

#### 5.4.4.4 Calculations

The densities of the specimens may be determined graphically or by calculation from the levels to which they settle, as follows :

a) Make a plot of float density *versus* float position on a chart large enough to be read accurately to $\pm 0,000\ 1$ g/ml and $\pm 1$ mm. Plot the positions of the specimens on the chart and read their corresponding densities, or

b) Calculate the density $\varrho_{S,x}$ of the specimen by interpolation, using the equation

$$\varrho_{S,x} = \varrho_{F_1} + \frac{(x - y)\,(\varrho_{F_2} - \varrho_{F_1})}{z - y}$$

where

$\varrho_{F_1}$ and $\varrho_{F_2}$ are the respective densities of the two standard floats bracketing the specimen;

$x$ is the distance of the specimen above an arbitrary level;

$y$ and $z$ are the distances (measured from the same arbitrary level) of the two standard floats.

NOTE — This method does not reveal calibration errors, which can be detected with the graphical method, and may be applied only when the calibration is known to be linear within the range under test.

## 6 Correction for buoyancy of air

Because the weighings are made in air, the values of the "apparent masses" obtained shall be corrected, if necessary to compensate for the different effect of the air buoyancy on the specimen and on the weights used, if required. (This will be the case if the accuracy of the results is to be between 0,2 % and 0,05 %.)

The true mass $m_T$ is calculated using the equation

$$m_T = m_{APP} \left( 1 + \frac{0,001\ 2}{\varrho_{S,t}} - \frac{0,001\ 2}{\varrho_L} \right)$$

where

$m_{APP}$ is the apparent mass;

$\varrho_{S,t}$ is the density of the specimen;

$\varrho_L$ is the density of the weight used.

## 7 Test report

The test report shall include the following particulars :

a) reference to this International Standard;

b) complete identification of the material tested;

c) the method used (A, B, C or D);

d) the individual values and the arithmetic mean of the density ($\varrho_{S,t}$ or $\varrho_{S,x}$), in kilograms per cubic metre, grams per cubic centimetre or grams per millilitre, where $t$ is the temperature of the test, or the relative density ($d_{t_2}^{t_1}$) and the identity of the reference substance.

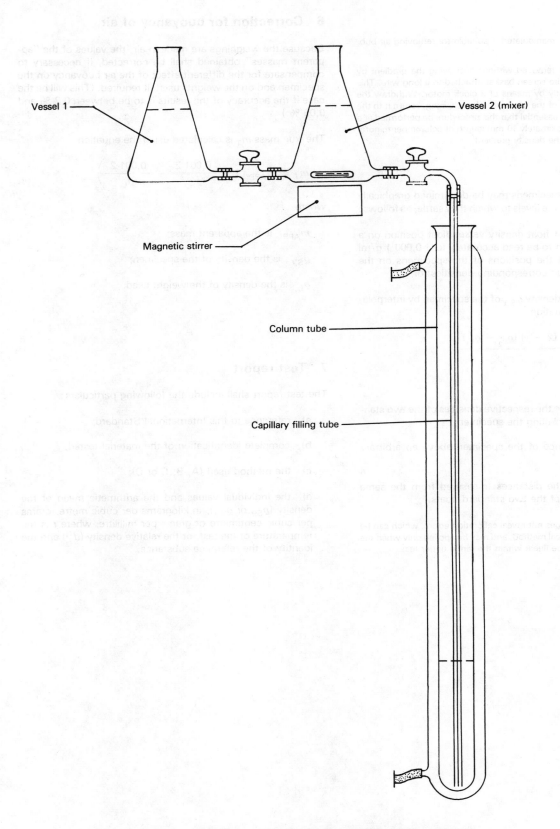

Vessel 1

Vessel 2 (mixer)

Magnetic stirrer

Column tube

Capillary filling tube

**Figure — Density gradient column filling devices**

# Annex

# Liquid systems for density gradient columns

(This annex does not form an integral part of the Standard.)

WARNING — Some of the following chemicals may be hazardous.

| System | Density range g/ml |
|---|---|
| Methanol/benzyl alcohol | 0,08 to 0,92 |
| Isopropanol/water | 0,79 to 1,00 |
| Isopropanol/diethylene glycol | 0,79 to 1,11 |
| Ethanol/carbon tetrachloride | 0,79 to 1,59 |
| Ethanol/water | 0,79 to 1,00 |
| Toluene/carbon tetrachloride | 0,87 to 1,59 |
| Water/aqueous solution of sodium bromide | 1,00 to 1,41 |
| Water/aqzeous solution of calcium nitrate | 1,00 to 1,60 |
| Ethanol/aqueous solution of zinc chloride | 0,80 to 1,70 |
| Carbon tetrachloride/1,3-dibromopropane | 1,60 to 1,99 |
| 1,3-Dibromopropane/ethylene bromide | 1,99 to 2,18 |
| Ethylene bromide/bromoform | 2,18 to 2,89 |
| Carbon tetrachloride/bromoform | 1,60 to 2,89 |
| Isopropanol/methylglycolacetate | 0,79 to 1,00 |

The following may also be used in various mixtures:

| | Density |
|---|---|
| n-Octane | 0,70 |
| Dimethylformamide | 0,94 |
| Tetrachlorethane | 1,59 |
| Ethyl iodide | 1,93 |
| Methylene iodide | 3,33 |

# INTERNATIONAL STANDARD

## ISO
## 1210

Second edition
1992-08-01

## Plastics — Determination of the burning behaviour of horizontal and vertical specimens in contact with a small-flame ignition source

*Plastiques — Détermination du comportement au feu d'éprouvettes horizontales et verticales au contact d'une petite flamme comme source d'allumage*

Reference number
ISO 1210:1992(E)

# Foreword

ISO (the International Organization for Standardization) is a worldwide federation of national standards bodies (ISO member bodies). The work of preparing International Standards is normally carried out through ISO technical committees. Each member body interested in a subject for which a technical committee has been established has the right to be represented on that committee. International organizations, governmental and non-governmental, in liaison with ISO, also take part in the work. ISO collaborates closely with the International Electrotechnical Commission (IEC) on all matters of electrotechnical standardization.

Draft International Standards adopted by the technical committees are circulated to the member bodies for voting. Publication as an International Standard requires approval by at least 75 % of the member bodies casting a vote.

International Standard ISO 1210 was prepared by Technical Committee ISO/TC 61, *Plastics*, Sub-Committee SC 4, *Burning behaviour*.

This second edition cancels and replaces the first edition (ISO 1210:1982), of which it constitutes a technical revision

Annex A of this International Standard is for information only.

International Organization for Standardization
Case Postale 56 • CH-1211 Genève 20 • Switzerland

Printed in Switzerland

# Plastics — Determination of the burning behaviour of horizontal and vertical specimens in contact with a small-flame ignition source

## 1 Scope

**1.1** This International Standard specifies a small-scale laboratory screening procedure for comparing the relative burning behaviour of vertically or horizontally oriented plastic specimens exposed to a small-flame ignition source.

**1.2** This method of test determines the afterflame/afterglow times and damaged length of specimens. It is applicable to solid and cellular materials having an apparent density of not less than 250 kg/m$^3$, determined in accordance with ISO 845. This method is not applicable to materials that shrink away from the applied flame without igniting.

**1.3** The classification system described in annex A is intended for quality assurance and the preselection of component materials for products.

This system is not intended for assessment of the fire behaviour of building materials or furnishings. The method of test described may be used for the preselection acceptance of a material, providing positive results are obtained at the thickness equal to the smallest thickness used in the application.

NOTE 1 Test results are influenced by material components, e.g. pigments, fillers and fire-retardants, and properties such as the direction of anisotropy and molecular mass.

## 2 Normative references

The following standards contain provisions which, through reference in this text, constitute provisions of this International Standard. At the time of publication, the editions indicated were valid. All standards are subject to revision, and parties to agreements based on this International Standard are encouraged to investigate the possibility of applying the most recent editions of the standards indicated below. Members of IEC and ISO maintain registers of currently valid International Standards.

ISO 291:1977, *Plastics — Standard atmospheres for conditioning and testing*.

ISO 293:1986, *Plastics — Compression moulding test specimens of thermoplastic materials*.

ISO 294:1975, *Plastics — Injection moulding test specimens of thermoplastic materials*.

ISO 295:1991, *Plastics — Compression moulding of test specimens of thermosetting materials*.

ISO 845:1988, *Cellular plastics and rubbers — Determination of apparent (bulk) density*.

ISO 1043-1:1987, *Plastics — Symbols — Part 1: Basic polymers and their special characteristics*.

ISO 5725:1986, *Precision of test methods — Determination of repeatability and reproducibility for a standard test method by inter-laboratory tests*.

ISO 10093:—[1]), *Plastics — Fire tests — Standard ignition sources*.

## 3 Definitions

For the purposes of this International Standard, the following definitions apply.

**3.1 afterflame:** Persistence of flaming of a material, under specified test conditions, after the ignition source has been removed.

**3.2 afterflame time:** The length of time for which a material continues to flame, under specified test

---

1) To be published.

conditions, after the ignition source has been removed.

**3.3 afterglow:** Persistence of glowing of a material, under specified test conditions, after cessation of flaming or, if no flaming occurs, after removal of the ignition source.

**3.4 afterglow time:** The time during which a material continues to glow, under specified test conditions, after cessation of flaming or after the ignition source has been removed.

## 4 Principle

A test specimen bar is supported horizontally or vertically by one end and the free end is exposed to a specified gas flame. The burning behaviour of the bar is assessed by measuring the linear burning rate (method A) or the afterflame/afterglow times (method B).

## 5 Significance of test

**5.1** Tests made on a material under the conditions specifed can be of considerable value in comparing the relative burning behaviour of different materials, in controlling manufacturing processes, or in assessing any change in burning characteristics prior to, or during use. The results obtained from this method are dependent on the shape, orientation and environment surrounding the specimen and on the conditions of ignition. Correlation with performance under actual service conditions is not implied.

**5.2** Results obtained in accordance with this International Standard shall not be used to describe or appraise the fire hazard presented by a particular material or shape under actual fire conditions. Assessment for fire hazard requires consideration of such factors as: fuel contribution, intensity of burning (rate of heat release), products of combustion and environmental factors such as the intensity of source, orientation of exposed material and ventilation conditions.

**5.3** Burning behaviour as measured by this test method is affected by such factors as density, any anisotropy of the material and the thickness of the specimen.

**5.4** Certain materials may shrink from the applied flame without igniting. In this event, test results are not valid and additional test specimens will be required to obtain valid tests. If the test specimens continue to shrink from the applied flame without igniting, these materials are not suitable for evaluation by this method of test.

**5.5** The burning behaviour of some plastic materials may change with time. It is accordingly advisable to make tests before and after ageing by an appropriate procedure. The preferred oven conditioning shall be 7 days at 70 °C. However, other ageing times and temperatures may be used by agreement between the interested parties and shall be noted in the test report.

**5.6** The effect on the burning behaviour of additives, deterioration, and possible loss of volatile components are measurable using this method. Results obtained using this method may serve for comparing the relative performance of materials and can be helpful in material assessment.

## 6 Apparatus

**6.1 Laboratory fume hood/cupboard**, having an inside volume of at least 0,5 m$^3$. The chamber shall permit observation and shall be draught-free, while permitting normal thermal circulation of air past the specimen during burning. For safety and convenience, it is desirable that this enclosure (which can be completely closed) be fitted with an evacuation device, such as an exhaust fan, to remove products of combustion, which may be toxic. However, it is important to be able to turn the device off during the actual test and start it again immediately after the test to remove the products of combustion.

NOTE 2    The amount of oxygen available to support combustion is naturally important for the conduct of these flame tests. For tests conducted by this method when burning times are protracted, chamber sizes less than 1 m$^3$ may not provide accurate results.

**6.2 Laboratory burner**, as specified in ISO 10093, (ignition source P/PF2), having a barrel length of 100 mm ± 10 mm and an internal diameter of 9,5 mm ± 0,3 mm. Do not equip the barrel with an end attachment such as a stabilizer.

**6.3 Ring stand**, with clamps or the equivalent, adjustable for positioning of the specimen.

**6.4 Timing device**, accurate to 1 s.

**6.5 Measuring scale**, graduated in millimetres.

**6.6 Supply of technical-grade methane gas**, with regulator and meter for uniform gas flow.

NOTE 3    Other gas mixtures having a heat content of approximately 37 MJ/m$^3$ have been found to provide similar results.

**6.7 Desiccator**, containing anhydrous calcium chloride or other drying agent.

**6.8 Conditioning room or chamber**, capable of being maintained at 23 °C ± 2 °C and a relative humidity of (50 ±5) %.

**6.9 Micrometer**, capable of being read to 0,01 mm.

**6.10 Air-circulating oven**, capable of being maintained at 70 °C ± 1 °C while providing not less than five air changes per hour.

## 7 Specimens

**7.1** All specimens shall be cut from a representative sample of the material (sheets or end-products), or shall be cast or injection- (see ISO 294), compression- (see ISO 293 or ISO 295) or transfer-moulded to the necessary shape. After any cutting operation, care shall be taken to remove all dust and any particles from the surface; cut edges shall have a smooth finish.

**7.2** Bar specimens should preferably be 125 mm ± 5 mm long, 13,0 mm ± 0,3 mm wide and 3,0 mm ± 0,2 mm thick.

Other thicknesses may be used by agreement between the interested parties and, if so, shall be noted in the test report. However, the maximum thickness shall not exceed 13 mm.

NOTE 4    Tests made on specimens of different thicknesses, densities, molecular masses, directions of anisotropy and types or levels of colour(s) or filler(s) and flame-retardant(s) may not be comparable.

**7.3** A minimum of 26 bar specimens shall be prepared.

NOTE 5    It is advisable to prepare additional specimens in the event that the situation described in 5.4 is encountered.

## 8 Method A — Determination of linear burning rate of horizontal specimens

### 8.1 Complementary apparatus (see figure 1)

**8.1.1 Wire gauze**, 20 mesh (approximately 20 openings per 25 mm), made with 0,40 mm to 0,45 mm diameter steel wire and cut into 125-mm squares.

**8.1.2 Support fixture**, for testing specimens that are not self-supporting (see figure 2).

### 8.2 Specimens

Three specimens shall be tested. Each specimen shall be marked with two lines perpendicular to the longitudinal axis of the bar, 25 mm and 100 mm from the end that is to be ignited.

Dimensions in millimetres,
unless stated otherwise

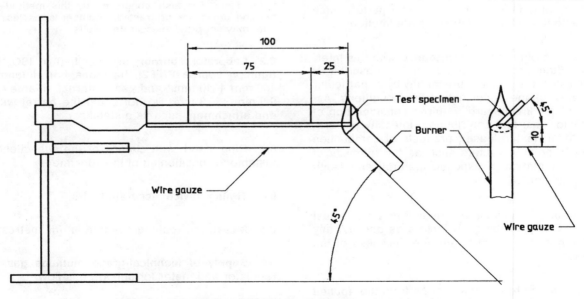

**Figure 1 — Horizontal burning test apparatus** (Method A)

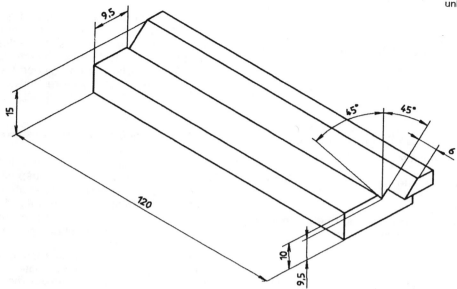

Dimensions In millimetres,
unless stated otherwise

**Figure 2 — Flexible-specimen support fixture** (Method A)

## 8.3  Conditioning

Unless otherwise required by the specification for the material being tested, two sets of three specimens shall be preconditioned in accordance with ISO 291, at 23 °C ± 2 °C and (50 ± 5) % relative humidity for 48 h. Testing shall be conducted at ambient room conditions within one hour of being conditioned.

## 8.4  Procedure

**8.4.1**  Clamp the specimen at the end farthest from the 25 mm mark, with its longitudinal axis horizontal and its transverse axis inclined at 45°. Clamp the wire gauze (8.1.1) horizontally beneath the specimen, with a distance of 10 mm between the lower edge of the specimen and the gauze, and with the free end of the specimen even with the edge of the gauze (see figure 1).

**8.4.2**  With the central axis of the burner tube vertical, set the burner (6.2) to produce a blue flame 20 mm ± 2 mm high by adjusting the gas supply (6.6) and air ports of the burner until an approximately 20-mm yellow-tipped blue flame is produced and then increase the air supply until the yellow tip disappears. Measure the height of the flame again and adjust it if necessary.

**8.4.3**  If the specimen sags at its free end during initial setting up, position the support fixture (8.1.2) illustrated in figure 2 under the specimen with the small extended portion of the support fixture approximately 20 mm from the free end of the speci-

men. Provide enough clearance at the clamped end of the specimen so that the support fixture can be moved freely sidewards. As the combustion front progresses along the specimen, withdraw the support fixture at the same approximate rate.

**8.4.4**  Apply the flame to the free end at the lower edge of the specimen so that the central axis of the burner tube is in the same vertical plane as the longitudinal bottom edge of the specimen and inclined towards the end of the specimen at an angle of approximately 45° to the horizontal (figure 1).

**8.4.5**  Position the burner so that the flame impinges on the free end of the specimen to a depth of 6 mm ± 1 mm. Apply the test flame for 30 s without changing its position; remove the burner after 30 s, or as soon as the combustion front of the specimen reaches the 25-mm mark (if less than 30 s). Restart the timing device (6.4) when the combustion front reaches the 25-mm mark.

**8.4.6**  If the specimen continues to burn (with or without a flame) after application of the test flame, record the time, in seconds, for the combustion front to travel from the 25-mm mark to the 100-mm mark and record the damaged length $L$, as 75 mm. If the combustion front passes the 25-mm mark but does not pass the 100-mm mark, record the elapsed time $t$, in seconds, and the damaged length $L$, in millimetres, between the 25-mm mark and where the combustion front stops.

**8.4.7**  Conduct the test procedure on at least three specimens.

201

## 8.5 Expression of results

**8.5.1** Calculate the linear burning rate $v$, in millimetres per minute, for each specimen, using the equation:

$$v = \frac{60L}{t}$$

where

$L$     is the damaged length, in millimetres, as defined in 8.4.6;

$t$     is the time, in seconds, as defined in 8.4.6.

NOTE 6    The SI unit of linear burning rate is the metre per second. In practice, the unit millimetre per minute is used.

**8.5.2** Calculate the average linear burning rate.

## 8.6 Precision

### 8.6.1 Interlaboratory trials

The precision data were determined from an interlaboratory experiment conducted in 1988 involving ten laboratories, three materials (levels) and three replicates, each using the average of three data points. All tests were conducted on 3-mm-thick specimens. The results were analysed in accordance with ISO 5725, and are summarized in table 1.

**Table 1 — Rate of burning**

Values in millimetres per minute

| Parameter | PE | ABS | Acrylic |
|-----------|------|------|---------|
| Average | 15,1 | 27,6 | 29,7 |
| Repeatability | 2,5 | 5,7 | 5,2 |
| Reproducibility | 3,6 | 11,4 | 6,4 |

NOTE — Material symbols are defined in ISO 1043-1.

### 8.6.2 Repeatability

In the normal and correct operation of the method, the difference between two averages determined from three specimens using identical test material and the same apparatus by one analyst within a short time interval will not exceed the repeatability value shown in table 1 more than once in 20 cases on average.

### 8.6.3 Reproducibility

In the normal and correct operation of the method, the difference between two independent averages (determined from three specimens) found by two operators working in different laboratories on identical test material will not exceed the reproducibility value shown in table 1 more than once in 20 cases on average.

### 8.6.4 Guidance on precision assessment

The two averages (determined from three specimens) shall be considered suspect and not equivalent if they differ by more than the repeatability and the reproducibility shown in table 1. Any judgement per 8.6.2 or 8.6.3 would have an approximately 95 % (0,95) probability of being correct.

NOTE 7    Table 1 is only intended to present a meaningful way of considering the approximate precision of this test method for a range of materials. These data should not be rigorously applied to acceptance or rejection of material, as the data are specific to the interlaboratory test and may not be representative of other lots, conditions, thicknesses, materials or laboratories.

## 8.7 Test report

The test report shall include the following particulars:

a) a reference to this International Standard;

b) all details necessary to identify the product tested, including the manufacturer's name, number or code;

c) the thickness, to the nearest 0,1 mm, of the test specimen;

d) the nominal apparent density (rigid cellular materials only);

e) the direction of any anisotropy relative to the test specimen dimensions;

f) any conditioning treatment;

g) any treatment before testing, other than cutting, trimming and conditioning;

h) whether or not the combustion front passed the 25-mm and 100-mm marks;

i) for specimens with which the combustion front passed the 100-mm mark, the average linear burning rate;

j) whether the flexible specimen support fixture was used.

## 9 Method B — Determination of afterflame and/or afterglow times on vertical specimens

### 9.1 Complementary apparatus (see figure 3)

**9.1.1 Supply of dry, absorbent surgical cotton.**

**9.1.2 Full-draught air-circulating oven**, minimum of five air changes per hour, capable of being maintained at 70 °C ± 1 °C or another agreed temperature.

### 9.2 Conditioning

Unless otherwise required by the material specification, the following shall apply:

**9.2.1** Two sets of 5 bar specimens shall be preconditioned for at least 48 h + 2 h at 23 °C ± 2 °C and (50 ± 5) % relative humidity.

**9.2.2** Two sets of 5 bar specimens shall be aged for 168 h ± 2 h at 70 °C ± 1 °C and then cooled in the desiccator (6.7) for at least 4 h at ambient temperature.

**9.2.3** All specimens shall be tested in a standard laboratory atmosphere of 23 °C ± 2 °C and (50 ± 5) % relative humidity in accordance with ISO 291.

### 9.3 Procedure

**9.3.1** Clamp the specimen from the upper 6 mm of its length with the longitudinal axis vertical so that the lower end of the specimen is 300 mm above a horizontal 50 mm × 50 mm layer of dry, absorbent surgical cotton (9.1.1) thinned to a maximum uncompressed thickness of 6 mm (see figure 3).

**9.3.2** With the central axis of the burner tube vertical, set the burner (6.2) to produce a blue flame 20 mm ± 2 mm high by adjusting the gas supply (6.6) and air ports of the burner until an approximately 20-mm yellow-tipped blue flame is produced and then increase the air supply until the yellow tip disappears. Measure the height of the flame again and adjust it if necessary.

**9.3.3** Apply the flame of the burner centrally to the middle point of the bottom edge of the specimen so that the top of the burner is 10 mm below that point, and maintain it at that distance for 10 s, moving the burner as necessary in response to any changes in the length or position of the specimen.

Dimensions in millimetres

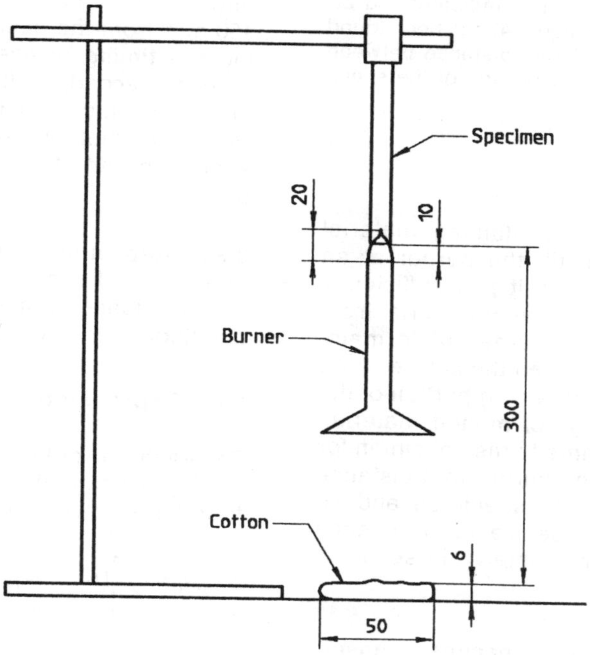

Figure 3 — Vertical-burning test apparatus (Method B)

Dimensions in millimetres

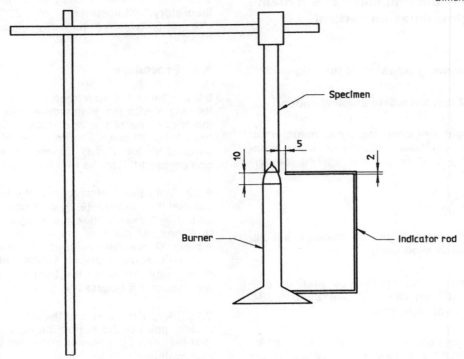

**Figure 4 — Optional indicator rod attachment** (Method B)

NOTE 8    For specimens which move under the influence of the burner flame, the use of a small indicator rod attached to the burner (as shown in figure 4) has been found to be helpful in maintaining the 10-mm distance between the top of the burner and the major portion of the specimen.

If the specimen drips molten or flaming material during the flame application, tilt the burner at an angle of up to 45° and withdraw it just sufficiently from beneath the specimen to prevent material from dropping into the barrel of the burner while maintaining the 10-mm spacing between the centre of the outlet of the burner and the remaining portion of the specimen, ignoring any strings of molten material. After the application of the flame to the specimen for 10 s, immediately withdraw the burner to a distance at least 150 mm away from the specimen and simultaneously use the timing device to commence measurement of the afterflame time $t_1$, in seconds. Note $t_1$.

**9.3.4**  When afterflaming of the specimen ceases, immediately place the flame and the burner again under the specimen and maintain the burner at a distance of 10 mm from the remaining lower edge of the specimen for 10 s while moving the burner clear of dropping material as necessary, as described in 9.3.3. After this application of the flame to

the specimen for 10 s, immediately extinguish the burner or remove it to a distance of at least 150 mm from the specimen and simultaneously, using the timing device, begin measurement, to the nearest second, of the afterflame time, $t_2$ and the afterglow time $t_3$ of the specimen. Note $t_2$ and $t_3$. Note also whether any particles or drips fall from the specimen and, if so, whether they ignite the cotton pad.

**9.3.5**  Repeat the procedure until at least five specimens have been tested which were preconditioned in accordance with 9.2.1, and five specimens preconditioned in accordance with 9.2.2.

## 9.4   Expression of results

For each set of five specimens from the two conditioning treatments, calculate the total afterflame time for the set $t_f$, in seconds, using the equation

$$t_f = \sum_{i=1}^{5} (t_{1,i} + t_{2,i})$$

where

$t_{1,i}$    is the first afterflame time, in seconds, of the $i$th specimen;

$t_{2,i}$    is the second afterflame time, in seconds, of the $i$th specimen.

## 9.5 Precision

### 9.5.1 Interlaboratory trials

The precision data were determined from an interlaboratory experiment conducted in 1978 involving four laboratories, four materials (levels) and two replicates each using the average of five data points. The results were analysed in accordance with ISO 5725, and are summarized in table 2.

### 9.5.2 Repeatability

In the normal and correct operation of the method, the difference between two averages determined from five specimens using identical test material and the same apparatus by one analyst within a short time interval will not exceed the reproducibility value shown in table 2 more than once in 20 cases on average.

### 9.5.3 Reproducibility

In the normal and correct operation of the method, the difference between two independent averages (determined from five specimens) found by two operators working in different laboratories on identical material will not exceed the reproducibility value shown in table 2 more than once in 20 cases on average.

### 9.5.4 Guidance on precision assessment

The two averages (determined from five specimens) are to be considered suspect and not equivalent if they differ by more than the repeatability and reproducibility shown in table 2. Any judgement per 9.5.2 or 9.5.3 would have an approximately 95 % (0,95) probability of being correct.

NOTE 9    Table 2 is only intended to present a meaningful way of considering the approximate precision of this test method for a range of materials. These data should not be rigorously applied to acceptance or rejection of material, as the data are specific to the interlaboratory test and may not be representative of other conditions, thicknesses, materials or laboratories.

## 9.6 Test report

The test report shall include the following particulars:

a) a reference to this International Standard;

b) all details necessary to identify the product tested, including the manufacturer's name, number or code;

c) the thickness, to the nearest 0,1 mm, of the test specimen;

d) the nominal apparent density (rigid cellular materials only);

e) the direction of any anisotropy relative to the test specimen dimensions;

f) any conditioning treatment;

g) any treatment before testing, other than cutting, trimming and conditioning;

h) the individual values of $t_1$, $t_2$ and $t_3$ for each specimen and the average values for each set;

i) the total afterflame time, $t_f$, for each set of five specimens from the two conditioning treatments (see 9.2.1 and 9.2.2);

j) whether any particles or drips fell from the specimen and whether they ignited the cotton.

### Table 2 — Afterflame and afterflame plus afterglow times

Values in seconds

| Stage | Time measured | Parameter | Material | | | |
|-------|---------------|-----------|------|------|------|------|
| | | | PC | PPO | ABS | PF |
| After first flame application | Afterflame | Average Repeatability Reproducibility | 1,70 1,10 1,76 | 10,05 10,79 12,19 | 0,43 0,83 1,26 | 0,78 0,87 1,80 |
| After second flame application | Afterflame plus afterglow | Average Repeatability Reproducibility | 3,55 1,52 2,65 | 15,95 14,49 13,24 | 1,13 2,16 1,89 | 49,33 45,73 50,77 |

NOTE — Symbols for plastics materials are defined in ISO 1043-1.

# Annex A
## (informative)

# Classification system for determining the burning behaviour of horizontal and vertical specimens in contact with a small-flame ignition source

## A.1 General

This annex describes a classification system that can be used to characterize the burning behaviour of rigid materials in response to a small-flame ignition source. The use of a category designation code is optional and is determined by examining the test results of materials tested by this method. Each category code represents a preferred range of performance levels that simplifies description in material designations or specifications and may assist certification bodies to determine compliance with applicable requirements.

## A.2 Category designations

### A.2.1 Method A

The behaviour of specimens can be classified in one of the following four categories (FH = flaming horizontal):

FH-1: No visible flame after the ignition source is removed, or if combustion front does not pass the 25-mm mark.

FH-2: Combustion front passes the 25-mm mark but does not pass the 100-mm mark. The length of the burnt area shall be added to the category designation (for example FH-2-70 mm).

FH-3: Combustion front passes the 100-mm mark and the linear burning rate does not exceed 40 mm/min for specimens having a thickness between 3 mm and 13 mm or 75 mm/min for specimens having a thickness less than 3 mm. The burning rate shall be added to the category designation (for example FH-3-30 mm/min).

FH-4: Same as FH-3 except the linear burning rate exceeds the values specified. The burning rate shall be added to the classification designation (for example FH-4-60 mm/min).

If all the specimens of a material which are tested do not have the same numerical category designation, the category with the highest number is reported as the designation for the material.

### A.2.2 Method B

The behaviour of specimens may be classified in one of the categories shown in table A.1 (FV = flaming vertical), by selecting the appropriate column using test results to answer the conditional questions posed.

## A.3 Test report

Recording the category designation in the test report is optional.

**Table A.1 — FV categories**

| Conditions | Category[1] [2] | | | |
|---|---|---|---|---|
| | Path 1 | Path 2 | Path 3 | Path 4 |
| If: every individual specimen afterflame time $t_1$ and $t_2$ is | $\leqslant$ 10 s | $\leqslant$ 30 s | $\leqslant$ 30 s | > 30 s |
| And: total set afterflame time $t_f$ for any conditioning is | $\leqslant$ 50 s | $\leqslant$ 250 s | $\leqslant$ 250 s | > 250 s |
| And: every individual specimen afterflame plus afterglow time after the second flame application ($t_2 + t_3$) is | $\leqslant$ 30 s | $\leqslant$ 60 s | $\leqslant$ 60 s | > 60 s |
| And: afterflame or afterglowing progresses up to the holding clamp | No | No | No | Yes |
| And: the cotton indicator pad is ignited by flaming particles or drops | No | No | Yes | Yes or no |
| Then: the category is | FV-0[3] | FV-1[3] | FV-2[3] | —[4] |

1) Select the highest-numbered path in which each individual element complies with the criterion specified.

2) If only one specimen from a set of five specimens for a given preconditioning treatment does not comply with the requirements for a category, another set of five specimens subjected to the same preconditioning shall be tested. All specimens from the second set shall comply with the appropriate requirements for the category.

3) Insert the minimum thickness, in millimetres, to the nearest 0,1 mm, which gives this category.

4) The material cannot be categorized by method B. Use method A (horizontal burning) to categorize the burning behaviour of the material.

# International Standard  2577

INTERNATIONAL ORGANIZATION FOR STANDARDIZATION•МЕЖДУНАРОДНАЯ ОРГАНИЗАЦИЯ ПО СТАНДАРТИЗАЦИИ•ORGANISATION INTERNATIONALE DE NORMALISATION

# Plastics — Thermosetting moulding materials — Determination of shrinkage

*Plastiques — Matières à mouler thermodurcissables — Détermination du retrait*

**Second edition — 1984-12-15**

UDC  678.072 : 678.019.252

Ref. No.  ISO 2577-1984 (E)

**Descriptors** : plastics,  thermosetting materials,  moulding materials,  tests,  determination,  shrinkage.

ISO 1984-1984 (E)

# Foreword

ISO (the International Organization for Standardization) is a worldwide federation of national standards bodies (ISO member bodies). The work of preparing International Standards is normally carried out through ISO technical committees. Each member body interested in a subject for which a technical committee has been established has the right to be represented on that committee. International organizations, governmental and non-governmental, in liaison with ISO, also take part in the work.

Draft International Standards adopted by the technical committees are circulated to the member bodies for approval before their acceptance as International Standards by the ISO Council. They are approved in accordance with ISO procedures requiring at least 75 % approval by the member bodies voting.

International Standard ISO 2577 was prepared by Technical Committee ISO/TC 61, *Plastics*.

ISO 2577 was first published in 1975. This second edition cancels and replaces the first edition, of which it constitutes a technical revision.

# Plastics — Thermosetting moulding materials — Determination of shrinkage

## 1 Scope and field of application

This International Standard specifies a method of determining the moulding shrinkage and the shrinkage after heat treatment of moulded test specimens of thermosetting moulding materials.

These characteristics are useful for the production control of thermosetting material and for checking uniformity of manufacture. Furthermore, knowledge of the initial shrinkage of thermosetting materials is important for the construction of moulds, and knowledge of post-shrinkage for establishing the suitability of the moulding material for the manufacture of moulded pieces with accurate dimensions.

## 2 Reference

ISO 291, *Plastics — Standard atmospheres for conditioning and testing*.

ISO 295, *Plastics — Compression moulding test specimens of thermosetting materials*.

## 3 Definitions

For the purpose of this International Standard, the following definitions apply :

**3.1 moulding shrinkage** : The difference in dimensions between a moulding and the mould cavity in which it was moulded, both the mould and the moulding being at normal temperature when measured.

**3.2 post-shrinkage** : Shrinkage of a plastic product after moulding, during post-treatment, storage or use.

## 4 Apparatus

**4.1 Mould, press, etc.,** suitable for moulding the test specimens specified in clause 6. For compression moulding, a positive or a semi-positive mould with single or multiple cavities shall be used. For injection moulding, the type of mould is under study.

If required, marks may be engraved in the mould near opposite ends of the specimen to facilitate the accurate measurement of the length of the cavity and the specimens.

NOTE — If multiple cavities are used with a positive mould, resulting variations in test specimen density may be sufficient to produce inconsistent shrinkage.

**4.2 Equipment,** suitable for measuring the lengths of the test specimen and the corresponding cavity of the mould to within 0,02 mm.

**4.3 Oven** (for post-shrinkage only).

## 5 Sampling

A representative sample shall be taken from the moulding material and be kept at room temperature in airtight containers, without any conditioning, until moulded into test specimens.

## 6 Test specimens

**6.1** The test specimen shall be

a) for compression moulding — bars of length 120 mm, width 15 mm and thickness 10 mm ;

b) for injection moulding — flat square plaques approximately 120 mm × 120 mm × 4 mm.

**6.2** The specimens shall be moulded to shape by compression or injection moulding using a mould with single or multiple cavities.

## 7 Procedure

**7.1** If not already known, measure the lengths of the cavities (or the distances between the engraved marks in the mould) to the nearest 0,02 mm at a temperature of 23 ± 2 °C (ISO 291 "atmosphere 23").

Record these measurements for use in the calculation of shrinkage.

NOTE — From time to time, moulds should be checked for wear, etc. As an alternate to measuring directly the lengths of the cold moulds, the gauge for the moulds may be obtained very precisely by cold-moulding specimens from lead and measuring their lengths.

**7.2** Mould at least two specimens from the sample to be tested, under the conditions given below.

a) For compression moulding :

Mould the specimens under the conditions of pressure, temperature, time, etc., specified in ISO 295 or in the relevant specification for the material.

b) For injection moulding :

Under study; to be added later.

NOTE — In the case of those fibrous materials that are to be injection-moulded as a plaque, at least four specimens should be tested.

**7.3** After removal from the mould, allow the test specimens to cool to room temperature by placing them on a material with low thermal conductivity and under an appropriate load to avoid warping. Store them at a temperature of 23 ± 2 °C and a relative humidity of 45 % to 55 % (ISO 291 ''atmosphere 23/50'') for between 16 and 72 h, or for such shorter time as can be shown to give the same test results.

**7.4** Before measuring the lengths of the test specimens, place them on a flat surface or against a straight edge in order to determine any warp or distortion. Any test specimen that has a warp exceeding 1 % of its length shall be discarded.

**7.5** For the determination of moulding shrinkage, measure, to the nearest 0,02 mm, the lengths of the bar specimens parallel to their major axis between opposite end faces or the distances between the gauge marks, at a temperature of 23 ± 2 °C (ISO 291 ''atmosphere 23''). Measurement of plaque specimens shall be made at distance of 20 mm from the corners, two measurements in the same direction.

NOTE — In order to measure the effect of orientation on the shrinkage of an injection-moulded specimen, shrinkages in two directions at right-angles (each of which is calculated from an average of two measurements in the same direction) are measured and calculated independently.

**7.6** For the determination of post-shrinkage, place the test specimens, measured as described in 7.5, in an oven maintained at the temperature given below. Support the specimens to avoid deformation and in such a way that they are separated from each other.

The heating temperatures shall be :

80 ± 2 °C for urea-formaldehyde moulding materials;

110 ± 3 °C for all other thermosetting moulding materials.

The times of exposure shall be :

48 ± 1 h for rapid determination;

168 ± 2 h for normal determination.

NOTE — Post-shrinkage depends strongly on the time of exposure. Therefore the exposure time should be noted [see 8.2 and 9 f)] and should be specified in the specification for the material.

At the end of the heating period, remove the test specimens from the oven and allow them to cool in a standard atmosphere of 23 ± 2 °C and a relative humidity of 45 % to 55 % (ISO 291 ''atmosphere 23/50'') for at least 3 h.

After the cooling period, measure the test specimens again, at a temperature of 23 ± 2 °C (ISO 291 ''atmosphere 23'') to the nearest 0,02 mm, as specified in 7.5.

## 8 Expression of results

**8.1** The moulding shrinkage (*MS*) is given, as a percentage, by the formula

$$MS = \frac{L_0 - L_1}{L_0} \times 100$$

where

$L_0$ is the length, in millimetres, of the dimension of the mould, determined as in 7.1;

$L_1$ is the length, in millimetres, of the corresponding dimension measured on the test specimen according to 7.5.

NOTE — When shrinkage is being determined using injection-moulded plaques, $L_0$ and $L_1$ are each the averages of two readings, measured in the same direction, taken 20 mm from the corners of the mould and the test specimen respectively.

**8.2** Post-shrinkage (*PS*) is given, as a percentage, by the formula

$$PS_{48\,h} \text{ or } PS_{168\,h} = \frac{L_1 - L_2}{L_1} \times 100$$

where

$L_1$ is as defined in 8.1;

$L_2$ is the length, in millimetres, of the same dimension of the test specimen, measured after heat treatment for 48 or 168 h according to 7.6.

NOTE — When post-shrinkage is being determined using injection-moulded plaques, $L_2$ is the average of two readings, measured in the same direction, taken 20 mm from the corners of the test specimen.

## 9 Test report

The test report shall include the following particulars :

a) reference to this International Standard;

b) the grade and designation of the moulding material;

c) the type of test specimen used (bar or plaque);

d) the method of moulding the specimens (compression or injection) and the moulding conditions;

e) the number of test specimens discarded because of excessive warping;

f) the conditions of heat treatment for the determination of post-shrinkage;

g) the moulding shrinkage ($MS$) and the post-shrinkage ($PS_{48\,h}$ and/or $PS_{168\,h}$), as a percentage, including the individual values, the arithmetic mean and, for injection-moulded plaques, the direction of measurement with respect to the direction of injection;

h) the dates of moulding the test specimens, measurement of moulding shrinkage, post-shrinkage heat treatment, and measurement of post-shrinkage.

# INTERNATIONAL STANDARD

## ISO
## 2818

Third edition
1994-08-15

# Plastics — Preparation of test specimens by machining

*Plastiques — Préparation des éprouvettes par usinage*

Reference number
ISO 2818:1994(E)

# Foreword

ISO (the International Organization for Standardization) is a worldwide federation of national standards bodies (ISO member bodies). The work of preparing International Standards is normally carried out through ISO technical committees. Each member body interested in a subject for which a technical committee has been established has the right to be represented on that committee. International organizations, governmental and non-governmental, in liaison with ISO, also take part in the work. ISO collaborates closely with the International Electrotechnical Commission (IEC) on all matters of electrotechnical standardization.

Draft International Standards adopted by the technical committees are circulated to the member bodies for voting. Publication as an International Standard requires approval by at least 75 % of the member bodies casting a vote.

International Standard ISO 2818 was prepared by Technical Committee ISO/TC 61, *Plastics*, Subcommittee SC 2, *Mechanical properties*.

This third edition cancels and replaces the second edition (ISO 2818:1980), which has been revised with respect to the following points:

— normative references for the geometry of cutting tools and abrasive tools and products;

— introduction of notching;

— extension of the table for recommended machining conditions.

Annex A of this International Standard is for information only.

International Organization for Standardization
Case Postale 56 • CH-1211 Genève 20 • Switzerland
Printed in Switzerland

# Introduction

The preparation of test specimens by machining influences the finished surfaces and, in some cases, even the internal structure of the specimens. Since test results are strongly dependent on both of these parameters, exact definitions of tools and machining conditions are required for reproducible test results with machined specimens.

# Plastics — Preparation of test specimens by machining

## 1 Scope

This International Standard establishes the general principles and procedures to be followed when machining and notching test specimens from compression-moulded and injection-moulded plastics, extruded sheets, plates and partially finished or wholly finished products.

In order to establish a basis for reproducible machining and notching conditions, the following general standardized conditions should be applied. It is assumed, however, that the exact procedures to be used will be selected or specified by the relevant material specification or by the standards on the particular test methods. If sufficiently detailed procedures are not thus specified, it is essential that the interested parties agree on the conditions to be used.

## 2 Normative references

The following standards contain provisions which, through reference in this text, constitute provisions of this International Standard. At the time of publication, the editions indicated were valid. All standards are subject to revision, and parties to agreements based on this International Standard are encouraged to investigate the possibility of applying the most recent editions of the standards indicated below. Members of IEC and ISO maintain registers of currently valid International Standards.

ISO 3002-1:1982, *Basic quantities in cutting and grinding — Part 1: Geometry of the active part of cutting tools — General terms, reference systems, tool and working angles, chip breakers.*

ISO 3017:1981, *Abrasive discs — Designation, dimensions and tolerances — Selection of disc outside diameter/centre hole diameter combinations.*

ISO 3855:1977, *Milling cutters — Nomenclature.*

ISO 6104:1979, *Abrasive products — Diamond or cubic boron nitride grinding wheels and saws — General survey, designation and multilingual nomenclature.*

ISO 6106:1979, *Abrasive products — Grain sizes of diamond or cubic boron nitride.*

ISO 6168:1980, *Abrasive products — Diamond or cubic boron nitride grinding wheels — Dimensions.*

## 3 Definitions

For the purposes of this International Standard, the following definitions apply:

### 3.1 Milling

In this machining operation, the tool has a circular primary motion and the workpiece a suitable feed motion. The axis of rotation of the primary motion retains its position with respect to the tool, independently of the feed motion (see ISO 3855). Complete dumb-bell and rectangular test specimens, as well as notches in finished specimens, may be prepared by milling.

#### 3.1.1 Geometry (see 3002-1 and figure 1)

Only a few details of the exact geometrical conditions of the milling tool and its position with respect to the workpiece given in ISO 3002-1 are relevant to this standard, as follows:

**3.1.1.1 tool-cutting-edge angle, $\alpha_r$:** The angle between the tool-cutting-edge plane $P_s$ and the assumed working plane $P_f$, measured in the tool back plane $P_r$.

**3.1.1.2 tool back clearance, $\alpha_p$:** The angle between the flank $A_\alpha$ of the cutter and the tool-cutting-edge plane $P_s$, measured in the tool back plane $P_p$.

**3.1.1.3 tool side clearance, $\alpha_f$:** The angle between the flank $A_\alpha$ of the cutter and the tool-cutting-edge plane $P_s$, measured in the assumed working plane $P_f$.

**3.1.1.4 tool radius, $R$:** The distance between the axis of the circular primary motion of the tool and its cutting edge.

**3.1.1.5 number of cutting teeth, $z$:** The number of cutting edges on the outer periphery of the rotating milling tool.

### 3.1.2 Tool and workpiece motions (see ISO 3002-1 and figure 2)

**3.1.2.1 rotational speed of tool, $n$:** The speed, in revolutions per minute, of the circular primary motion of the tool.

**3.1.2.2 cutting speed, $v_c$:** The instantaneous velocity, in metres per minute, of the primary motion of a selected point on the cutting edge relative to the workpiece. The relationship between $v_c$ and $n$ is given by the equation $v_c = n \cdot 2\pi R$.

**3.1.2.3 feed speed, $v_f$:** The instantaneous velocity, in metres per minute, of the feed motion of a selected point on the cutting edge relative to the workpiece.

**3.1.2.4 feed path, $\lambda$:** The distance, in millimetres, at any given point on the surface of the workpiece covered during the time between two successive cutting operations. The feed path is given by the equation $\lambda = v_f / z \cdot n$.

**3.1.2.5 cutting depth, $a$:** The (mean) distance, in millimetres, between the surfaces of the workpiece before and after one complete milling run.

## 3.2 Cutting of rectangular test specimens

In this machining operation, rectangular test specimens are cut by means of a circular or band saw, made from hardened steel or coated with diamond or cubic boron nitride powder, or cut with the aid of an abrasive disc of which the cutting edge may be coated with diamond or boron nitride powder. For further details on abrasive discs and abrasive products, see ISO 3017 and ISO 6104.

### 3.2.1 Geometry

**3.2.1.1 tool radius, $R$:** The distance, in millimetres, between the rotary axis of a circular saw or an abrasive disc and the cutting edges of the tool.

**3.2.1.2 number of cutting teeth, $z$:** The number of cutting teeth on the periphery of a circular saw.

### 3.2.2 Tool and workpiece motions

**3.2.2.1 rotational speed of tool, $n$:** The speed of rotation, in revolutions per minute, of a circular saw or an abrasive disc.

**3.2.2.2 cutting speed, $v_c$:** The instantaneous velocity, in metres per minute, of the cutting tip of a saw tooth, or of a selected point on the cutting edge of an abrasive disc, relative to the workpiece. For a circular saw or an abrasive disc, the relationship between $v_c$ and $n$ is given by the equation $v_c = n \cdot 2\pi R$.

**3.2.2.3 feed speed, $v_f$:** The instantaneous velocity, in metres per minute, of the tool feed parallel to the saw or disc plane and perpendicular to the cutting direction relative to the workpiece.

## 3.3 Cutting of disc-shaped test specimens (see figure 4)

In this machining operation, disc-shaped test specimens are cut from sheet material with the aid of a circular cutter with a saw-toothed edge of hardened steel or which may be coated with diamond or cubic boron nitride powder. The test specimens may also be cut by means of a milling cutter with one or more teeth, as described in 3.1, which moves in a circular orbit. Furthermore, the test specimens may also be cut from a roughly preshaped pack of individual sheets with the aid of a turning lathe.

### 3.3.1 Geometry

**3.3.1.1 tool radius, $R$:** The distance, in millimetres, between the rotary axis of the circular cutter and the inner limit of the cutting edge. The tool radius is equal to the radius of the finished test specimen.

**3.3.1.2 number of cutting teeth, $z$:** The number of teeth on the sawtooth cutting edge of a circular cutter. If a lathe is used for cutting circular test specimens, the geometrical definitions of the cutting tool are the same as those given in 3.1.

### 3.3.2 Tool and workpiece motions

**3.3.2.1 rotational speed of tool, $n$:** The speed of rotation, in revolutions per minute, of a circular cutter.

**3.3.2.2 cutting speed, $v_c$:** The instantaneous velocity, in metres per minute, of a selected point on the cutting edge relative to the workpiece. The relationship between $v_c$ and $n$ is given by the equation $v_c = n \cdot 2\pi R$.

**3.3.2.3   feed speed, $v_f$:** The instantaneous velocity, in metres per minute, of the tool feed parallel to the rotary axis of the circular cutter and perpendicular to the cutting direction relative to the workpiece.

## 3.4   Planing of rectangular bars and planing or broaching of notches in finished test specimens

In this machining operation, sawed or sliced rectangular bars are finished by planing. Also, notches in finished specimens can be cut by planing or broaching.

### 3.4.1   Geometry

**3.4.1.1   tool-cutting-edge angle, $\alpha_r$:** As defined in 3.1.1.1.

**3.4.1.2   tool back clearance, $\alpha_p$:** As defined in 3.1.1.2.

**3.4.1.3   tool side clearance, $\alpha_f$:** As defined in 3.1.1.3.

### 3.4.2   Tool and workpiece motions

**3.4.2.1   cutting speed, $v_c$:** The instantaneous velocity, in metres per minute, of the primary motion of a selected point on the cutting edge relative to the workpiece.

**3.4.2.2   cutting depth, $a$:** The (mean) distance, in millimetres, between the surfaces of the workpiece before and after one planing run.

## 3.5   Stamping of arbitrarily shaped test specimens fabricated from thin sheets

In this operation, arbitrarily shaped test specimens are stamped under high pressure from thin sheets by means of a tool with a sharp edge made from hardened steel and located in a plane parallel to the plane of the sheet.

### 3.5.1   Geometry

**3.5.1.1   shape of the stamping tool:** The geometric shape of the stamping edge in a plane parallel to the sheet plane. The shape of the stamping tool depends on the shape of the test specimen to be stamped, along with its required dimensions and tolerances.

### 3.5.2   Forces on the tool and tool motion

**3.5.2.1   contact force, $F_c$:** The force, in newtons, applied to the stamping tool in the direction perpendicular to the sheet plane.

**3.5.2.2   feed speed, $v_f$:** The instantaneous velocity, in metres per minute, of the feed motion of the edge plane of the stamping tool in a direction perpendicular to the sheet plane.

## 4   Test specimens

### 4.1   Shape and state of the test specimens

The following types of test specimen can be prepared by the machining processes described in this International Standard:

— rectangular bars;

— notched rectangular bars;

— rectangular plates;

— curvilinear test specimens (e.g. dumb-bells);

— discs.

The exact shape, dimensions and tolerances of the test specimens shall conform to the standard for the particular test method in question. The machined surfaces and edges of the finished specimens shall be free of visible flaws, scratches or other imperfections when viewed with a low-power magnifying glass (approximately ×5 magnification).

Rectangular bars shall be free of twist and shall have perpendicular pairs of parallel surfaces. The surfaces and edges shall be free from scratches, pits, sink marks and flashes. Each specimen shall be checked for conformity with these requirements by visual observation against straight-edges, squares and flat plates, and by measuring with micrometer callipers.

The requirements on the quality of the edges of disc-shaped specimens used for impact-penetration tests are less rigorous than those for tensile-test specimens.

Any specimen showing a measurable or observable departure from the requirements given above shall be rejected or machined to proper size and shape before testing.

## 4.2 Preparation of test specimens

The test specimens shall be machined from plates or sheets made from the material to be tested by compression moulding, injection moulding, casting, polymerizing *in situ*, extrusion or other processing operations to produce semifinished products. Plates may also be obtained in an appropriate manner from finished products. If the sample from which the specimens are prepared is not isotropic, prepare test specimens with their main axis parallel to and perpendicular to the main orientation axis. In all cases, the exact conditions for producing the test specimens, and the position and orientation of the specimens within the samples, shall be agreed upon by the interested parties, and such details shall be described in the test report.

NOTE 1    Attention is drawn to the fact that the room temperature and the temperature of the material during the machining may influence the properties of the specimen.

## 5  Machinery and tools

For preparing test specimens from plastics materials and for notching of finished specimens, the machines mentioned in 5.1 to 5.5 can be used (see also clause 3). Recommended machining conditions for various specimen shapes and specimen materials are given in table 1. Any required conditions for preparation of test specimens by machining will be specified for each material in part 2 of the appropriate ISO material standard. The conditions given in table 1 for machining notches have also been found to give satisfactory results for numerous materials; however, because of the wide variety of materials tested, other conditions may also be appropriate.

### 5.1  Milling cutters

These can be used to prepare dumb-bell test specimens and rectangular bars. They may contain one tooth or a number of teeth arranged in a manner described in ISO 3855 and may cut at variable speeds (at high speeds, for instance, in the case of copy milling machines). They can also be used to cut notches in rectangular specimens. In this case, more than one tooth shall be used only if the notches can be made with the same quality as with one tooth.

### 5.2  Slicing or sawing machines

These may be used to prepare rectangular bar or plate test specimens. They can be equipped with a circular or a band saw, or with a circular disc, the edge of which is coated with an abrasive material such as diamond or cubic boron nitride.

### 5.3  Tubular cutting machines

These are used to prepare disc test specimens from flat plates or sheet material. The cutting edge of this kind of tool may be saw-toothed or coated with an abrasive material.

### 5.4  Lathes

These can be used for the same purpose as indicated in 5.3, i.e. for cutting disc test specimens from roughly pre-shaped packs of individual sheets.

### 5.5  Planing machines

These can be used to cut finished sawed or sliced rectangular bars and to cut notches.

### 5.6  Stamping tools

Stamping tools are suitable for preparing test specimens of any shape from thin sheets made of materials of adequate ductility.

### 5.7  Broaching tools

These can be used for notching. They may be hand-operated or machine-driven.

## 6  Procedure

The machining speed is dependent on the material being tested and shall be such that overheating of the material is avoided. This is particularly important in the case of thermoplastic materials. If the use of a cooling agent is necessary, this will be stated in part 2 of the appropriate ISO material standard. The use of a cooling agent shall have no deleterious effect on the material being machined (see also table 1). Fine abrasives may be used to achieve a smooth finish. In the case of tools with edges coated with diamond, cubic boron nitride or another abrasive material, ISO 3017, ISO 6104, ISO 6106 and ISO 6168 should be considered.

NOTE 2    When machining specimens, care should be taken to avoid skin contact and inhalation of dust, as dust may cause irritation.

### 6.1  Preparation of dumb-bell specimens

Prepare such specimens by low-speed milling with a hand-controlled milling tool or, preferably, by high-speed copy milling using the conditions given in table 1.

Examine the milled surfaces and edges of the finished test specimens with a magnifying glass having an approximately ×5 magnification for the presence of flaws, scratches and other imperfections. After cutting a maximum of 500 specimens, examine the cutting edge with the aid of a microscope or profile projector with a ×50 to ×100 magnification.

## 6.2 Preparation of rectangular test specimens by sawing or cutting with an abrasive disc

The detailed conditions used in these methods are given in table 1. Prepare the test specimens by sawing only if there are no particular requirements regarding the quality of the specimen surfaces, or if the surfaces are to be subsequently finished by another method such as milling or planing. In the latter case, examine the surfaces as specified in 6.1.

## 6.3 Preparation of disc-shaped test specimens

In general, disc specimens are used to perform impact-penetration tests. In such cases, imperfections in the machined surfaces have no serious influence on the test results. Prepare such specimens using the conditions given in table 1 and ensure that the plane surfaces of the specimens are smooth and free from flaws.

## 6.4 Stamping out test specimens of any shape

Use this method for the preparation of test specimens only if the material in question is sufficiently soft and the specimens are to be made from sufficiently thin sheets. The specimen is stamped from the sheet using a single stroke of a knife-edged punch of appropriate shape and dimensions. The cutting edge of the punch shall be sufficiently sharp and free from notches. The sheet shall be supported on a slightly yielding material with a smooth surface (for example leather, rubber or good-quality cardboard) on a flat, rigid base.

The criterion for the applicability of this method is the quality of the specimen edges and surfaces as revealed by examination using the method outlined in 6.1.

## 6.5 Notching finished test specimens by milling or broaching

Notching may be carried out with the aid of a milling or broaching machine or a lathe, preferably with a single-tooth cutter. Use a tool having a cutting edge made of high-speed steel, hardened steel or diamond. Use a multi-tooth cutter only if notches can be prepared having the same quality as notches made with a single-tooth cutter. For specimens prepared by stamping, machine the notch in a secondary operation (i.e. the notch shall not be stamped).

In preparing notches, the use of abrasives is not permissible.

For milling, choose the feed rate so that the thickness $d_s$ of the shavings is from 0,003 mm to 0,07 mm (see figures 2 and 3). The thickness $d_s$ is given, in millimetres, by the equation

$$d_s = v_f^2 \left( n^2 \cdot R \right)$$

where

$v_f$    is the feed speed, in millimetres per minute;

$n$    is the rotary speed of the tool, in revolutions per minute;

$R$    is the distance, in millimetres, between the axis of the milling machine and the tip of the cutter.

It is essential that close tolerances are established on the contour and radius of the notch because these parameters largely determine the degree of stress concentration at the base of the notch. To obtain reproducible results, carefully grind and hone the cutting edge to ensure sharpness and freedom from nicks and burrs.

Before the first use and after cutting about 500 notches, or more often if the cutter has been used to notch a hard abrasive material, inspect the cutter for sharpness, absence of nicks, correct tip radius and correct tip contour. If the radius and contour do not fall within the specified limits, replace the cutter by a newly sharpened and honed one.

A microscope or profile projector with a ×50 to ×100 magnification is suitable for checking the cutter and the notch. In the case of single-tooth cutters, the contour of the tip of the cutting tool may be checked instead of the contour of the notch in the specimen, provided that, for the type of notch produced, the two correspond or that a definite relationship exists between them. There is some evidence that notches cut by the same cutter in widely differing materials may differ in contour.

In the case of transparent materials, it is often possible to detect undesirable changes in the specimen by means of photoelastic effects. For example, undesirable heating or melting caused by machining, especially of injection-moulded specimens, becomes visible by virtue of distinct changes in coloured interference lines or areas within the zone near the machined surface.

NOTE 3   Experience using notched specimens has shown that there are materials (e.g. PMMA, PC) for which the measured values obtained in tests using such specimens decrease gradually in spite of the fact that the cutter is optically satisfactory. In such cases, it is recommended that the cutter be checked using a reference material.

# 7   Test report

The test report shall include the following information:

a)   a reference to this International Standard;

b)   a description of the material tested and of the sample from which the test specimens were machined (shape, method of preparation, orientation, etc.);

c)   a precise description of the position and orientation of the test specimens as taken from a semifinished or finished product;

d)   the dimensions of the test specimens;

e)   the method of machining used;

f)   the machining conditions used (see table 1);

g)   any other relevant details.

## Table 1 — Recommended machining conditions for four types of test specimen and for notches

| Material | Method of machining | Rotational speed $n$ tr/min | Diameter $2R$ mm | Cutting edge angle $\alpha_r$ | Back clearance $\alpha_p$ | Side clearance $\alpha_f$ | Number of teeth $z$ | Cutting speed $v_c$ m/min | Feed speed $v_f$ m/min | Feed path $\lambda$ mm | Cutting depth $a$ mm | Coolant |
|---|---|---|---|---|---|---|---|---|---|---|---|---|
| **1) Dumb-bell specimen (see 6.1)** | | | | | | | | | | | | |
| Thermoplastics | Medium-speed milling | 180 to 500 | 125 to 150 | 5 to 15 | 5 to 20 | — | 10 to 16 | 70 to 250 | Slowly | — | 1 to 5 | None, air or water |
| Thermosets | | — | 125 to 150 | — | — | — | — | 70 to 250 | Slowly | — | 1 to 5 | None, air or water |
| Thermoplastics | High-speed copy milling | 8 000 to 30 000 | 5 to 20 | 10 to 15 | 5 to 20 | — | 4 to 8 | 125 to 2 000 | Slowly | — | 0,2 | Air or water |
| Thermosets | | 20 000 | 15 to 20 | 10 to 15 | 5 to 20 | — | 4 to 8 | 100 to 1 500 | Slowly | — | 0,5 | Air or water |
| **2) Rectangular specimen (see 6.2)** | | | | | | | | | | | | |
| Thermoplastics | Sawing with a circular saw | 1 000 to 2 000 | 50 to 150 | — | — | — | 30 to 100 | 150 to 1 000 | Medium | — | — | None or air |
| Thermosets | | 1 000 to 2 000 | 50 to 150 | — | — | — | 50 to 150 | 150 to 1 000 | Medium | — | — | None or air |
| Thermoplastics | Sawing with a band saw | — | — | — | — | — | as for circular saw | 3 to 15 | Medium | — | — | None or air |
| Thermosets | | — | — | — | — | — | | 3 to 15 | Medium | — | — | None or air |
| Thermoplastics | Cutting with an abrasive disc | 2 000 to 13 000 | 50 to 150 | — | — | — | — | 1 000 to 2 000 | Slowly | — | — | Air or water |
| Thermosets | | 2 000 to 13 000 | 50 to 150 | — | — | — | — | 1 000 to 2 000 | Slowly | — | — | Air or water |
| **3) Disc-shaped specimen (see 6.3)** | | | | | | | | | | | | |
| Thermoplastics | Cutting with a circular-saw-like cutter | 100 to 200 | 40 to 100 | — | — | — | 30 to 100 | 10 to 100 | Medium | — | — | None or air |
| Thermosets | | 100 to 200 | 40 to 100 | — | — | — | 30 to 100 | 10 to 100 | Medium | — | — | None or air |
| Thermoplastics | Cutting with a circular abrasive cutter | 300 to 1 500 | 40 to 100 | — | — | — | — | 100 to 200 | Slowly | — | — | Air or water |
| Thermosets | | 300 to 1 500 | 40 to 100 | — | — | — | — | 100 to 200 | Slowly | — | — | Air or water |
| Thermoplastics | Cutting with a single-tooth milling cutter | 100 to 200 | 40 to 100 | 5 to 15 | 5 to 20 | — | 1 | 10 to 100 | Slowly | — | — | None or air |
| Thermosets | | 100 to 200 | 40 to 100 | 5 to 15 | 5 to 20 | — | 1 | 10 to 100 | Slowly | — | — | None or air |
| Thermoplastics | Turning with a lathe | 500 to 1 000 | 20 to 100 | 5 to 15 | 5 to 20 | — | 1 | 30 to 300 | Slowly | — | — | None or air |
| Thermosets | | 500 to 1 000 | 20 to 100 | 5 to 15 | 5 to 20 | — | 1 | 30 to 300 | Slowly | — | — | None or air |
| **4) Stamped specimens of any shape (see 6.4)** | | | | | | | | | | | | |
| Thermoplastics | Stamping from thin sheets | — | — | — | — | — | — | — | Slowly by pressure | — | — | None |
| Thermosets | | — | — | — | — | — | — | — | | — | — | None |
| **5) Cutting notches (see 6.5)** | | | | | | | | | | | | |
| Thermoplastics | Medium-speed milling | 200 to 1 000 | 60 to 80 | 2 to 7 | 2 to 7 | 2 to 7 | 1 | 50 to 250 | 0,07 to 2 | 1 to 2 | 0,2 to 2 | Air or water |
| Thermosets | | 200 to 1 000 | 60 to 80 | 2 to 7 | 2 to 7 | 2 to 7 | 1 | 50 to 250 | 0,07 to 2 | 1 to 2 | 0,2 to 2 | Air or water |
| Thermoplastics | Broaching | — | — | 2 to 7 | 2 to 7 | 2 to 7 | 1 | 12 to 20 | Slowly | — | 0,1 to 0,3 | Air or water |
| Thermosets | | — | — | 2 to 7 | 2 to 7 | 2 to 7 | 1 | 12 to 20 | Slowly | — | 0,1 to 0,3 | Air or water |

NOTE — These machining conditions may vary, depending on the specific materials and tools used. The machining conditions used should be those which provide specimens conforming to the specified dimensions and free of flaws when examined under the specified magnification.

Particular machining conditions are reported in part 2 of the designation standard for the material concerned.

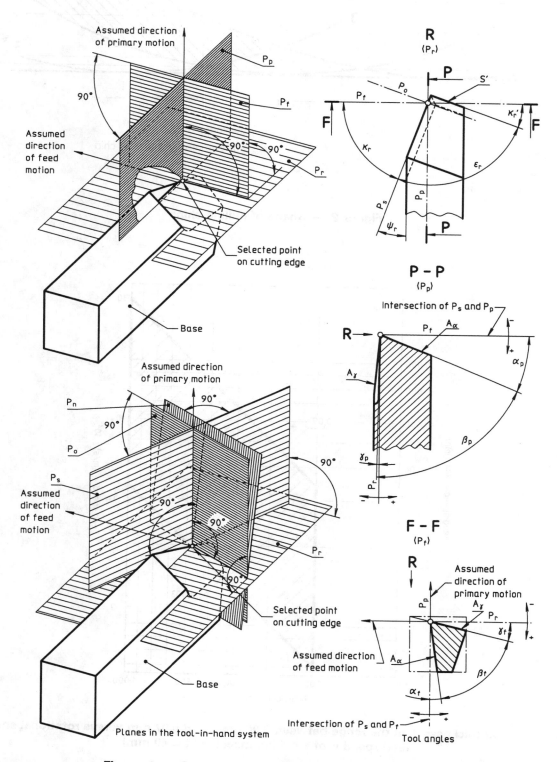

**Figure 1 — Geometry of the active part of cutting tools**

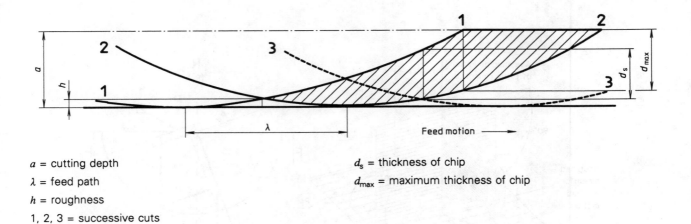

$a$ = cutting depth  
$\lambda$ = feed path  
$h$ = roughness  
1, 2, 3 = successive cuts

$d_s$ = thickness of chip  
$d_{max}$ = maximum thickness of chip

**Figure 2 — Shape of milling chips**

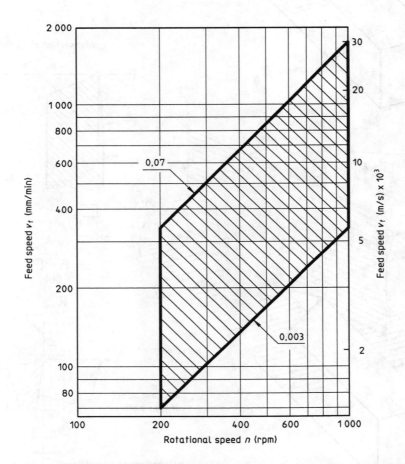

**Figure 3 — Chip thickness $d_s$ in the range between 0,003 mm and 0,07 mm versus rotational speed $n$ and feed speed $v_f$ of a milling machine ($R$ = 40 mm)**

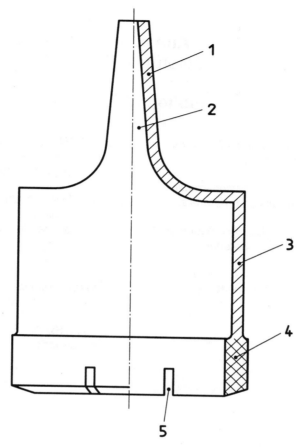

1  Cylindrical or conical fixing shank

2  Axial cooling-water feed channel

3  Body of the cutter

4  Abrasive sleeve (slightly thicker than the body of the cutter)

5  Drainage holes for cooling water and dust

**Figure 4 — Example of a circular cutter (half-section of a diamond-edged cutter)**

# Annex A
(informative)

# Bibliography

[1] ISO 291:1977, *Plastics — Standard atmospheres for conditioning and testing.*

[2] ISO 293:1986, *Plastics — Compression moulding test specimens of thermoplastic materials.*

[3] ISO 294:1975, *Plastics — Injection moulding test specimens of thermoplastic materials.*

[4] ISO 295:1991, *Plastics — Compression moulding of test specimens of thermosetting materials.*

[5] ISO 468:1982, *Surface roughness — Parameters, their values and general rules for specifying requirements.*

[6] ISO 2557-1:1989, *Plastics — Amorphous thermoplastics — Preparation of test specimens with a specified maximum reversion — Part 1: Bars.*

[7] ISO 2557-2:1986, *Plastics — Amorphous thermoplastics — Preparation of test specimens with a specified reversion — Part 2: Plates.*

[8] ISO 3167:1993, *Plastics — Multipurpose test specimens.*

# International Standard

ISO 3146

INTERNATIONAL ORGANIZATION FOR STANDARDIZATION●МЕЖДУНАРОДНАЯ ОРГАНИЗАЦИЯ ПО СТАНДАРТИЗАЦИИ●ORGANISATION INTERNATIONALE DE NORMALISATION

# Plastics — Determination of melting behaviour (melting temperature or melting range) of semi-crystalline polymers

*Plastiques — Détermination du comportement à la fusion (température de fusion ou plage de température de fusion) des polymères semi-cristallins*

**Second edition — 1985-12-15**

UDC 678.7 : 620.1 : 536.421.1

Ref. No. ISO 3146-1985 (E)

Descriptors : plastics, polymers, tests, determination, melting points, test equipment.

# Foreword

ISO (the International Organization for Standardization) is a worldwide federation of national standards bodies (ISO member bodies). The work of preparing International Standards is normally carried out through ISO technical committees. Each member body interested in a subject for which a technical committee has been established has the right to be represented on that committee. International organizations, governmental and non-governmental, in liaison with ISO, also take part in the work.

Draft International Standards adopted by the technical committees are circulated to the member bodies for approval before their acceptance as International Standards by the ISO Council. They are approved in accordance with ISO procedures requiring at least 75 % approval by the member bodies voting.

International Standard ISO 3146 was prepared by Technical Committee ISO/TC 61, *Plastics*.

This second edition cancels and replaces the first edition (ISO 3146-1974), of which it constitutes a minor revision.

Users should note that all International Standards undergo revision from time to time and that any reference made herein to any other International Standard implies its latest edition, unless otherwise stated.

# Plastics — Determination of melting behaviour (melting temperature or melting range) of semi-crystalline polymers

## 0 Introduction

The melting behaviour of a crystalline or partly crystalline polymer is a structure-sensitive property.

In polymers a sharp melting point, such as is observed for low molecular mass substances, usually does not occur; instead a melting temperature range is observed on heating, from the first change of shape of the solid particles to the transformation into a highly viscous or viscoelastic liquid, with accompanying disappearance of the crystalline phase, if present. The melting range depends upon a number of parameters, such as molecular mass, molecular mass distribution, per cent crystallinity, and thermodynamic properties.

It may also depend on the previous thermal history of the specimens. The lower or upper limit of the melting range, or its average value, is sometimes conventionally referred to as the "melting temperature".

## 1 Scope and field of application

This International Standard specifies three methods for evaluating the melting behaviour of semi-crystalline polymers.

**Section one** specifies a capillary tube method (method A), which is based on the changes in shape of the polymer. This method is applicable to all polymers and their compounds, even if there is no crystalline phase.

**Section two** specifies a polarizing microscope method (method B), which is based on changes in the optical properties of the polymer.

This method is applicable to polymers containing a birefringent crystalline phase; it may not be suitable for plastics compounds containing pigments and/or other additives which could interfere with the birefringence of the polymeric crystalline zone.

**Section three** specifies a thermal analytical method (method C), having two variants :

— method C1, which uses Differential Thermal Analysis (DTA);

— method C2, which uses Differential Scanning Calorimetry (DSC).

Both are applicable to all polymers containing a crystalline phase and their compounds.

The melting temperatures determined by the different methods usually differ by several kelvins for the reasons explained in the Introduction.

Of the methods given above, experiments have indicated DSC (Differential Scanning Calorimetry) to be the method of choice as having the best reproducibility of results.

## 2 Definitions

**2.1 semi-crystalline polymers :** Polymers containing a crystalline phase surrounded by amorphous materials.

**2.2 melting range :** The temperature range over which crystalline polymers lose their crystallinity when heated.

NOTE — The conventional "melting temperatures" determined by methods A and B are defined in clauses 3 and 8.

# Section one : Method A — Capillary tube

## 3 Principle

Heating of a specimen, at a controlled rate, and observation for change in shape.

Reporting of the temperature of the specimen at the first visible deformation as the melting temperature.

NOTE — This method may also be used for non-crystalline materials according to the relevant specifications or by agreement between the interested parties.

## 4 Apparatus (see figure 1)

**4.1 Melting apparatus**, consisting of the following items :

a) cylindrical metal block, the upper part of which is hollow and forms a chamber;

b) metal plug, with two or more holes, allowing a thermometer and one or more capillary tubes to be mounted into the metal block a);

c) heating system for the metal block a) provided, for example, by an electrical resistance enclosed in the block;

d) rheostat for regulation of the power input, if electrical heating is used;

e) four windows of heat-resistant glass on the lateral walls of the chamber, diametrically disposed at right angles to each other. In front of one of these windows is mounted an eyepiece for observing the capillary tube. The other three windows are used for illuminating the inside of the enclosure by means of lamps.

NOTE — Other suitable melting apparatuses may be used, provided that they give the same results.

**4.2 Capillary tube**, of heat-resistant glass, closed at one end.

NOTE — The maximum external diameter should preferably be 1,5 mm.

**4.3 Calibrated thermometer**, graduated in divisions of 1 K. The thermometer probe shall be positioned in such a way that heat dispersion in the apparatus is not impeded.

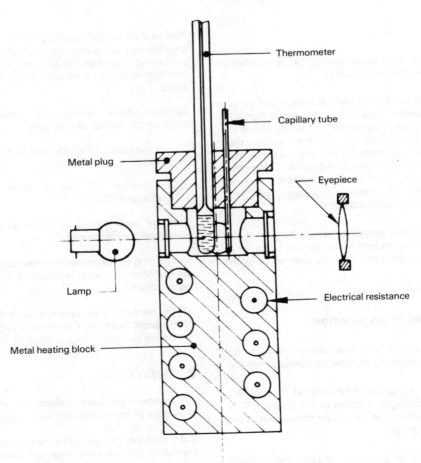

Figure 1 — Apparatus for method A

NOTE — Other suitable temperature-measuring devices may be used.

## 5 Test specimens

The specimens used shall be representative of the sample of material to be tested.

### 5.1 Characteristics

Powder of particle size up to 100 μm or cut pieces of films of thickness 10 to 20 μm should preferably be used. Comparison tests shall be carried out on specimens of the same or similar particle size, or similar thickness in the case of layers or films.

### 5.2 Conditioning

If not otherwise specified or agreed to by the interested parties, the sample shall be conditioned at 23 ± 2 °C and relative humidity of 50 ± 5 % for 3 h prior to the measurement.

## 6 Procedure

### 6.1 Calibration

Calibrate the temperature-measuring system periodically over the temperature range used for the test, with reagent grade or certified chemicals.

Chemicals recommended for calibration are listed in table 1.

### 6.2 Determination

**6.2.1** Insert the thermometer (4.3) and the capillary tube (4.2) containing the specimen into the heating chamber [4.1a)] and start the heating. When the temperature of the specimen is about 20 K below the expected melting temperature, regulate the rate of temperature increase to 2 ± 0,5 K/min. Record the temperature at which the specimen begins to change shape.

**6.2.2** Repeat the operations specified in 6.2.1 with a second specimen. If the two results obtained by the same operator on the same sample differ by more than 3 K, repeat the procedure on two new specimens.

Insufficient data are available for establishing reproducibility.

**Table 1 — Calibration standards**

| Chemical | Melting temperature[1] (°C) |
|---|---|
| L-Menthol-1 | 42,5 |
| Azobenzene | 69,0 |
| 8-Hydroxyquinoline | 75,5 |
| Naphthalene | 80,2 |
| Benzyl | 96,0 |
| Acetanilide | 113,5 |
| Benzoic acid | 121,7 |
| Phenacetin [N (4-ethoxyphenyl) acetamide] | 136,0 |
| Adipic acid | 151,5 |
| Indium | 156,4 |
| Sulfanilamide | 165,7 |
| Hydroquinone | 170,3 |
| Succinic acid | 189,5 |
| 2-Chloroanthraquinone | 208,0 |
| Anthracene | 217,0 |
| Saccharin | 229,4 |
| Tin | 231,9 |
| Tin(II) chloride | 247,0 |
| Phenolphthalein | 261,5 |

[1] The temperatures indicated refer to theoretically pure chemicals; the values of the actual melting point for the standard materials used should be certified by the supplier.

## 7 Test report

The test report shall include the following information :

a) reference to this International Standard;

b) reference of the method used (method A);

c) complete identification and description of the material tested;

d) shape and size (or mass) of the specimens;

e) previous thermal history of the specimens;

f) conditioning;

g) heating rate;

h) temperatures, in degrees Celsius or in kelvins, of two successive individual measurements, and their arithmetic mean;

i) any operational details not specified in this International Standard or regarded as optional, as well as any incidents liable to have affected the results.

# Section two : Method B — Polarizing microscope

## 8   Principle

Heating of a specimen, positioned between the polarizer and analyser of a microscope, at a controlled rate.

Measurement of the temperature at which the crystalline polymer loses its optical anisotropy, as detected by the disappearance of birefringence, as the melting temperature.

## 9   Apparatus

Ordinary laboratory apparatus and

**9.1   Microscope**, with a disk polarizer and a cap analyser, or a polarizing microscope with built-in analyser, with magnification from X 50 to X 100.

**9.2   Micro hot-stage**, consisting of an insulated metal block that can be mounted slightly above the microscope stage. This block shall be

a)   provided with a hole for light passage;

b)   electrically heated, with adequate controls for adjustment of heating and cooling rates;

c)   constructed to provide a chamber with a heat baffle and a glass cover, for carrying out measurements in an inert atmosphere;

d)   provided with a hole for insertion of a temperature-measuring device near the light hole.

**9.3   Thermometers**, calibrated, or equivalent **temperature-measuring devices**, for the test temperature ranges.

## 10   Test specimens

### 10.1   Powdered materials

Place a 2 to 3 mg portion of the powder (particle size not more than 100 μm) on a clean slide and cover with a cover glass.

Heat the specimen, the slide and the cover on a hot-plate slightly above the melting temperature of the polymer. By a slight pressure on the cover glass, form a thin film of thickness 0,01 to 0,04 mm and allow it to cool slowly by switching off the hot-plate.

### 10.2   Moulded or pelleted materials

Cut from the sample, with a microtome, a film of thickness approximately 0,02 mm, place it on a clean slide and cover with a cover glass. Heat and melt it as specified in 10.1.

### 10.3   Film or sheet materials

Cut a 2 to 3 mg portion of the film or sheet, place it on a clean slide, cover with a cover glass and proceed as specified in 10.1.

NOTE — The preliminary melting of the specimens between slide and cover presents the advantage of destroying any birefringence due to orientation or internal stresses, and also of reducing the danger of oxidation during the test. The need for an inert gas stream — as described in 11.2 — is thus limited to very special cases. The reproducibility of the measurements is also increased. However, by agreement between the interested parties, the determination may be carried out directly on the powder or cut film piece without preliminary melting. This deviation should be stated in the test report.

### 10.4   Conditioning

See 5.2.

## 11   Procedure

### 11.1   Calibration

See 6.1.

### 11.2   Determination

Place the glass microscope slide with the specimen on the micro hot-stage (9.2). Adjust the light source to maximum light intensity and focus the microscope (9.1).

For specimens that are degradable by air, adjust the gas inlet to the stage so that a slight stream of inert gas blankets the stage, keeping it under slight positive pressure to prevent ingress of air. Rotate the analyser to obtain a dark field; the crystalline material will appear bright on a dark field. Adjust the controller to heat the stage gradually (at a rate not higher than 10 K/min) to a temperature that is lower than the melting temperature, $\theta_m$, as determined approximately by previous test, by the following amounts :

10 K for $\theta_m \leqslant 150\ °C$

15 K for $150\ °C < \theta_m \leqslant 200\ °C$

20 K for $\theta_m > 200\ °C$

Then adjust the controller so that the temperature rises at a rate of 1 to 2 K/min.

Observe the temperature at which birefringence disappears, leaving a totally dark field. Record this temperature as the melting temperature of the sample.

Turn off the heating and remove the glass cover, heat baffle and specimen slide.

Repeat the procedure with another specimen. If the two results obtained by the same operator on the same sample differ by more than 1 K, repeat the procedure on two new specimens.

According to the results of round robins, the repeatability was 2 K. Insufficient data are available for establishing reproducibility.

## 12 Test report

The test report shall include the following information :

a) reference to this International Standard;

b) reference of the method used (method B);

c) complete identification and description of the material tested;

d) shape and size (or mass) of the specimens;

e) previous thermal history of the specimens;

f) conditioning;

g) description of preliminary heating on the slide, if applicable;

h) presence and type of inert gas, if applicable;

i) heating rate;

j) temperatures, in degrees Celsius or kelvins, of two successive individual measurements and their arithmetic mean;

k) any operational details not specified in this International Standard or regarded as optional, as well as any incidents liable to have affected the results.

# Section three : Method C — Thermal analysis (DTA or DSC)

## 13   Additional definitions

**13.1   Differential Thermal Analysis; DTA :**[1] A technique in which the temperature difference between a substance and a reference material is measured as a function of temperature while the substance and reference material are subjected to a controlled temperature programme.

NOTE — The record is the differential thermal or DTA curve; the temperature difference, $\Delta T$, should be plotted on the ordinate with endothermic reactions downwards and temperature, $T$, or time, $t$, on the abscissa, increasing from left to right.

**13.2   Differential Scanning Calorimetry; DSC :**[1] A technique in which the difference in energy inputs into a substance and a reference material is measured as a function of temperature while the substance and reference material are subjected to a controlled temperature programme.

NOTE — Two modes, power-compensation differential scanning calorimetry and heat-flux differential scanning calorimetry, can be distinguished, depending on the method of measurement used.

**13.3   baseline :**[2] The portion or portions of the DTA or DSC curve for which $\Delta T$ or the heat flux is approximately constant (approximately zero in DTA).

For example, see AB and DE in figure 2.

**13.4   peak :**[2] That portion of a DTA or DSC curve which departs from, and subsequently returns to, the baseline.

For example, see BCD in figure 2.

NOTE — A peak is attributable to the occurrence of some single process. It is normally characterized by a deviation from the established baseline, a maximum deflection, and a re-establishment of a baseline, not necessarily identical to that before the peak.

**13.5   endothermic peak; endotherm :**[2]

(1)   In DTA, a peak where the temperature of the sample falls below that of the reference material; that is $\Delta T$ is negative. (The melting phenomenon is an endothermic change.)

(2)   In DSC, a peak where the energy input to the sample is larger than that to the reference material.

**13.6   peak height :**[2] The distance, vertical to the temperature axis, between the interpolated baseline and the peak tip.

For example, see CF in figure 2.

NOTE — There are several ways of interpolating the baseline; that shown in figure 2 is only an example. Locations of points B and D depend on the method of interpolation of the baseline. Other examples of interpolation of the baseline are shown in figure 3.

**13.7   peak area :**[2] The area enclosed between the peak and the interpolated baseline(s). (See the note to 13.6.)

For example, see BCDB in figure 2.

**13.8   extrapolated onset temperature :**[2] The temperature determined by the point of intersection of the tangent drawn at the point of greatest slope on the leading edge of the peak (for example, tangent t in figure 2) with the extrapolated baseline (for example BG in figure 2).

For example, see G in figure 2, and figure 3.

NOTE — For polymers having a wide melting range, the extrapoled onset temperature indicates the initial point of the rapid rise of the melting curve, but not necessarily the initial melting.

**13.9   peak temperature :**[2] The temperature at the time at which the differential temperature or heat flux during that peak has the maximum value.

For example, see C in figure 2.

**13.10   sample :**[2] The actual material to be tested, whether diluted with an inert material or undiluted.

**13.11   specimens :**[2] Portions of the sample to be tested and the reference material.

**13.12   reference material :**[2] A substance known to be thermally inactive over the temperature range of interest, for example $\alpha$-aluminium oxide ($\alpha$-Al$_2$O$_3$).

**13.13   sample holder :**[2] The container or support for the test portion of the sample.

**13.14   reference holder :**[2] The container or support for the reference material.

**13.15   specimen holder :**[2] The complete assembly in which the specimens are housed. When the heating or cooling source is incorporated in one unit with the containers or supports for the sample and reference material, this is regarded as part of the specimen-holder assembly.

**13.16   block :**[2] A type of specimen-holder assembly in which a relatively large mass of materials is in intimate contact with the specimens or specimen holders.

---

1)   Definition taken from : Nomenclature Committee of the International Confederation for Thermal Analysis. Nomenclature in thermal analysis : Part IV. *J. Thermal Anal.* **13** 1978 : 387-392.

2)   Definition taken from : Nomenclature Committee of the International Confederation for Thermal Analysis. Nomenclature in thermal analysis : Part II. *Talanta* **19** 1972 : 1079-1081.

**13.17  differential thermocouple; $\Delta T$ thermocouple :**[1]
The thermocouple system used to measure temperature or heat-flux differences.

## 14  Principle

Heating of a sample and an appropriate reference material at a controlled rate in a suitable DTA (method C1) or DSC (method C2) apparatus.

Recording of a DTA or DSC curve from which one or several characteristic points, related to the melting behaviour, are determined.

## 15  Apparatus

**15.1  Differential thermal analyser or differential scanning calorimeter.**

Most commercially available and custom-built instruments may be used.

The principal design characteristics of such instruments are

   a)  a heated block with two holders for the specimens identically positioned and guaranteeing the same heat transfer conditions for both;

   b)  a temperature-recording system;

   c)  an electrical heater in the block or a furnace with adequate controls for adjustment of heating rate to linear conditions within ± 0,5 K/min.

Synchronous recordings of the DTA curve and the temperature difference $\Delta T$ shall permit temperature recording with a sensitivity not less than 2 K per millimetre on the recorder scale.

The DSC apparatus is constructed similarly to the DTA apparatus but with separate compensating devices for the two specimen holders and with electronic equipment maintaining the temperatures of the specimens of the sample and reference material at the same level by variation of the required power input.

The peak area shall be not less than 2 cm², the peak height shall be at least 10 times the height of the noise level.

**15.2  Thermocouples,** rigidly fixed in the specimen compartments, possessing identical characteristics within the graduation precision limits.

**15.3  Equipment for filling the specimen containers with inert gas,** or for passage of a constant measured flow of inert gas through the containers.

## 16  Test specimens

### 16.1  Characteristics

Recommended specimens are given in table 2.

Since milligram quantities of material are used, it is essential to ensure that the specimens are homogeneous and representative.

**Table 2 — Characteristics of test specimens**

| Sample form | Particle size | Mass of test portion |
|---|---|---|
| Powder | diameter : up to 0,5 mm | }up to 50 mg |
| Cut film | thickness : 0,05 to 0,50 mm <br> area : 0,25 to 4,00 mm² | |
| Fibre | diameter : up to 0,50 mm <br> length : up to 2,00 mm | |

NOTE — Since the results are affected by the mass of the specimen and by the particle size and shape, the specimens to be compared should have approximately the same particle size and shape and the same mass.

### 16.2  Conditioning

See 5.2.

## 17  Procedure

### 17.1  Calibration

Calibrate the temperature measuring system periodically over the temperature range used for the test.

Certified reference materials are available and are listed in table 3.

**Table 3 — DTA reference materials for transition temperatures in the range 125 to 435 °C**

| Reference material[1] | DTA mean values, (°C) | |
|---|---|---|
| | Extrapolated onset temperature, $T_o$ | Peak temperature, $T_p$ |
| Potassium nitrate | 128 | 135 |
| Indium metal | 154 | 159 |
| Tin metal | 230 | 237 |
| Silver sulfate | 424 | 433 |

1) ICTA-NBS Certified Reference Materials for Differential Thermal Analysis, available from the US National Bureau of Standards, Washington, DC 20234, USA. The indicated values refer only to a specific certified batch.

NOTE — The characteristic points ($T_o$ and $T_p$) are not to be confused with the true melting temperatures as reported in the literature.

---

1) Definition taken from : Nomenclature Committee of the International Confederation for Thermal Analysis. Nomenclature in thermal analysis : Part II. *Talanta* **19** 1972 : 1079-1081.

## 17.2 Determination (methods C1 and C2)

**17.2.1** Weigh the specimen into the sample holder of the instrument (15.1) immediately after removal of the sample from the conditioning area. If applicable, start the passage of a stream of an inert gas (freed from oxygen and dried, for example by bubbling through an alkaline pyrogallol solution) through the sample holder. Operate the instrument according to the manufacturer's instructions. The maximum temperature recorded shall be at least 50 K above the peak temperature. The recommended heating rate is 10 K/min.

From the DTA or DSC curve recorded during this first thermal cycle, determine the melting bevaviour of the material "as received", which may include thermal memories from the manufacturing process.

**17.2.2** If it is desired to erase the effects of the previous thermal history of the specimen in order to obtain an unambiguous material identification, a second thermal cycle shall be performed, as follows:

— at the end of the first cycle, hold the specimen under inert gas atmosphere, about 30 K above the melting peak ($T_p$) for 10 min, then cool it at a rate of 10 K/min to 50 K below the peak crystallization temperature;

— immediately repeat the heating cycle at a rate of 10 K/min and record the heating curve.

## 18 Expression of results

The following values, in degrees Celsius or in kelvins, are read from the endothermic DTA or DSC curves recorded by the instrument :

a) $T_o$: extrapolated onset melting temperature;

b) $T_p$: peak melting temperature.

According to the results of a round robin carried out in 1984, the repeatability was within 2 K and the reproducibility within 4 K.

## 19 Test report

The test report shall include the following information :

a) reference to this International Standard;

b) reference of the method used (method C1 or C2);

c) complete identification and description of the material tested;

d) shape and size (or mass) of the specimens;

e) previous thermal history of the specimens;

f) conditioning;

g) kind of instrument used;

h) type of the sample holder (including its shape, material, etc.);

i) type of temperature-measuring system (including type of thermocouple);

j) location of the temperature-measuring system (inside or outside the sample holder);

k) kind, size and form of the reference material;

l) if necessary, composition and physical parameters (pressure, flow rate, moisture content, etc.) of the inert gas;

m) heating rate;

n) onset temperature $T_o$ and peak temperature $T_p$ in degrees Celsius or in kelvins, of the endothermic curve representing the initial and final melting temperatures, respectively, of the sample, both corrected by the calibration data; the results of both thermal cycles shall be reported; if not otherwise specified, the reference results are those of the second cycle;

o) any operational details not specified in this International Standard or regarded as optional, as well as any incidents liable to have affected the results.

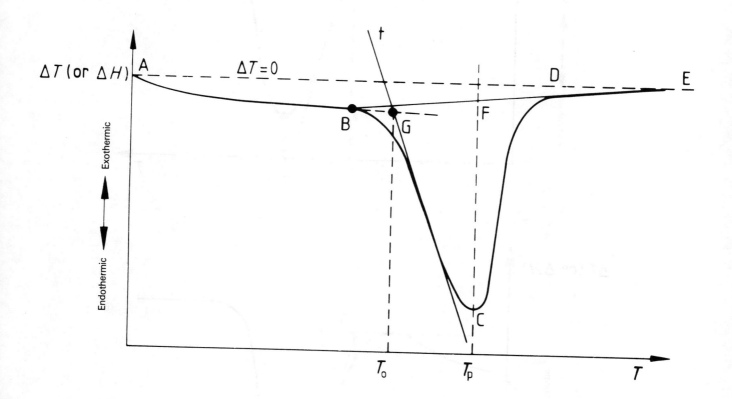

Figure 2 — Formalized DTA (or DSC) curve

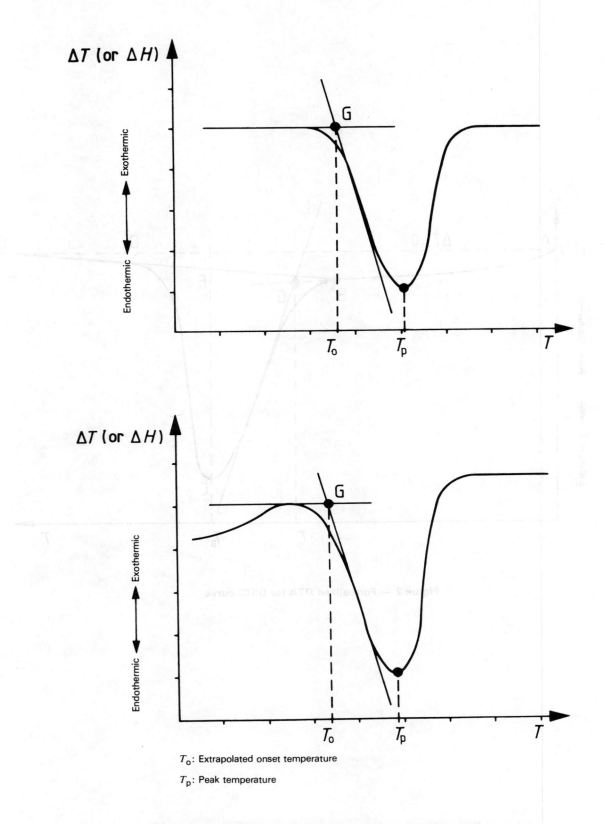

$T_o$: Extrapolated onset temperature

$T_p$: Peak temperature

Figure 3 — Examples of evaluation of $T_o$ and $T_p$ on DTA (or DSC) curves

# INTERNATIONAL STANDARD

**ISO
3167**

Third edition
1993-06-15

# Plastics — Multipurpose test specimens

*Plastiques — Éprouvettes à usages multiples*

Reference number
ISO 3167:1993(E)

# Foreword

ISO (the International Organization for Standardization) is a worldwide federation of national standards bodies (ISO member bodies). The work of preparing International Standards is normally carried out through ISO technical committees. Each member body interested in a subject for which a technical committee has been established has the right to be represented on that committee. International organizations, governmental and non-governmental, in liaison with ISO, also take part in the work. ISO collaborates closely with the International Electrotechnical Commission (IEC) on all matters of electrotechnical standardization.

Draft International Standards adopted by the technical committees are circulated to the member bodies for voting. Publication as an International Standard requires approval by at least 75 % of the member bodies casting a vote.

International Standard ISO 3167 was prepared by Technical Committee ISO/TC 61, *Plastics*, Sub-Committee SC 2, *Mechanical properties*.

This third edition cancels and replaces the second edition (ISO 3167:1983), which has been extended to introduce the preferred specimen type A with a smaller radius, in order to facilitate the testing of parts with simple machining for a variety of tests.

With respect to existing moulds, the specimen type described in the second edition is included as type B. It may be possible to eliminate type B at the next revision of this International Standard.

The designations of dimensions are harmonized with those of the International Standards for testing which relate to a multipurpose test specimen, in accordance with ISO 31.

Annexes A and B of this International Standard are for information only.

# Plastics — Multipurpose test specimens

## 1 Scope

**1.1** This International Standard specifies requirements relating to multipurpose test specimens for plastic moulding materials intended for processing by injection or direct compression moulding.

**1.2** Specimens of types A and B are tensile test specimens, from which with simple machining, specimens for a variety of other tests can be taken (see annex A). Because they have such wide utility, these tensile specimens are referred to in this International Standard as multipurpose test specimens.

**1.3** The principal advantage of a multipurpose test specimen is that it allows all the test methods mentioned in annex A to be carried out on the basis of comparable mouldings. Consequently, the properties measured are coherent as all are measured with specimens in the same state. In other words, it can be expected that test results for a given set of specimens will not vary appreciably due to unintentionally different moulding conditions. On the other hand, if desired, the influence of moulding conditions and/or different states of the specimens can be assessed without difficulty for all of the properties measured.

**1.4** For quality-control purposes, the multipurpose test specimen may serve as a convenient source of further specimens not readily available. Furthermore, the fact that only one mould is required may be advantageous.

**1.5** The use of multipurpose test specimens shall be agreed upon by the interested parties, because there may be significant differences between properties of the multipurpose test specimens and those specified in the relevant test methods.

## 2 Normative references

The following standards contain provisions which, through reference in this text, constitute provisions of this International Standard. At the time of publication, the editions indicated were valid. All standards are subject to revision, and parties to agreements based on this International Standard are encouraged to investigate the possibility of applying the most recent editions of the standards indicated below. Members of IEC and ISO maintain registers of currently valid International Standards.

ISO 293:1986, *Plastics — Compression moulding test specimens of thermoplastic materials.*

ISO 294:—[1], *Plastics — Injection moulding of test specimens of thermoplastic materials.*

ISO 295:1991, *Plastics — Compression moulding of test specimens of thermosetting materials.*

ISO 2818:—[2], *Plastics — Preparation of test specimens by machining.*

## 3 Dimensions of test specimens

For the purposes of this International Standard, the preferred multipurpose test specimen is the tensile specimen type A according to figure 1. This can be made suitable for a variety of other tests by simple cutting, because the length $l_1$ of its narrow parallel portion is 80 mm ± 2 mm.

## 4 Preparation of test specimens

### 4.1 Moulding of multipurpose test specimens

The specimens shall be moulded as specified in ISO 293, ISO 294 and ISO 295, as appropriate, and

---

1) To be published. (Revision of ISO 294:1975)
2) To be published. (Revision of ISO 2818:1980)

under conditions defined for the material under examination.

Strict control of moulding conditions is essential to ensure that all test specimens in a set are actually in the same state.

## 4.2 Machining of test specimens

**4.2.1** Machining of specimens from the multipurpose test specimens shall be performed either as specified in ISO 2818, or as agreed upon by the interested parties. The surface of the central parallel-sided portion of the test specimens shall remain as moulded.

**4.2.2** Test specimens having a width of 10 mm shall be cut symmetrically from the central parallel-sided portion of the multipurpose test specimen.

**4.2.3** For test specimens longer than 80 mm, the broad ends of the multipurpose test specimen type A (or type B for test specimens longer than 60 mm) shall be machined to the width of the central parallel-sided portion. During the machining operation, care shall be taken to avoid any damage to the moulded surfaces of the central portion. The width of the machined portions of the specimen shall be not less than

that of the central parallel-sided portion, but may exceed the width of the latter by not more than 0,2 mm.

## 5 Report on preparation of test specimens

The report shall contain the following information:

a) reference to this International Standard;

b) indication of specimen type (A or B);

c) type, source, manufacturer's code, grade and form, including history, etc. if known;

d) method of moulding and the conditions used;

e) method of machining and the conditions used;

f) number of test specimens;

g) the standard atmosphere for conditioning, plus any special conditioning treatment if required by the standard for the material or product concerned;

h) date of preparation.

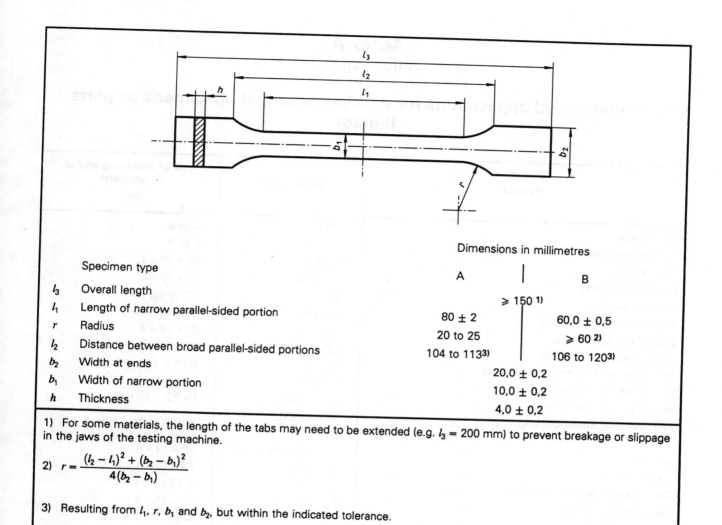

Dimensions in millimetres

| Specimen type | A | | B |
|---|---|---|---|
| $l_3$ Overall length | | $\geqslant 150$ [1] | |
| $l_1$ Length of narrow parallel-sided portion | $80 \pm 2$ | | $60{,}0 \pm 0{,}5$ |
| $r$ Radius | 20 to 25 | | $\geqslant 60$ [2] |
| $l_2$ Distance between broad parallel-sided portions | 104 to 113 [3] | | 106 to 120 [3] |
| $b_2$ Width at ends | | $20{,}0 \pm 0{,}2$ | |
| $b_1$ Width of narrow portion | | $10{,}0 \pm 0{,}2$ | |
| $h$ Thickness | | $4{,}0 \pm 0{,}2$ | |

[1] For some materials, the length of the tabs may need to be extended (e.g. $l_3 = 200$ mm) to prevent breakage or slippage in the jaws of the testing machine.

[2] $r = \dfrac{(l_2 - l_1)^2 + (b_2 - b_1)^2}{4(b_2 - b_1)}$

[3] Resulting from $l_1$, $r$, $b_1$ and $b_2$, but within the indicated tolerance.

**Figure 1 — Multipurpose test specimen types A and B**

# Annex A
(informative)

# Recommended applications for multipurpose test specimens or parts thereof

| Method | Reference[1] | Type of specimen and/or dimensions<br>mm |
|---|---|---|
| Tensile test | ISO 527-2 | A or B |
| Tensile creep test | ISO 899 | A or B |
| Flexural test | ISO 178 | 80 × 10 × 4 |
| Flexural creep test | ISO 6602 | 80 × 10 × 4 |
| Compressive test | ISO 604 | (10 to 40) × 10 × 4 |
| Impact strength — Charpy | ISO 179 | 80 × 10 × 4 |
| Impact strength — Izod | ISO 180 | 80 × 10 × 4 |
| Impact strength — tensile | ISO 8256 | 80 × 10 × 4 |
| Temperature of deflection under load | ISO 75 | (110 or 80) × 10 × 4 |
| Vicat softening temperature | ISO 306 | 10 × 10 × 4 |
| Hardness, ball indentation | ISO 2039-1 | (⩾ 20) × 20 × 4 |
| Environmental stress cracking | ISO 4599 ⎫<br>ISO 4600 ⎭ | A or B or 80 × 10 × 4 |
| Density | ISO 1183, method A | 30 × 10 × 4 |
| Oxygen index | ISO 4589 | 80 × 10 × 4 |
| Comparative tracking index (CTI) | IEC 112 | > 15 × 15 × 4 |
| Electrolytic corrosion | IEC 426 | 30 × 10 × 4 |
| Linear expansion | | > 30 × 10 × 4 |

1) See annex B.

# Annex B

(informative)

# Bibliography

[1] ISO 75:1987, *Plastics and ebonite — Determination of temperature of deflection under load.*

[2] ISO 178:1993, *Plastics — Determination of flexural properties.*

[3] ISO 179:1993, *Plastics — Determination of Charpy impact strength.*

[4] ISO 180:1993, *Plastics — Determination of Izod impact strength.*

[5] ISO 306:1987, *Plastics — Thermoplastic materials — Determination of Vicat softening temperature.*

[6] ISO 527-2:1993, *Plastics — Determination of tensile properties — Part 2: Test conditions for moulding and extrusion plastics.*

[7] ISO 604:1973, *Plastics — Determination of compressive properties.*

[8] ISO 899:1981, *Plastics — Determination of tensile creep.*

[9] ISO 1183:1987, *Plastics — Methods for determining the density and relative density of non-cellular plastics.*

[10] ISO 2039-1:1987, *Plastics — Determination of hardness — Part 1: Ball indentation method.*

[11] ISO 4589:1984, *Plastics — Determination of flammability by oxygen index.*

[12] ISO 4599:1986, *Plastics — Determination of resistance to environmental stress cracking (ESC) — Bent strip method.*

[13] ISO 4600:1992, *Plastics — Determination of environmental stress cracking (ESC) — Ball or pin impression method.*

[14] ISO 6602:1985, *Plastics — Determination of flexural creep by three-point loading.*

[15] ISO 8256:1990, *Plastics — Determination of tensile-impact strength.*

[16] ISO 10350:—[3], *Plastics — Acquisition and presentation of comparable single-point data.*

[17] IEC 112:1979, *Method for determining the comparative and the proof tracking indices of solid insulating materials under moist conditions.*

[18] IEC 426:1973, *Determining electrolytic corrosion with insulating materials.*

---

3) To be published.

# International Standard

# ISO 4589

INTERNATIONAL ORGANIZATION FOR STANDARDIZATION•МЕЖДУНАРОДНАЯ ОРГАНИЗАЦИЯ ПО СТАНДАРТИЗАЦИИ•ORGANISATION INTERNATIONALE DE NORMALISATION

# Plastics — Determination of flammability by oxygen index

*Plastiques — Essais de réaction au feu — Détermination de l'indice d'oxygène*

**First edition — 1984-12-15**

UDC  678.5/.8 : 620.1 : 536.468

**Ref. No.  ISO 4589-1984 (E)**

**Descriptors :** plastics, tests, fire tests, determination, flammability, test equipment, test specimens, dimensions, marking.

# Foreword

ISO (the International Organization for Standardization) is a worldwide federation of national standards bodies (ISO member bodies). The work of preparing International Standards is normally carried out through ISO technical committees. Each member body interested in a subject for which a technical committee has been established has the right to be represented on that committee. International organizations, governmental and non-governmental, in liaison with ISO, also take part in the work.

Draft International Standards adopted by the technical committees are circulated to the member bodies for approval before their acceptance as International Standards by the ISO Council. They are approved in accordance with ISO procedures requiring at least 75 % approval by the member bodies voting.

International Standard ISO 4589 was prepared by Technical Committee ISO/TC 61, *Plastics*.

# Plastics — Determination of flammability by oxygen index

## 0 Introduction

This International Standard has been prepared to rationalize procedures developed independently for similar methods of test by various national standards bodies.

## 1 Scope and field of application

This International Standard specifies methods for determining the minimum concentration of oxygen, in admixture with nitrogen, that will support combustion of small vertical test specimens under specified test conditions. The results are defined as oxygen index values.

Methods are provided for testing materials that are self-supporting in the form of vertical bars or sheet up to 10,5 mm thick. These methods are suitable for solid, laminated or cellular materials characterized by an apparent density greater than 100 kg/m$^3$. The methods may also be applicable to some cellular materials having an apparent density of less than 100 kg/m$^3$. A method is provided for testing flexible sheet or film materials while supported vertically.

Oxygen index results obtained using the methods described in this International Standard can provide a sensitive measure of the burning characteristics of materials under certain controlled laboratory conditions, and hence may be useful for quality control purposes. The results obtained are dependent upon the shape, orientation and isolation of the test specimen and the conditions of ignition. For particular materials or applications, it may be necessary or appropriate to specify different test conditions. Such requirements should be referred to in other standards. Results obtained from test specimens of differing thickness or by using different ignition procedures may not be comparable and no correlation with flammability behaviour under other fire conditions is implied.

Results obtained in accordance with this International Standard must not be used to describe or appraise the fire hazard presented by a particular material or shape under actual fire conditions, unless used as one element of a fire risk assessment that takes into account all of the factors pertinent to the assessment of the fire hazard of a particular application for the material.

NOTES

1  It may not be possible to apply these methods satisfactorily to materials that exhibit high levels of shrinkage when heated, e.g. highly oriented thin film.

2  For assessing the flame propagation properties of cellular materials of density < 100 kg/m$^3$, attention is drawn to the method of ISO 3582 for testing horizontal burning characteristics.

## 2 References

ISO 291, *Plastics — Standard atmospheres for conditioning and testing.*

ISO 293, *Plastics — Compression moulding test specimens of thermoplastic materials.*

ISO 294, *Plastics — Injection moulding test specimens of thermoplastic materials.*

ISO 295, *Plastics — Compression moulding test specimens of thermosetting materials.*

ISO 845, *Cellular plastics and rubbers — Determination of density.*

ISO 2818, *Plastics — Preparation of test specimens by machining.*

ISO 2859, *Applications of statistical methods — Sampling procedures and tables of inspection by attributes.*

ISO 3167, *Plastics — Preparation and use of multi-purpose test specimens.*

ISO 3582, *Cellular plastic and cellular rubber materials — Laboratory assessment of horizontal burning characteristics of small specimens subjected to a small flame.*

ISO 4893, *Plastics — Preparation of test specimens from thermosetting materials by injection moulding.*[1]

ISO 5725, *Precision of test methods — Determination of repeatability and reproducibility.*

ISO 6400, *Plastics — Test specimens from semi-crystalline thermoplastic moulding materials by compression moulding — Preparation of reference test specimens with a reproducible state.*[1]

---

1)  At present at the stage of draft.

## 3  Principle

A small test specimen is supported vertically in a mixture of oxygen and nitrogen flowing upwards through a transparent chimney. The upper end of the specimen is ignited and the subsequent burning behaviour of the specimen is observed to compare the period for which burning continues, or the length of specimen burnt, with specified limits for such burning. By testing a series of specimens in different oxygen concentrations, the minimum oxygen concentration is estimated at which the burning behaviour of 50 % of specimens representing the material under test would exceed at least one of the specified limits for burning.

## 4  Definition

For the purposes of this International Standard, the following definition applies.

**oxygen index**: The minimum concentration of oxygen by percentage volume in a mixture of oxygen and nitrogen introduced at 23 ± 2 °C that will just support combustion of a material under specified test conditions.

## 5  Apparatus

The following apparatus shall be arranged as indicated by the diagrams in figures 1 and 2 as appropriate:

**5.1  Test chimney**: a heat-resistant glass tube supported vertically on a base through which oxygen-containing gas mixtures can be introduced.

The preferred dimensions of the chimney are 450 mm minimum height and 75 mm minimum diameter cylindrical bore. The upper outlet shall be restricted as necessary by an overhead cap having an outlet small enough to produce an exhaust velocity of at least 90 mm/s from a flow rate within the chimney of 30 mm/s (see the note). Chimneys of other dimensions, with or without restricted outlets, may be used, if shown to give equivalent results.

The bottom of the chimney, or the base upon which the chimney is supported, shall incorporate a means for distributing evenly the gas mixture entering the chimney. The preferred means comprises solid glass beads of between 3 and 5 mm diameter, in a layer between 80 and 100 mm deep. Other means, such as radial manifolds, may be used, if shown to give equivalent results. A porous screen may be mounted below the level of the specimen holder, to prevent falling combustion debris from fouling the gas entry and distribution paths.

The chimney support may incorporate a levelling device and indicator, to facilitate vertical alignment of the chimney and a test specimen supported therein. A dark background may be provided to facilitate observation of flames within the chimney.

NOTE — For tubes of 75 to 100 mm diameter, a cap converging to an outlet of 40 mm diameter at a level at least 10 mm above the top of the cylindrical chimney has been found satisfactory.

**5.2  Test specimen holder**, suitable for supporting a specimen vertically in the centre of the chimney.

For self-supporting materials, the specimen shall be held by a small clamp which is at least 15 mm away from the nearest point at which the specimen may burn before the extent-of-burning criterion is exceeded. For supported film or sheet test specimens, the specimen shall be supported by both vertical edges in a frame equivalent to that illustrated by figure 2, with reference marks at 20 mm and 100 mm below the top of the frame.

The profile of the holder and its support should be smooth to minimize induction of turbulence in the rising flow of gas.

**5.3  Gas supplies**, comprising pressurized sources of oxygen and/or nitrogen not less than 98 % ($m/m$) pure and/or clean air (containing 20,9 % oxygen), as appropriate.

The moisture content of the gas mixture entering the chimney shall be < 0,1 % ($m/m$), unless the results have been shown to be insensitive to higher moisture levels in the gas mixture. The gas supply system shall incorporate a drying device, or provision for monitoring or sampling the gas supply for moisture content, unless the moisture content of the gas supplies is known to be acceptable.

The constituent gas supply lines shall be linked in a manner which thoroughly mixes the gases, before they enter the gas distribution device at the base of the chimney, so that the variation in oxygen concentration in the gas mixture rising in the chimney, below the level of the test specimen, is < 0,2 % ($V/V$).

NOTE — It should not be assumed that bottled oxygen or nitrogen will always contain < 0,1 % ($m/m$) of water; moisture contents of 0,003 % to 0,01 % ($m/m$) are typical for commercial supplies as filled bottles ≥ 98 % ($m/m$) pure, but as such bottled gases are depressured to below about 1 MPa, the moisture content of the gas drawn off may rise above 0,1 % ($m/m$).

**5.4  Gas measurement and control devices**, suitable for establishing the concentration of oxygen in the gas mixture entering the chimney with an accuracy of ± 0,5 % ($V/V$) of the mixture and for adjusting the concentration with a precision of ± 0,1 % ($V/V$) of the mixture when the gas velocity through the chimney is 40 ± 10 mm/s at 23 ± 2 °C.

Means shall be provided for checking or ensuring that the temperature of the gas mixture entering the chimney is 23 ± 2 °C. If this involves an internal probe, its position and profile shall be designed to minimize induction of turbulence within the chimney.

NOTE — Systems of measurement and control that have proved satisfactory include the following:

a) needle valves on individual and mixed gas supply lines, a paramagnetic oxygen analyser that continuously samples the mixed gas, and a flowmeter to indicate when the gas flow through the chimney is within the required limits;

b) calibrated orifices, gas pressure regulators and pressure gauges on the individual gas supply lines; or

c) needle valves and calibrated flowmeters on the individual gas supply lines.

Systems b) and c) may require calibration after assembly to ensure that the compounded errors of the component parts do not exceed the requirements of 5.4.

249

**5.5   Flame igniter,** comprising a tube that can be inserted into the chimney to apply to the test specimen a flame issuing from an outlet of 2 ± 1 mm diameter at the end of the tube.

The flame fuel shall be propane, without premixed air. The fuel supply shall be adjusted so that the flame will project 16 ± 4 mm vertically downwards from the outlet when the tube is vertical within the chimney and the flame is burning within the chimney atmosphere.

**5.6   Timing device,** capable of measuring periods up to 5 min with an accuracy of ± 0,2 s.

**5.7   Fume extraction system,** providing sufficient ventilation or exhaust to remove fumes or soot expelled from the chimney without disrupting the gas-flow rate or temperatures in the chimney.

NOTE — If soot-generating materials are being tested, the glass chimney may require cleaning to maintain good visibility, and the gas inlets, or inlet screen, and temperature sensor (if fitted) may also require cleaning to function properly. Suitable precautions should be taken to protect personnel from noxious materials or burns during testing or cleaning operations.

# 6   Calibration of equipment

For compliance with this method, calibrate the equipment periodically in accordance with the instructions given in annex A so that the maximum interval between recalibration and use complies with the periods stated in table 1.

**Table 1 — Equipment calibration frequencies**

| Item | Maximum period |
|---|---|
| Gas-flow rate controls | 6 months |
| Oxygen concentration controls | 6 months |
| Gas system joints (as required by clause A.2 in annex A) | |
| a) for joints disturbed during use or cleaning of the apparatus | 24 h |
| b) for undisturbed joints | 6 months |

# 7   Preparation of test specimens

## 7.1   Sampling

Obtain a sample sufficient for preparation of at least 15 test specimens. The sample shall be taken, if relevant, in accordance with the material specification, otherwise in accordance with ISO 2859.

NOTE — For a material for which the oxygen index is known to within ± 2, 15 test specimens may be sufficient. For materials of unknown oxygen index, or which exhibit erratic burning characteristics, between 15 and 30 test specimens may be required.

## 7.2   Test specimen dimensions and their preparation

Using, if applicable, procedures that comply with the appropriate material specification or ISO methods for specimen preparation, mould or cut test specimens that satisfy the dimensions specified for the most appropriate specimen form given in table 2.

Ensure that the surfaces of the specimens are clean and free from flaws that could affect burning behaviour, for example peripheral moulding flash or burrs from machining.

Note the position and orientation of test specimens with respect to any asymmetry in the sample material.

NOTES

1   Some material specifications may require choice and identification of the ''state of the test specimen'' used; for example, in a ''defined state'' or a ''basic state'' for a styrene-based polymer or copolymer.

2   In the absence of a relevant material specification, one or more procedures from ISO 293, ISO 294, ISO 295, ISO 2818, ISO 3167, ISO 4893 or ISO 6400 may be used.

3   Oxygen index results may be significantly affected by differences in ease of ignition or of burning behaviour, due to material inhomogeneity (for example, different levels of shrinkage when heated for specimens cut in different directions from asymmetrically-oriented thermoplastic film).

**Table 2 — Test specimen dimensions**

| Test specimen form [1] | Dimensions | | | Typical use |
|---|---|---|---|---|
| | Length mm | Width mm | Thickness mm | |
| I | 80 to 150 | 10 ± 0,5 | 4 ± 0,25 | For moulding materials |
| II | 80 to 150 | 10 ± 0,5 | 10 ± 0,5 | For cellular materials |
| III [2] | 80 to 150 | 10 ± 0,5 | < 10,5 | For sheet materials ''as received'' |
| IV | 70 to 150 | 6,5 ± 0,5 | 3 ± 0,25 | Alternative size for self-supporting moulding or sheet materials, for electrical purposes |
| V [2] | $140\,^{0}_{-5}$ | 52 ± 0,5 | < 10,5 | For flexible film or sheet |

1)   Test specimens of forms I, II, III and IV are suitable for materials that are self-supporting at these dimensions.

Test specimens of form V are suitable for materials that require support during testing.

2)   Results obtained using form III or form V test specimens may only be comparable for specimens of the same form and thickness. It is assumed that the amount of variation in thickness for such materials will be controlled by other standards.

## 7.3 Marking of test specimens

For monitoring the distance over which a specimen burns, it may be marked with transverse lines at one or more levels which are dependent upon the specimen form and the ignition procedure to be used. Self-supporting specimens are preferably marked on at least two adjacent faces. If wet inks are used, the marks shall be dry before the specimen is ignited.

### 7.3.1 Marks for testing by top surface ignition

Test specimens of form I, II, III or IV, to be tested in accordance with procedure A (see 8.2.1), shall be marked 50 mm from the end to be ignited.

### 7.3.2 Marks for testing by propagating ignition

The reference marks for testing specimens of form V are carried by the supporting frame (see figure 2), but such specimens may be marked at 20 mm and at 100 mm from the end to be ignited, for convenience when testing heat-stable materials.

If specimens of form I, II, III and IV are to be tested in accordance with procedure B (see 8.2.1 and 8.2.3), they shall be marked at 10 mm and at 60 mm from the end to be ignited.

## 7.4 Conditioning

Unless otherwise specified in other established standards, each test specimen shall be conditioned for at least 88 h at $23 \pm 2$ °C and $(50 \pm 5)$ % RH immediately prior to use.

NOTE — Samples of cellular materials that may contain volatile flammable material should be purged of such volatile material prior to conditioning at 23 °C and 50 % RH. Test specimens may be purged satisfactorily by pre-conditioning in suitable ventilated ovens for 168 h. Larger blocks of such materials may require longer pre-treatment. Facilities for cutting specimens from cellular material that may contain volatile flammable material must be suitable for the hazards involved.

## 8 Procedure

## 8.1 Setting up the apparatus and test specimen

**8.1.1** Maintain the ambient temperature for the test apparatus at $23 \pm 2$ °C. If necessary, keep the test specimens in an enclosure at $23 \pm 2$ °C and $(50 \pm 5)$ % RH from which each test specimen may be taken when required.

**8.1.2** Recalibrate equipment components, if necessary (see clause 6 and annex A).

**8.1.3** Select an initial concentration of oxygen to be used. When possible, this may be based on experience of results for similar materials. Alternatively, try to ignite a test specimen in air, and note the burning behaviour. If the specimen burns rapidly, select an initial concentration of about 18 % ($V/V$) of oxygen; if the test specimen burns gently or unsteadily, select an initial oxygen concentration of about 21 %; if the specimen

does not continue to burn in air, select an initial concentration of at least 25 %, depending upon the difficulty of ignition or the period of burning before extinguishment in air.

**8.1.4** Ensure that the test chimney is vertical (see figure 1). Mount a specimen vertically in the centre of the chimney so that the top of the specimen is at least 100 mm below the open top of the chimney and the lowest exposed part of the specimen is at least 100 mm above the top of the gas distribution device at the base of the chimney (see figure 1 or figure 2 as appropriate).

**8.1.5** Set the gas mixing and flow controls so that an oxygen/nitrogen mixture at $23 \pm 2$ °C, containing the desired concentration of oxygen, is flowing through the chimney at a rate of $40 \pm 10$ mm/s. Let the gas flow purge the chimney for at least 30 s prior to ignition of each specimen, and maintain the flow without change during ignition and combustion of each specimen.

Record the oxygen concentration used as the volume per cent calculated according to the equations given in annex C.

## 8.2 Igniting the test specimen

**8.2.1** Select one of two alternative ignition procedures which are dependent upon the specimen form as follows:

a) for specimen forms I, II, III and IV (see table 2), use procedure A, top surface ignition, as described in 8.2.2;

b) for specimen form V, use procedure B, propagating ignition, as described in 8.2.3.

Ignition shall imply, for the purposes of this International Standard, the initiation of flaming combustion.

NOTES

1 For tests on materials that exhibit steady burning and spread of combustion in oxygen concentrations at, or close to, their oxygen index value, or for self-supporting specimens of ≤3 mm thickness, procedure B (with specimens marked in accordance with 7.3.2) may be found to give more consistent results than procedure A. Procedure B may then be used for specimens of form I, II, III or IV.

2 Some materials may exhibit a non-flaming type of combustion (for example, glowing combustion) instead of, or at a lower oxygen concentration than that required for, flaming combustion. When testing such materials, it is necessary to identify the type of combustion for which the oxygen index is required or measured.

### 8.2.2 Procedure A — Top surface ignition

For top surface ignition, the igniter is used to initiate burning only on the top surface of the upper end of the specimen.

Apply the lowest visible part of the flame to the top of the specimen using a sweeping motion, if necessary, to cover the whole surface, but taking care not to maintain the flame against the vertical faces or edges of the specimen. Apply the flame for up to 30 s, removing it every 5 s for just sufficient time to observe whether or not the entire top surface of the specimen is burning.

251

Consider the specimen to be ignited, and commence measurement of the period and distance of burning, as soon as removal of the igniter, after a contact period increment of 5 s, reveals burning supported by the whole of the top end surface of the specimen.

### 8.2.3 Procedure B — Propagating ignition

For propagating ignition, the igniter is used to produce burning across the top and partially down the vertical faces of the specimen.

Lower and move the igniter sufficiently to apply the visible flame to the end face of the specimen and also, to a depth of approximately 6 mm, to its vertical faces. Continue to apply the igniter for up to 30 s, with interruptions for inspection of the specimen every 5 s, until its vertical faces are burning steadily or until the visibly burning portion first reaches the level of the upper reference mark on the support frame or, if used for specimens of form I, II, III or IV, on the specimen.

Consider the specimen to be ignited, for the purpose of measuring the period and extent of burning, as soon as any part of the visibly burning portion reaches the level of the upper reference mark.

NOTE — The burning portion includes any burning drips that may run down the surface of the specimen.

### 8.3 Assessing burning behaviour

For the purposes of 8.4 to 8.6 inclusive, observe and terminate the burning of individual test specimens as follows.

**8.3.1** Commence measurement of the period of burning as soon as the specimen has been ignited in accordance with 8.2.2 or 8.2.3, as applicable, and observe its burning behaviour. If burning ceases but spontaneous re-ignition occurs in <1 s, continue the observation and measurements.

**8.3.2** If neither the period nor the extent of burning exceeds the relevant limit specified in table 3 for the applicable specimen, note the duration and extent of burning. This is recorded as an "O" response.

Alternatively, if either the period or extent of burning exceeds the relevant limit specified in table 3, note the burning behaviour accordingly, and extinguish the flame. This is recorded as an "X" response.

Note also the burning characteristics of the material, for example dripping, charring, erratic burning, glowing combustion or after-glow.

**8.3.3** Remove the specimen and clean, as necessary, any surfaces within the chimney or on the igniter that have become contaminated with soot, etc. Allow the chimney to regain a temperature of $23 \pm 2$ °C, or replace it with another so conditioned.

NOTE — If sufficiently long, the specimen may be inverted, or trimmed to remove the burnt end, and re-used. Results from such specimens can save material when establishing an approximate value for the minimum oxygen concentration required for combustion, but cannot be included among those used for estimation of the oxygen index, unless the specimen is reconditioned at the temperature and humidity appropriate for the material involved.

### 8.4 Selecting successive oxygen concentrations

The procedure described in 8.5 and 8.6 is based upon the "Up-and-down method for small samples" [1], using the specific case where $N_T - N_L = 5$ (see 8.6.2 and 8.6.3), with an arbitrary step size for certain changes to be made in the oxygen concentration used.

During the testing, select the oxygen concentration to be used for testing the next test specimen as follows:

a) decrease the oxygen concentration if the burning behaviour of the preceding specimen gave an "X" response,

### Table 3 — Criteria for oxygen index measurements[1]

| Test specimen form (see table 2) | Ignition procedure | Period of burning after ignition, s | Extent of burning [2] |
|---|---|---|---|
| I, II, III and IV | A Top surface ignition | 180 | 50 mm below the top of the specimen |
| | B Propagating ignition | 180 | 50 mm below the upper reference mark |
| V | B Propagating ignition | 180 | 80 mm below the upper reference mark (on the frame) |

1) These criteria do not necessarily produce equivalent oxygen index results for specimens of differing shape or tested using different ignition conditions or procedures.

2) The extent of burning is exceeded when any part of the visibly burning portion of a specimen, including burning drips descending the vertical faces, passes the level defined in the fourth column of table 3.

[1] Dixon, W.J. *American Statistical Association Journal*, pp. 967-970 (1965).

otherwise

b) increase the oxygen concentration if the preceding specimen gave an "O" response.

Choose the size of the change in oxygen concentration in accordance with 8.5 and 8.6, as appropriate.

## 8.5 Determining the preliminary oxygen concentration

Repeat the procedures specified in 8.1.4 to 8.4 inclusive, using oxygen concentration changes of any convenient step size, until two oxygen concentrations, in per cent volume, have been found that differ by $\leqslant 1,0$ and of which one gave an "O" response and the other an "X" response. From this pair of oxygen concentrations, note that which gave the "O" response as the preliminary oxygen concentration level and then proceed in accordance with 8.6.

NOTES

1 The two results, at oxygen concentrations $\leqslant 1,0$ apart, which give opposite responses, do not have to be from successive specimens.

2 That concentration which gave the "O" response does not have to be lower than that which gave the "X" response.

3 A format convenient for recording the information required by this and subsequent clauses is illustrated in annex D.

## 8.6 Oxygen concentration changes

**8.6.1** Using, again, the preliminary oxygen concentration (8.5), test one specimen by repeating 8.1.4 to 8.3 inclusive. Record both the oxygen concentration ($c_O$) used and the response, "X" or "O", as the first of the $N_L$ and of the $N_T$ series of results.

**8.6.2** Change the oxygen concentration, in accordance with 8.4, using concentration changes ($d$) of 0,2 % ($V/V$) (see note) of the total gas mixture to test further specimens in accordance with 8.1.4 to 8.4 inclusive, noting the values of $c_O$ and corresponding responses until a different response to that obtained in 8.6.1 is recorded.

The result from 8.6.1 plus those, if any, of like response from 8.6.2 constitute the $N_L$ series of results. (See example in annex D, Part 2.)

NOTE — Where experience has shown that the requirements of 8.6.4 are usually satisfied by a value of $d$ other than 0,2 %, that value may be selected as the initial value of $d$.

**8.6.3** Test four more specimens, in accordance with 8.1.4 to 8.4 inclusive, maintaining $d = 0,2$ %, and note the $c_O$ used for, and response of, each specimen. Designate the oxygen concentration used for the last specimen as $c_F$.

These four results together with the last result from 8.6.2 (i.e. that which differed in response from that of 8.6.1) constitute the remainder of the $N_T$ series, so that

$$N_T = N_L + 5$$

(See example in annex D, Part 2.)

**8.6.4** Calculate the estimated standard deviation, $\hat{\sigma}$, of the oxygen concentration measurements from the last six responses in the $N_T$ series (including $c_F$), in accordance with 9.3. If the condition

$$\frac{2\hat{\sigma}}{3} < d < 1,5\hat{\sigma}$$

is satisfied, calculate the oxygen index in accordance with 9.1, otherwise

a) if $d < 2\hat{\sigma}/3$, repeat steps 8.6.2 to 8.6.4, using increased values for $d$, until the condition is satisfied, or

b) if $d > 1,5\hat{\sigma}$, repeat steps 8.6.2 to 8.6.4, using decreased values for $d$, until the condition is satisfied, except that $d$ shall not be reduced below 0,2 unless so required by the relevant material specification.

## 9 Calculations and expression of results

### 9.1 Oxygen index

Calculate the oxygen index OI, expressed as a percentage by volume, from the relationship

$$OI = c_F + kd$$

where

$c_F$ is the final value of oxygen concentration, in per cent volume to one decimal place, used in the series of $N_T$ measurements performed in accordance with 8.6, and noted in accordance with 8.6.3;

$d$ is the interval, in per cent volume to at least one decimal place, between oxygen concentration levels used and controlled in accordance with 8.6;

$k$ is a factor to be obtained from table 4, as described in 9.2.

For the purpose of calculation of $\hat{\sigma}$, as required by 8.6.4 and 9.3, the OI shall be calculated to two decimal places.

For the purpose of reporting OI results, express OI values to the nearest 0,1, with exactly intermediate results being rounded downwards.

### 9.2 Determination of $k$

The value and sign of $k$ are dependent upon the pattern of the responses of specimens tested in accordance with 8.6, and may be determined from table 4 as follows:

a) if the response of the specimen tested according to 8.6.1 was "O", so that the first contrary response (see 8.6.2) was an "X", refer to column 1 of table 4 to select the row for which the last four response symbols correspond to those found when testing in accordance with 8.6.3. The value and sign of $k$ will be that shown in column 2, 3, 4 or 5 for which the number of "O"s shown in row (a) of the table corresponds to the number of "O" responses found for the $N_L$ series, in accordance with 8.6.1 and 8.6.2.

**Table 4 — Values of $k$ for calculating oxygen index concentration from determinations made by Dixon's "Up-and-down" method**

| 1 | 2 | 3 | 4 | 5 | 6 |
|---|---|---|---|---|---|
| Responses for the last five measurements | Values of $k$ for which the first $N_L$ determinations are : | | | | |
| | (a)  O | OO | OOO | OOOO | |
| XOOOO | − 0,55 | − 0,55 | − 0,55 | − 0,55 | OXXXX |
| XOOOX | − 1,25 | − 1,25 | − 1,25 | − 1,25 | OXXXO |
| XOOXO | 0,37 | 0,38 | 0,38 | 0,38 | OXXOX |
| XOOXX | − 0,17 | − 0,14 | − 0,14 | − 0,14 | OXXOO |
| XOXOO | 0,02 | 0,04 | 0,04 | 0,04 | OXOXO |
| XOXOX | − 0,50 | − 0,46 | − 0,45 | − 0,45 | OXOXO |
| XOXXO | 1,17 | 1,24 | 1,25 | 1,25 | OXOOX |
| XOXXX | 0,61 | 0,73 | 0,76 | 0,76 | OXOOO |
| XXOOO | − 0,30 | − 0,27 | − 0,26 | − 0,26 | OOXXX |
| XXOOX | − 0,83 | − 0,76 | − 0,75 | − 0,75 | OOXXO |
| XXOXO | 0,83 | 0,94 | 0,95 | 0,95 | OOXOX |
| XXOXX | 0,30 | 0,46 | 0,50 | 0,50 | OOXOO |
| XXXOO | 0,50 | 0,65 | 0,68 | 0,68 | OOOXX |
| XXXOX | − 0,04 | 0,19 | 0,24 | 0,25 | OOOXO |
| XXXXO | 1,60 | 1,92 | 2,00 | 2,01 | OOOOX |
| XXXXX | 0,89 | 1,33 | 1,47 | 1,50 | OOOOO |
| | Values of $k$ for which the first $N_L$ determinations are : | | | | Responses for the last five measurements |
| | (b)  X | XX | XXX | XXXX | |
| | are as given in the above table opposite the appropriate response in column 6, but with the sign of $k$ reversed, i.e. OI = $c_F$ − $kd$ (see 9.1). | | | | |

or

b) if the response of the specimen tested according to 8.6.1 was "X", so that the first contrary response was an "O", refer to the sixth column of table 4 to select the row for which the last four response symbols correspond to those found when testing in accordance with 8.6.3. The value of $k$ will be that shown in column 2, 3, 4 or 5 for which the number of "X"'s shown in row (b) of the table corresponds to the number of "X" responses found for the $N_L$ series, in accordance with 8.6.1 and 8.6.2, but the sign of $k$ must be reversed, so that negative values shown in table 4 for $k$ become positive, and vice versa.

NOTE — An example of the determination of $k$ and the calculation of an OI is given in annex D.

### 9.3  Standard deviation of oxygen concentration measurements

For the purposes of 8.6.4, calculate the estimated standard deviation, $\hat{\sigma}$, of oxygen concentration measurements from the relationship

$$\hat{\sigma} = \left[ \frac{\sum (c_i - \text{OI})^2}{n - 1} \right]^{1/2}$$

where

$c_i$ represents, in turn, each of the per cent oxygen concentrations used during measurement of the last six responses in the $N_T$ series of measurements ;

OI is the oxygen index value, calculated in accordance with 9.1 ;

$n$ is the number of measurements of oxygen concentration contributing to $\sum (c_i - \text{OI})^2$.

NOTE — For this method, $n = 6$, in accordance with 8.6.4. For $n < 6$, the method loses precision. For $n > 6$, alternative statistical criteria would apply.

### 9.4  Precision of results

This method may be expected to be capable of the limits given in table 5 for materials that ignite without difficulty and burn steadily.

**Table 5 — Estimated precision limits[1]**

| Approximate values for 95 % confidence | Within laboratories | Between laboratories |
|---|---|---|
| Standard deviation | 0,2 | 0,5 |
| Repeatability ($r$) | 0,5 | — |
| Reproducibility ($R$) | — | 1,4 |

1) The precision data were determined from an international interlaboratory trial in 1978/1980 involving 16 laboratories and 12 samples.

NOTE — Materials that exhibit erratic combustion behaviour may increase the limits in table 5 by a factor up to 5. On the other hand, it may be found that, for materials that exhibit very consistent burning behaviour, $d > 1,5\hat{\sigma}$ even if $d$ is reduced to 0,1, indicating that greater precision is possible. For practical purposes, the accuracy and precision requirements specified for apparatus by this International Standard are inadequate for significant discrimination if using $d < 0,1$, and results obtained using this method have not been found to be significantly different for $d < 0,2$. More precise determination of the minimum oxygen concentration to just support combustion would require different apparatus and the use of different statistical relationships and factors to determine the value from a longer series of measurements.

# 10 Test report

The test report shall include the following information:

a) a reference to this International Standard;

b) a statement that test results relate only to the behaviour of the test specimens under the conditions of this test and that these results must not be used to infer the fire hazards of the material in other forms or under other fire conditions;

c) identification of the material tested, including, where relevant, the type of material, density, previous history, and the specimen orientation with respect to any anisotropy in the material or sample;

d) the test specimen form or dimensions;

e) the ignition procedure used (A or B), and the igniter used, if other than the standard propane flame;

f) the oxygen index;

g) the estimated standard deviation and the oxygen concentration increment used, if other than 0,2 %;

h) a description of any relevant ancillary characteristics or behaviour, such as charring, dripping, severe shrinkage, erratic burning, after-glow;

j) any variations from the requirements of this International Standard.

# Annex A

# Calibration of equipment

(This annex forms an integral part of the Standard.)

## A.1 Calibration of gas-flow rate controls

The system for indicating the gas-flow rate through the chimney, to satisfy 5.4 and 8.1.5, shall be checked using a water-sealed rotating drum meter (wet test meter), or an equivalent device, with an accuracy equivalent to $\pm 2$ mm/s flow rate through the chimney.

The flow rate shall be estimated by dividing the total gas-flow rate through the chimney by the cross-sectional area of the bore of the chimney, for example by using the equation

$$F = 1,27 \times 10^6 \frac{q_V}{D^2}$$

where

$F$ is the flow rate through the chimney, in millimetres per second;

$q_V$ is the total gas-flow at $23 \pm 2$ °C through the chimney, in litres per second;

$D$ is the diameter of the bore of the chimney, in millimetres.

## A.2 Calibration of oxygen concentration controls

The concentration of oxygen in the mixture of gases flowing into the chimney shall be checked to an accuracy of 0,1 % ($V/V$) of mixture, either by sampling the chimney atmosphere for analysis or by using an independently calibrated oxygen analyser in situ. Integral oxygen analysers may be calibrated using standard oxygen/nitrogen mixtures. The checks should be carried out for at least three different nominal concentrations, representing respectively maximum, minimum and intermediate levels for the oxygen concentration range for which the equipment is to be used.

Leak tests shall be carried out on all joints where leaks could change the oxygen concentration levels in the chimney from the concentration levels set or indicated.

## A.3 Calibration of complete equipment

The performance of the equipment may be checked, for a specific test procedure, by testing a calibrated material and comparing the measured results with the expected result for the calibrated material. For information on the availability and use of calibrated materials, see annex B.

# Annex B

# Materials calibrated for oxygen index values

(This annex does not form an integral part of the Standard.)

It may be found desirable periodically to test materials characterized for oxygen index, to check the overall performance of a particular combination of oxygen index measurement equipment and operator(s).

The results obtained for particular samples of certain materials tested using the procedures of this method in an interlaboratory experiment involving 16 laboratories from seven different countries have been expressed in table 6 as the range within which a single test result should be found, with 95 % confidence, for each particular material/igniter/test procedure combination.

While stocks of surplus materials from the 1978/1980 interlaboratory experiment last, samples with the oxygen index levels given in table 6 will be available only from the Rubber and Plastics Research Association, Shawbury, Shrewsbury, Shropshire, United Kingdom.

### Table 6 — Reference material oxygen index values [1]

| Material | Procedure A Top surface ignition | | Procedure B Propagating ignition | |
|---|---|---|---|---|
| | Flame | Incandescent | Flame | Incandescent |
| Polypropylene | 18,3 to 19,0 | 17,7 to 18,1 | 17,7 to 18,2 | 17,3 to 18,1 |
| Melamine-formaldehyde (MF) | 41,0 to 43,6 | 40,8 to 44,3 | 39,6 to 42,5 | 40,4 to 43,8 |
| PMMA 3 mm thick | 17,3 to 18,1 | 17,5 to 18,1 | 17,2 to 18,0 | 17,2 to 17,9 |
| PMMA 10 mm thick | 17,9 to 19,0 | 17,8 to 18,7 | 17,5 to 18,5 | 17,5 to 18,6 |
| Phenolic foam 10,5 mm thick | 39,1 to 40,7 | 38,9 to 40,8 | 39,6 to 40,9 | 39,1 to 40,9 |
| PVC film 0,02 mm thick | | | 22,4 to 23,6 | 22,6 to 23,5 |

[1]  Results for "Incandescent" ignition given in table 6 were obtained using equipment and procedure described in ISO/DP 4589.4 (1980-12-05) available from the ISO Central Secretariat or from BSI as Doc. 81/50253.

# Annex C

# Calculation of oxygen concentration

(This annex forms an integral part of the Standard.)

Oxygen concentrations required for the purposes of clause 8 shall be calculated according to the equation

$$c_O = \frac{100 V_O}{V_O + V_N}$$

where

$c_O$  is the oxygen concentration, in per cent by volume;

$V_O$  is the volume of oxygen per volume of mixture, at 23 °C;

$V_N$  is the volume of nitrogen per volume of mixture, at 23 °C.

NOTES

1   If an oxygen analyser is used, the oxygen concentration should be determined using the readout from the particular instrument used.

2   If the result is calculated from flow or pressure data for individual gas streams contributing to the mixture, it is necessary to allow for the proportion of oxygen present in streams other than a pure oxygen supply. For example, for mixtures made using air mixed with oxygen of 98,5 % ($V/V$) purity or with nitrogen containing 0,5 % ($V/V$) of oxygen, the oxygen concentration, in per cent by volume, should be calculated using the relationship

$$c_O = \frac{98,5 V_O' + 20,9 V_A' + 0,5 V_N'}{V_O' + V_A' + V_N'}$$

where

$V_O'$  is the volume of oxygen stream used, per volume of mixture;

$V_A'$  is the volume of air stream used, per volume of mixture;

$V_N'$  is the volume of nitrogen stream used, per volume of mixture;

assuming that the streams are at the same pressure at 23 °C.

For mixtures based on two gas streams, $V_O'$, $V_A'$ or $V_N'$ becomes zero, as appropriate.

# Annex D

# Typical test results sheet

(This annex does not form an integral part of the Standard.)

## Test results sheet for oxygen index according to *ISO 4589*

Material: *Phenolic laminate*
Specimen form: *III (4mm thick)*
Ignition procedure: (A)  B
Conditioning procedure: 23 — (23/50)
Oxygen concentration increment ($d$): 0,2

Oxygen Index (concentration, %): *29,5*
  (rounded to 0,1 %)
$\hat{\sigma}$: *0,152*
Date of test: *13/10/80*
Laboratory No. *19*   Test No. *1*

### Part 1: Determination of oxygen concentration for one pair of "X" and "O" response at ≤1 % $O_2$ concentration interval (in accordance with 8.5)

| Oxygen concentration (%) | 25,0 | 35,0 | 30,0 | 32,0 | 31,0 | | | | |
|---|---|---|---|---|---|---|---|---|---|
| Burning period (s) | 10 | >180 | 140 | >180 | >180 | | | | |
| Length burnt (mm) | | | | | | | | | |
| Response ("X" or "O") | O | X | O | X | X | | | | |

Oxygen concentration of the "O" response for the pair = **30,0**
(being the concentration to be used again for the first measurement in Part 2)

### Part 2: Determination of the oxygen index value (in accordance with 8.6)

Step size to be used for successive changes in oxygen concentration of $d$ % =  **0,2**  %
(initially to be 0,2 %, unless otherwise instructed)

| | $N_T$ series measurements | | | | | | | | | | | |
|---|---|---|---|---|---|---|---|---|---|---|---|---|
| | $N_L$ series measurements (8.6.1 + 8.6.2) | | | | | | | (8.6.3) | | | | $c_F$ |
| Oxygen concentration (%) | 30,0 | 29,8 | 29,6 | 29,4 | | | | 29,4 | 29,6 | 29,4 | 29,6 | 29,8 |
| Burning period (s) | >180 | >180 | >180 | 150 | | | | 150 | >180 | 110 | 165 | >180 |
| Length burnt (mm) | | | | | | | | | | | | |
| Response ("X" or "O") | X | X | X | | | | → O | X | O | O | X |
| Column (2, 3, 4 or 5): **4** | | | | | | | Row (1 to 16): **7** | | | | |
| $k$ value from table 4: **1,25** | | | | | | | | | | | |
| | | | | | | | Hence, $k$ = **— 1,25** | | | | |

$$OI = c_F + kd = 29,8 + (-1,25 \times 0,2)$$
    = **29,5** % (to one decimal place, for reporting OI)
    = **29,55** % (to 2 decimal places, for calculation of and verification of $d$ as required in Part 3)

## Part 3: Verification of step size $d$ % oxygen concentration
## (in accordance with 8.6.4 and 9.3)

| Last six results | | Oxygen concentration (%) | | | | Estimation of standard deviation |
|---|---|---|---|---|---|---|
| | | $c_i$[1] | OI | $c_i - $ OI | $(c_i - $ OI$)^2$ | |
| $c_F$ | 1 | 29,8 | 29,55 | 0,25 | 0,062 5 | |
| | 2 | 29,6 | 29,55 | 0,05 | 0,002 5 | |
| | 3 | 29,4 | 29,55 | −0,15 | 0,022 5 | |
| | 4 | 29,6 | 29,55 | 0,05 | 0,002 5 | |
| | 5 | 29,4 | 29,55 | −0,15 | 0,022 5 | |
| $n$ | 6 | 29,6 | 29,55 | 0,15 | 0,002 5 | |
| | Total $\sum (c_i - $ OI$)^2$ | | | | 0,115 0 | |

$$\hat{\sigma} = \left[ \frac{\sum (c_i - \text{OI})^2}{n - 1} \right]^{1/2}$$

$$= \left( \frac{0,115}{5} \right)^{1/2} = 0,152$$

$$\frac{2}{3}\hat{\sigma} = 0,101$$

$$d = 0,2$$

$$\frac{3}{2}\hat{\sigma} = 0,227$$

1) Column $c_i$ contains the oxygen concentrations used for the measurements of $c_F$ and for each of the 5 preceding measurements, for $n = 6$.

If $\frac{2}{3}\hat{\sigma} < d < \frac{3}{2}\hat{\sigma}$ or if $0,2 = d > \frac{3}{2}\hat{\sigma}$, OI is valid.

Otherwise

if $\frac{2}{3}\hat{\sigma} > d$, repeat Part 2 using a larger value for $d$; or

if $\frac{3}{2}\hat{\sigma} < d$, repeat Part 2 using a smaller value for $d$.

Then again verify the step size, making further changes to the step size if necessary until one of the verification relationships is satisfied.

## Part 4: Ancillary information

a) These test results relate only to the behaviour of the specimens under the conditions of this test. These results must not be used to infer the relative hazards presented by differing materials or shapes under these or other fire conditions.

b) Special material history/characteristics, if applicable:

c) Variations from standard procedure, if applicable:

d) Description of observed burning behaviour:

e) Results measured/reported by

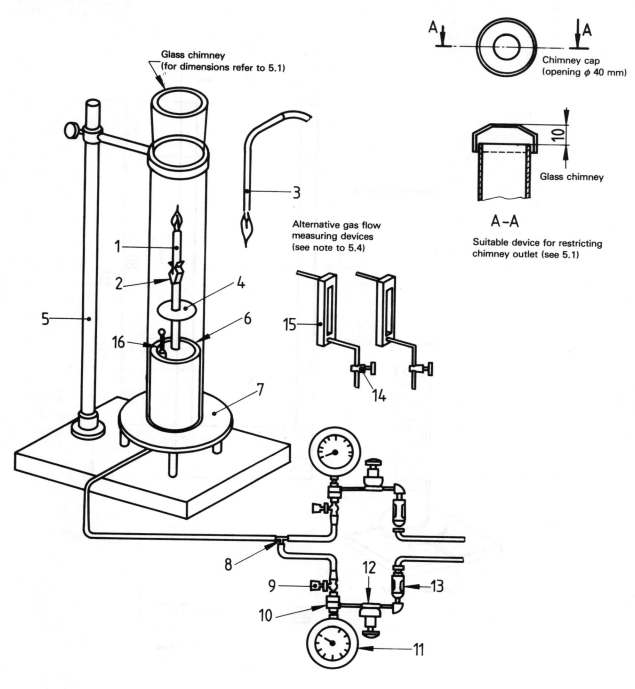

**Figure 1 — Diagram of typical apparatus for determination of oxygen index**

| | | | | |
|---|---|---|---|---|
| 1 | Burning test specimen | 7 | Base plate | |
| 2 | Specimen holder | 8 | Gas premixing point | |
| 3 | Igniter | 9 | Cut-off valve | |
| 4 | Debris screen of wire mesh | 10 | Orifice in holder | |
| 5 | Chimney support | 11 | Pressure gauge | |
| 6 | Bead bed | | | |

| | |
|---|---|
| 12 | Precision pressure regulator |
| 13 | Filter |
| 14 | Needle valve |
| 15 | Gas flow meter |
| 16 | Temperature sensor |

Dimensions in millimetres
with tolerances ±0,25 mm

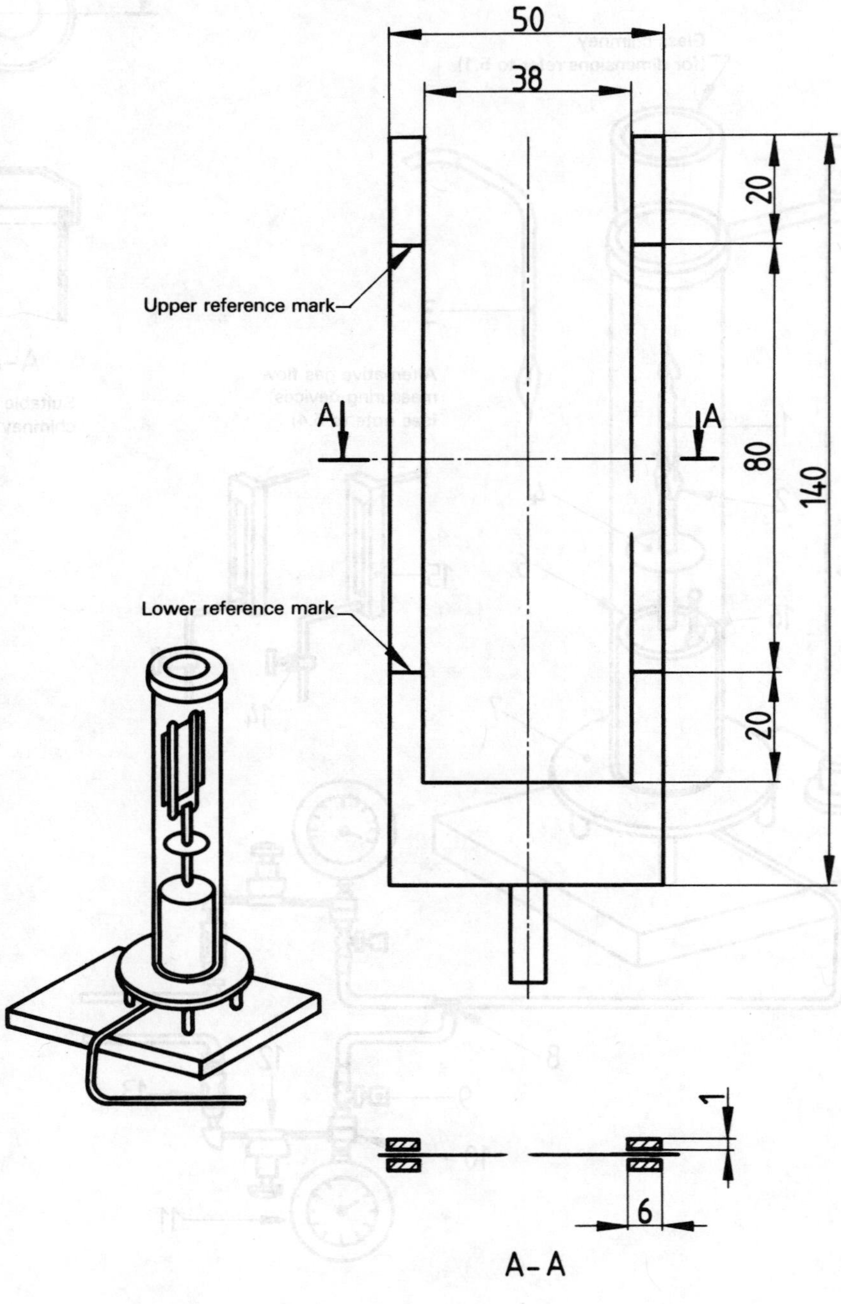

Two frame sections must hold the test specimen securely along both upright edges between forks each made of stainless steel.

**Figure 2 — Frame design for supporting non-self-supporting test specimens**

# INTERNATIONAL STANDARD

ISO
8256

First edition
1990-12-15

# Plastics — Determination of tensile-impact strength

*Plastiques — Détermination de la résistance au choc-traction*

Reference number
ISO 8256:1990(E)

# Foreword

ISO (the International Organization for Standardization) is a worldwide federation of national standards bodies (ISO member bodies). The work of preparing International Standards is normally carried out through ISO technical committees. Each member body interested in a subject for which a technical committee has been established has the right to be represented on that committee. International organizations, governmental and non-governmental, in liaison with ISO, also take part in the work. ISO collaborates closely with the International Electrotechnical Commission (IEC) on all matters of electrotechnical standardization.

Draft International Standards adopted by the technical committees are circulated to the member bodies for voting. Publication as an International Standard requires approval by at least 75 % of the member bodies casting a vote.

International Standard ISO 8256 was prepared by Technical Committee ISO/TC 61, *Plastics*.

Annexes A and B form an integral part of this International Standard. Annex C is for information only.

International Organization for Standardization
Case Postale 56 • CH-1211 Genève 20 • Switzerland

Printed in Switzerland

# Plastics — Determination of tensile-impact strength

## 1 Scope

**1.1** This International Standard specifies two methods for the determination of the energy required to rupture test specimens of plastics under a specified tensile-impact velocity. The tests can be described as tensile tests at comparatively high rates of straining. These methods can be used for materials too flexible or too thin to be tested with impact tests conforming to ISO 179 and ISO 180, and for more rigid materials. Different parameters are specified depending on the type of test specimen (see 6.1 and figure 3).

**1.2** These methods are used for investigating the behaviour of specified specimens under specified impact velocities, and for estimating the brittleness or the toughness of specimens within the limitations inherent in the test conditions. The response of plastics to comparatively high rates of straining is useful to describe, for example, the behaviour of materials when subjected to weathering or thermal ageing, as well as to assess their properties under corresponding service conditions.

**1.3** These methods are applicable to specimens prepared from moulding materials or to specimens taken from finished or semi-finished products (for example mouldings, films, laminates or extruded or cast sheets). The methods are suitable for production control as well as for quality control. Test results gained on test specimens obtained from moulding compounds cannot be applied directly to mouldings of any given shape, because values may depend on the design of the moulding and the moulding conditions.

**1.4** Results obtained by testing moulded specimens of different dimensions may not necessarily be the same. Equally, specimens cut from moulded products may not give the same results as specimens of the same dimensions moulded directly from the material. Results obtained by method A and method B may or may not be comparable.

**1.5** These methods are not suitable for use as a source of data for design calculations on components. Information on the typical behaviour of a material can be obtained, however, by testing different types of test specimen prepared under different conditions, and testing at different temperatures.

## 2 Normative references

The following standards contain provisions which, through reference in this text, constitute provisions of this International Standard. At the time of publication, the editions indicated were valid. All standards are subject to revision, and parties to agreements based on this International Standard are encouraged to investigate the possibility of applying the most recent editions of the standards indicated below. Members of IEC and ISO maintain registers of currently valid International Standards.

ISO 179:1982, *Plastics — Determination of Charpy impact strength of rigid materials.*

ISO 180:1982, *Plastics — Determination of Izod impact strength of rigid materials.*

ISO 291:1977, *Plastics — Standard atmospheres for conditioning and testing.*

ISO 293:1986, *Plastics — Compression moulding test specimens of thermoplastic materials.*

ISO 294:1975, *Plastics — Injection moulding test specimens of thermoplastic materials.*

ISO 295:1974, *Plastics — Compression moulding test specimens of thermosetting materials.*

ISO 1268:1974, *Plastics — Preparation of glass fibre reinforced, resin bonded, low-pressure laminated plates or panels for test purposes.*

ISO 2557-1:1989, *Plastics — Amorphous thermoplastics — Preparation of test specimens with a specified maximum reversion — Part 1: Bars.*

ISO 2557-2:1986, *Plastics — Amorphous thermoplastics — Preparation of test specimens with a specified reversion — Part 2: Plates.*

ISO 2818:1980, *Plastics — Preparation of test specimens by machining.*

ISO 3167:1983, *Plastics — Preparation and use of multipurpose test specimens.*

## 3 Definitions

For the purposes of this International Standard, the following definitions apply.

**3.1 tensile-impact strength of unnotched specimens:** The energy absorbed in breaking an unnotched specimen under specified conditions, referred to the original cross-sectional area of the specimen.

It is expressed in kilojoules per square metre (kJ/m$^2$).

**3.2 tensile-impact strength of notched specimens:** The energy absorbed in breaking a notched specimen under specified conditions, referred to the original cross-sectional area of the specimen at the notch.

It is expressed in kilojoules per square metre (kJ/m$^2$).

## 4 Principle

The energy utilized in this test method is delivered by a single swing of the pendulum of a tensile-impact machine. The energy to fracture is determined by the kinetic energy extracted from the pendulum in the process of breaking the specimen. Corrections are made for the energy to toss or bounce the crosshead.

The specimen is impacted at the bottom of the swing of the pendulum. The specimen is horizontal at rupture. One end of the specimen, at impact, is held either by the frame or the pendulum and the other end by the crosshead. The crosshead may be either mounted stationary on the support frame (method A) or carried downward together with the pendulum (method B).

## 5 Apparatus

### 5.1 Test machine

**5.1.1** The test machine shall be the pendulum type and shall be of rigid construction. It shall be capable of measuring the impact energy expended in breaking a test specimen. The value of the impact energy shall be taken as equal to the difference between the initial potential energy in the pendulum and the energy remaining in the pendulum after breaking the test specimen. The energy reading shall be accurately corrected for friction and air-resistance losses and for scale errors.

**5.1.2** The machine shall have the characteristics shown in table 1. The frictional loss shall be periodically checked.

NOTE 1    In order to apply the test to the full range of materials specified in 1.3, it is necessary to use more than one machine or to use a set of interchangeable pendulums. It is not advisable to compare results obtained with different pendulums.

**5.1.3** The machine shall be securely fixed to a foundation having a mass of at least 20 times that of the heaviest pendulum in use. It shall be adjusted so that the orientations of the striker and supports are as specified in 5.2 and 5.3.

**5.1.4** The distance between the axis of rotation and the centre of impact of the pendulum shall be within $\pm$ 1 % of the distance from the axis of rotation to the centre of the test specimen.

**5.1.5** The dial, or other indicator of the energy consumed, shall be capable of being read to an accuracy of $\pm$ 1 % of full-scale deflection.

**5.1.6** The machine shall be of the type shown schematically in figure 1 for method A, or of the type shown in figure 2 for method B.

### 5.2 Pendulum

**5.2.1** The pendulum shall be constructed of a single- or multiple-membered arm holding the head, in which the greatest mass is concentrated. A rigid pendulum is essential to maintain the proper clearances and geometric relationships between related parts and to minimize energy losses, which are always included in the measured impact-energy value.

**5.2.2** Accurate means shall be available to determine and minimize energy losses due to windage and friction (see annex B).

### 5.3 Crosshead

**5.3.1** The crosshead, which acts as a specimen clamp for method A, shall be made from a material which guarantees a substantially inelastic impact (e.g. aluminium).

The mass of the crosshead shall be selected from the values given in table 1.

**Table 1 — Characteristics of pendulum impact-testing machine**

| Initial potential energy | Velocity at impact | Maximum permissible frictional loss | Crosshead mass[1] | |
|---|---|---|---|---|
| | | | Method A | Method B |
| J | m/s | % | g | g |
| 2,0 | 2,6 to 3,2 | 1 | 15 ± 1 or 30 ± 1 | 15 ± 1 |
| 4,0 | 2,6 to 3,2 | 0,5 | 15 ± 1 or 30 ± 1 | 15 ± 1 |
| 7,5 | 3,4 to 4,1 | 0,5 | 30 ± 1 or 60 ± 1 | 30 ± 1 |
| 15,0 | 3,4 to 4,1 | 0,5 | 30 ± 1 or 60 ± 1 | 120 ± 1 |
| 25,0 | 3,4 to 4,1 | 0,5 | 60 ± 1 or 120 ± 1 | 120 ± 1 |
| 50,0 | 3,4 to 4,1 | 0,5 | 60 ± 1 or 120 ± 1 | 120 ± 1 |

1) For method A, use the lighter crosshead wherever possible.

**5.3.2** A jig shall be used to assist in clamping the crosshead in the specified position, at right angles to the longitudinal axis of the specimen.

## 5.4 Clamping devices/jaws

**5.4.1** For specimen types 1, 2, 3 and 4 (see table 2 and figure 3), the surfaces between which the specimen is clamped shall be clamped such that there is no slippage when the blow is struck. The same applies to the jaw faces of the clamping device attached to the frame. The clamping device shall be such as to ensure that it does not contribute to failure of the specimen.

Jaws may have file-like serrations, and the size of serrations shall be selected, according to experience, to suit the hardness and toughness of the specimen material and the thickness of the specimen. The edges of the serrated jaws in close proximity to the test region shall have a radius such that they cut across the edges of the first serrations.

**5.4.2** For specimen type 5, held only by embedding, a notched pair of jaws with different heights is necessary. The pair of jaws chosen for the test shall be the one whose height is greater than the thickness of the specimen but lower than 120 % of its thickness.

## 5.5 Micrometers and gauges

Micrometers and gauges suitable for measuring the dimensions of test specimens to an accuracy of 0,01 mm are required. For measuring the thickness of film and sheeting with thicknesses below 1 mm, use an instrument reading to an accuracy of not less than 5 % of the nominal thickness. In measuring the thickness of the specimen, the measuring face shall apply a load of 0,01 MPa to 0,05 MPa.

For notched specimens, see the requirements of 7.4.

## 6 Test specimens

### 6.1 Dimensions and notches

Five types of test specimen, as specified in table 2 and shown in figure 3, may be used. For method A, the preferred specimen types are type 1 (notched) and type 3 (unnotched), but type 2, 4 or 5 may also be used if required. For method B, the preferred specimen types are type 2 and type 4.

The test result depends on the type of specimen used and its thickness. For reproducible results, or in case of dispute, therefore, the type of test specimen and its thickness shall be agreed upon.

Specimens are tested at their original thickness up to and including 4 mm. The preferred specimen thickness is 4 mm ± 0,2 mm. Within the gauge area, the thickness shall be maintained to within a tolerance of ± 5 %. Above 4 mm, the test methods described in this International Standard are inapplicable, and use shall be made of ISO 179 or ISO 180.

NOTE 2   Specimen type 1 can be prepared from the multi-purpose test specimen described in ISO 3167.

### 6.2 Preparation

#### 6.2.1 Moulding or extrusion compounds

Specimens shall be prepared in accordance with the relevant material specification. When none exists, or when otherwise specified, specimens shall be directly extruded, compression or injection moulded from the material in accordance with ISO 293, ISO 294, ISO 295, ISO 2557-1 or ISO 2557-2 as appropriate, or machined in accordance with ISO 2818 from sheet that has been compression or injection moulded from the compound.

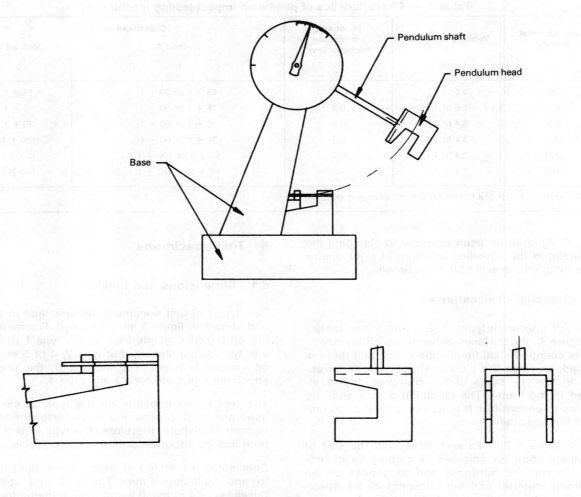

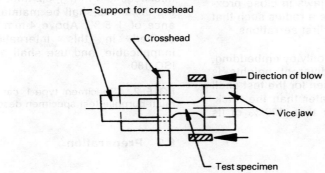

**Figure 1 — Diagram showing relationship of pendulum to specimen clamps for method A**

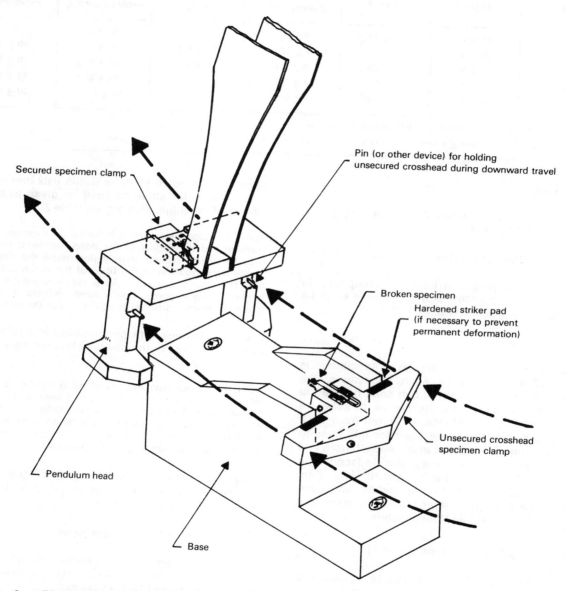

Secured specimen clamp

Pin (or other device) for holding
unsecured crosshead during downward travel

Broken specimen

Hardened striker pad
(if necessary to prevent
permanent deformation)

Unsecured crosshead
specimen clamp

Pendulum head

Base

**Figure 2 — Diagram showing relationship of pendulum to specimen clamps for method B after specimen
rupture**

**Table 2 — Specimen types and dimensions and distances between supports** (see also figure 3)

Dimensions in millimetres

| Specimen type | Length, $l$ | Width, $b$ | Preferred value of dimension, $x$ | Preferred value for $l_o$ | Free length between grips, $l_e$ | Radius of curvature, $r$ |
|---|---|---|---|---|---|---|
| 1[1] | 80 ± 2 | 10 ± 0,5 | 6 ± 0,2 | — | 30 ± 2 | — |
| 2 | 60 ± 1 | 10 ± 0,2 | 3 ± 0,05 | 10 ± 0,2 | 25 ± 2 | 10 ± 1 |
| 3 | 80 ± 2 | 15 ± 0,5 | 10 ± 0,5 | 10 ± 0,2 | 30 ± 2 | 20 ± 1 |
| 4 | 60 ± 1 | 10 ± 0,2 | 3 ± 0,1 | — | 25 ± 2 | 15 ± 1 |
| 5[2] | 80 ± 2 | 15 ± 0,5 | 5 ± 0,5 | 10 ± 0,2 | 50 ± 0,5 | 20 ± 1 |

1) Notch angle is 45° ± 1°, radius of notch 1,0 mm ± 0,02 mm.

2) For type 5:  $b' = 23$ mm ± 2 mm  $r' = 6$ mm ± 0,5 mm  $l' = 11$ mm ± 1 mm

### 6.2.2 Sheets

Specimens shall be machined from sheets in accordance with ISO 2818.

### 6.2.3 Fibre-reinforced resins

A panel shall be prepared from the compound in accordance with ISO 1268, and specimens shall be machined in accordance with ISO 2818.

### 6.2.4 Thin films

For thin films, the use of multi-layer specimens is recommended. To prepare such specimens, the necessary number of layers of film shall be fixed in place before stamping, e.g. by means of adhesive tapes applied over a distance of 30 mm from each end. The adhesive tapes are used for thin films to hold the specimens together, at the ends, before and after stamping. In other cases, double-faced tape may be used between each layer of film. Film specimens shall be free from scratches, and each layer shall be free from either tension or slackness relative to other layers in the specimen.

## 6.3 Notching of specimens (type 1)

**6.3.1** Notches shall be machined in accordance with ISO 2818 on unnotched specimens prepared in accordance with 6.2.

**6.3.2** The radius of the notch base shall be 1,0 mm ± 0,02 mm, its angle 45° ± 1° (see figure 3). The profile of the cutting tooth (teeth) shall be such as to produce in the specimen, at right angles to its principal axis, two notches of the contour and depth shown in figure 3. The two lines drawn perpendicular to the length direction of the specimen through the

apex of each notch shall be within 0,02 mm of each other. Particular attention shall be given to the accuracy of the dimension $x$ (see table 2).

NOTE 3    Close tolerances have to be imposed on the contour and the radius of the notch for most materials because these factors largely determine the degree of stress concentration at the base of the notch during the test. The maintenance of a sharp, clean-edged cutting tool is particularly important since minor defects at the base of the notch can cause large deviations in the test results.

**6.3.3** Specimens with moulded-in notches may be used if specified in the specification for the material being tested.

NOTE 4    Specimens with moulded-in notches generally do not give the same results as specimens with machined notches, and allowance should be made for this difference in interpreting the results. Specimens with machined notches are generally preferred because skin effects and/or localized anisotropy are minimized.

**6.3.4** For samples prepared by cutting with a puncher, the notch shall not be punched but shall be machined in a second step.

## 6.4 Number of test specimens

**6.4.1** Unless otherwise specified in the specification for the material being tested, a minimum of 10 specimens shall be tested.

**6.4.2** The impact properties of certain types of sheet material may differ depending on the direction of measurement in the plane of the sheet. In such cases, it is customary to prepare two groups of test specimens with their major axes respectively parallel and perpendicular to the direction of some feature of the sheet which is either visible or inferred from a knowledge of the method of its manufacture.

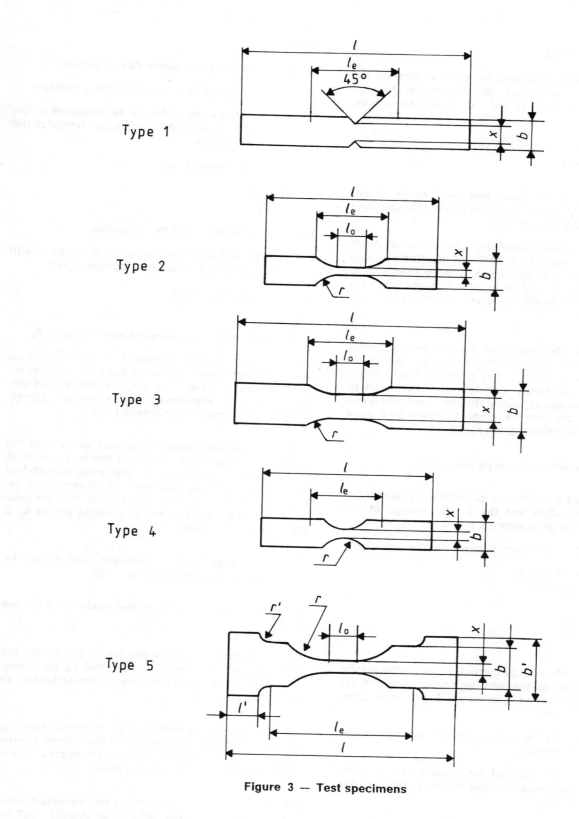

Figure 3 — Test specimens

## 6.5 Conditioning

Unless otherwise specified in the specification for the material being tested, the specimens shall be preconditioned and tested in accordance with ISO 291.

## 7 Procedure

**7.1** Check that the pendulum machine is of the correct energy range and that it has the specified striking velocity (see table 1).

The selected pendulum shall consume at least 20 %, but not more than 80 %, of its stored energy in breaking the test specimen. If more than one of the pendulums described in table 1 meets these requirements, the pendulum having the higher energy shall be used.

**7.2** If applicable, adjust the pointer on the energy scale so that it touches the driving pin when the pendulum is in the starting position. Carry out a blank test three times (i.e. without a specimen or crosshead in place), calculate the mean frictional loss and ensure that the mean frictional loss does not exceed the values given in table 1.

### 7.3 Determination of energy corrections

**7.3.1 Method A — Energy correction $E_q$ due to the plastic deformation and the kinetic energy of the crosshead** (see also annex C)

The correction $E_q$ is determined from the equation

$$E_q = \frac{E_{max}\mu(3 + \mu)}{2(1 + \mu)}$$

$$\approx \frac{3}{2}E_{max}\mu$$

where

$E_q$    is the energy correction due to the plastic deformation and the kinetic energy of the crosshead (see annex B or annex C for details);

$E_{max}$    is the maximum impact energy of the pendulum;

$\mu$    is the mass of the crosshead divided by the reduced mass of the pendulum (i.e. $m_{cr}/m_p$).

The reduced mass of the pendulum is given by the equation

$$m_p = \frac{E_{max}}{gL(1 - \cos \alpha)}$$

where

$g$    is the acceleration due to gravity;

$L$    is the reduced length of the pendulum;

$\alpha$    is the angle between the positions of the maximum and the minimum height of the pendulum.

$L$ is given by the equation

$$L = \frac{gT^2}{4\pi^2}$$

where $T$ is the period of the pendulum.

If $\alpha$ is 160° and $m_p$ is measured in kilograms, with $E_{max}$ in joules and $L$ in metres, it follows that

$$m_p = 5,3 \times 10^{-2}\frac{E_{max}}{L}$$

### 7.3.2 Method B — Crosshead-bounce energy $E_b$

The crosshead-bounce energy $E_b$ is determined for each specimen and pendulum from the crosshead-bounce energy curve. This curve is determined only once for each crosshead and pendulum combination. (See annex A for details.)

**7.4** Measure the thickness $d$ and width $x$ of the central, parallel-sided section of the test specimen to the nearest 0,02 mm. In the case of notched specimens, carefully measure the dimension $x$ using, for example, a micrometer fitted with an anvil of width 2 mm to 3 mm and of suitable profile to fit the shape of the notch.

**7.5** Lift and arrest the pendulum, and adjust the pointer in accordance with 7.2.

**7.6** Insert the specimen in the holder and tighten firmly.

**7.7** Carefully release the pendulum. Read from the scale the impact energy $E_s$ absorbed by the specimen and apply corrections for frictional losses as necessary (see 7.2).

**7.8** If the resulting corrected tensile-impact energy is below 20 % of the capacity of the 2-joule pendulum (see table 1), multi-layer specimens prepared in accordance with 6.2.4 shall be used.

**7.9** If various materials are to be compared, pendulums with the same velocity at impact shall be used for each. In cases of dispute, it is recommended that test results be compared only with results obtained with pendulums of identical nominal energy and specimens of the same geometry.

**7.10** Immediately after the test has been completed, a check shall be made to ensure that the specimen was firmly clamped or whether it had slipped in one of the two grips, and that the failure occurred in the narrow, parallel-sided part of the specimen. If any of the specimens tested do not meet these requirements, the results for these specimens shall be discarded and additional specimens tested.

# 8 Expression of results

In order to calculate the tensile-impact strength, the consumed energy $E_s$ must first be corrected for the toss energy $E_q$.

## 8.1 Calculation of energy correction

### 8.1.1 Energy correction for method A

The corrected tensile-impact energy $E_c$, in joules, is calculated using the equation

$$E_c = E_s - E_q$$

where

$E_s$ is the consumed energy (non-corrected) in joules;

$E_q$ is the elastic toss energy, in joules, of the crosshead, calculated as specified in 7.3.1.

### 8.1.2 Energy correction for method B

The corrected tensile-impact energy $E_c$, in joules, is calculated using the equation

$$E_c = E_s + E_b$$

where

$E_s$ is the consumed energy (non-corrected), in joules;

$E_b$ is the crosshead-bounce energy, in joules, of the crosshead, as determined from the measured value of $E_s$ and the graph prepared for the particular impact tester used, as specified in 7.3.2 and annex A.

## 8.2 Calculation of tensile-impact strength

The tensile-impact strength $E$ or the tensile-impact strength (notched) $E_n$, expressed in kilojoules per square metre, is calculated using the equation

$$E \text{ or } E_n = \frac{E_c}{x \cdot d} \times 10^3$$

where

$E_c$ is the corrected impact energy, in joules, calculated in accordance with 8.1;

$x$ is the width, in millimetres, of the narrow, parallel-sided section of the specimen or the distance between the notches (see figure 3);

$d$ is the thickness, in millimetres, of the narrow, parallel-sided section of the specimen (or for plied film, the total thickness).

Calculate the arithmetic mean, the standard deviation and the coefficient of variation of the ten results as required.

Report all calculated values to two significant figures.

# 9 Precision

The precision of this test method is not known because inter-laboratory data are not available. When inter-laboratory data are obtained, a precision statement will be added with the following revision.

# 10 Test report

The test report shall include the following information:

a) a reference to this International Standard;

b) the method used (A or B);

c) full identification of the material tested, including manufacturer's code, material grade and material form;

d) the type of test specimen used or the dimensions of the test specimens;

e) the method of preparing the test specimens;

f) the thickness of moulded specimens or, for sheets, the thickness of the sheet and, if applicable, the directions of the major axes of the specimens in relation to some feature of the sheet;

g) details of preconditioning and the test conditions;

h) the maximum energy of the pendulum used;

i) the mass of the crosshead used;

j) the tensile-impact strength $E$ or $E_n$ of the material, expressed in kilojoules per square metre, reported as the arithmetic mean of the results on notched and/or unnotched test specimens, as applicable;

k) the individual test results, if required;

l) if required, the standard deviation and the coefficient of variation of the results;

m) the type of fracture exhibited by the test specimens.

# Annex A
## (normative)

## Determination of bounce-correction factor

After impact and rebound of the crosshead, the specimen is pulled by two moving bodies, the pendulum with an energy of $0,5MV^2$, and the crosshead with an energy of $0,5mv^2$. When the specimen breaks, only that energy is recorded on the pendulum dial which is lost by the pendulum. Therefore, one must add the incremental energy contributed by the crosshead to determine the true energy used to break the specimen. The correction (i.e. the incremental energy contributed by the crosshead) can be calculated as follows:

By definition,

$$E = \frac{1}{2} M(V^2 - V_2^2) \qquad \qquad \dots \text{(A.1)}$$

and

$$e = \frac{1}{2} m(v_1^2 - v_2^2) \qquad \qquad \dots \text{(A.2)}$$

where

$M$      is the mass, in kilograms, of pendulum;

$m$      is the mass, in kilograms, of crosshead;

$V$      is the maximum velocity, in metres per second, of centre of impact of crosshead of pendulum;

$V_2$      is the velocity, in metres per second, of centre of impact of pendulum at time when specimen breaks;

$v_1$      is the crosshead velocity, in metres per second, immediately after bounce;

$v_2$      is the crosshead velocity, in metres per second, at time when specimen breaks;

$E$      is the energy, in joules, read on pendulum dial;

$e$      is the energy contribution, in joules, of crosshead, i.e. bounce-correction factor to be added to pendulum reading.

Once the crosshead has rebounded, the momentum of the system (in the horizontal direction) remains constant. Neglecting vertical components, the momentum equation for the impact can be written as follows:

$$MV - mv_1 = MV_2 - mv_2 \qquad \qquad \dots \text{(A.3)}$$

Equations (A.1), (A.2) and (A.3) can be combined to give:

$$e = \frac{1}{2} m \left\{ v_1^2 - \left[ v_1 - \frac{M}{m} \left( V - \sqrt{V^2 - \frac{2E}{M}} \right) \right]^2 \right\} \qquad \qquad \dots \text{(A.4)}$$

If $e$ is plotted as a function of $E$ (for fixed values of $V$, $M$, $m$ and $v_1$), $e$ will increase from zero, pass through a maximum (equal to $0,5mv_1^2$) and then decrease, passing again through zero before becoming negative. The only part of this curve for which a reasonably accurate analysis has been made is the initial portion between $e = 0$ and $e = 0,5mv_1^2$. Once the crosshead reverses its direction of travel, the correction becomes less clearly defined and, after it contacts the anvil a second time, the correction becomes much more difficult to evaluate. It is assumed, therefore, for the sake of simplicity, that once $e$ has reached its maximum value the correction factor will remain constant at a value of $0,5mv_1^2$. It should be clearly realised that the use of that portion of the curve in figure A.1 where $e$ is constant does not give an accurate correction. However, as $E$ grows larger, the correction factor becomes relatively less important and no great sacrifice of overall accuracy results from the assumption that the maximum correction is $0,5mv_1^2$.

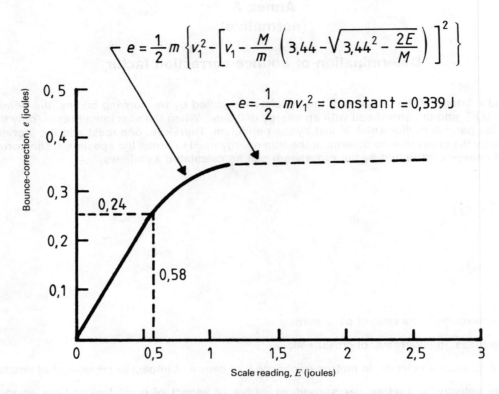

$$e = \frac{1}{2} m \left\{ v_1^2 - \left[ v_1 - \frac{M}{m} \left( 3,44 - \sqrt{3,44^2 - \frac{2E}{M}} \right) \right]^2 \right\}$$

$$e = \frac{1}{2} m v_1^2 = \text{constant} = 0,339 \text{ J}$$

**Figure A.1 — Typical correction-factor curve for single bounce of crosshead specimen-in-head tensile-impact machine (6,8 J hammer, steel crosshead)**

# Annex B

(normative)

## Instructions for the construction and use of a windage and friction correction chart for method B

**B.1** The construction and use of the chart described is based upon the assumption that the friction and windage losses are proportional to the angle through which these forces act on the pendulum. Figure B.1 shows a plot of the assumed energy loss versus the pendulum position (i.e. the angle through which the pendulum has swung) during the pendulum swing. The correction chart to be described is essentially the left-hand half of figure B.1.

**B.2** Start the construction of the correction chart (figure B.2) by laying off to some convenient linear scale on the abscissa the pendulum position for the portion of the swing beyond the free-hanging position. Place the free-hanging reference point at the right-hand end of the abscissa, with the angular displacement increasing linearly to the left. The abscissa is referred to as scale C. Although angular displacement is the quantity to be represented linearly on the abscissa, this displacement is more conveniently expressed in terms of indicated energy read from the machine dial. This yields a non-linear scale C with indicated pendulum energy increasing to the right.

**B.3** On the right-hand ordinate, lay off a linear scale B, starting with zero at the bottom and stopping at the maximum expected pendulum friction and windage value at the top.

**B.4** On the left-hand ordinate construct a linear scale D, ranging from zero at the bottom to 1,2 times the maximum ordinate value appearing on scale B, but make the scale twice the scale used in the construction of scale B.

**B.5** Adjoining scale D draw a curve OA which is the locus of the points representing equal values of the energy correction on scale D and the indicated energy on scale C. This curve is referred to as scale A and utilizes the same divisions and numbering system as the adjoining scale D.

**B.6** The chart is used as follows:

**B.6.1** Locate and mark on scale A the reading A obtained from the free swing of the pendulum with the pointer prepositioned on the dial in the free-hanging or maximum-indicated-energy position.

**B.6.2** Locate and mark on scale B the reading B obtained after several free swings of the pendulum with the pointer pushed up close to the zero-indicated-energy position on the dial.

**B.6.3** Join the two points thus obtained by a straight line.

**B.6.4** From the indicated impact energy on scale C, project up to the constructed line and across to the left to obtain the correction for windage and friction from scale D.

**B.6.5** Subtract this correction from the indicated impact-energy reading to obtain the energy delivered to the specimen. (See 7.7.)

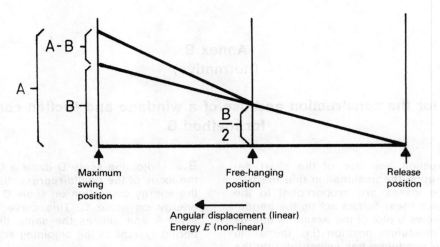

**Figure B.1 — Method of construction of a windage and friction correction chart**

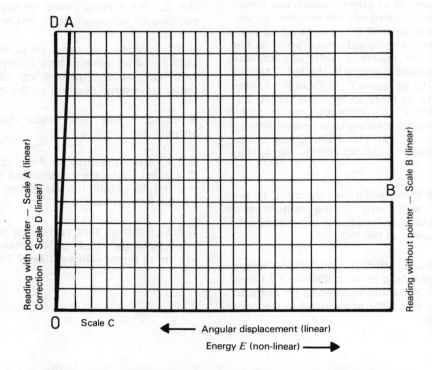

**Figure B.2 — Sample windage and friction correction chart**

# Annex C
## (informative)

## Determination of correction factor

**C.1** In calculating the correction $E_q$, the following energy terms are used:

$$E_{max} = \frac{1}{2} m_p v_0^2 = \text{the maximum impact energy of the pendulum} \qquad \ldots \text{(C.1)}$$

$$E_p = \frac{1}{2} m_p v_p^2 = \text{the residual energy of the pendulum after the impact} \qquad \ldots \text{(C.2)}$$

$$E_s = E_{max} - E_p = \text{the measured (consumed) energy} \qquad \ldots \text{(C.3)}$$

$E_c$ = the energy needed for deformation and fracture of the specimen (to be calculated) $\qquad \ldots$ (C.4)

$E_{cr,pl}$ = the energy consumed by the plastic deformation of the crosshead $\qquad \ldots$ (C.5)

$$E_{cr,kin} = \frac{1}{2} m_{cr} v_{cr}^2 = \text{the kinetic energy of the crosshead lost by the pendulum} \qquad \ldots \text{(C.6)}$$

$$= \frac{1}{2} m_{cr} v_p^2 \quad \text{if the elastic energy of the impact can be neglected}$$

where

$m_p$      is the reduced mass of the pendulum;

$v_0$      is the velocity of the pendulum immediately before impact;

$v_p$      is the velocity of the pendulum immediately after impact;

$m_{cr}$      is the mass of the crosshead;

$v_{cr}$      is the velocity of the crosshead immediately after impact.

**C.2** The energy equation for the impact is

$$E_s = E_c + E_{cr,pl} + E_{cr,kin} \qquad \ldots \text{(C.7)}$$

Furthermore, from (C.2) and (C.6), it follows that

$$E_{cr,kin} = \frac{m_{cr}}{m_p} \times E_p \qquad \ldots \text{(C.8)}$$

Combining this equation with (C.3) gives

$$E_{cr,kin} = \mu(E_{max} - E_s) \qquad \ldots \text{(C.9)}$$

where

$$\mu = \frac{m_{cr}}{m_p}$$

In order to obtain $E_c$ (the energy needed for deformation and fracture of the specimen), it is necessary to subtract the energy correction $E_q$ given by

$$E_q = E_s - E_c = E_{cr,pl} + E_{cr,kin} = E_{cr,pl} + \mu(E_{max} - E_s) \qquad \ldots \text{(C.10)}$$

**C.3** To calculate the energy $E_{cr,pl}$ consumed by the plastic deformation of the crosshead, it is necessary to consider the momentum equation at impact without a specimen (i.e. $E_c = 0$).

This case is indicated by an asterisk (*).

Since the crosshead is plastically deformed by the same amount with and without a specimen,

$$E_{cr,pl} = E_{cr,pl}^*$$ 
...(C.11)

The momentum equation can be written

$$m_p v_0 = m_p v_p^* + m_{cr} v_{cr}^*$$ 
...(C.12)

If the elastic energy of the impact is neglected,

$$v_{cr}^* = v_p^*$$ 
...(C.13)

and hence

$$v_p^* = \frac{v_0}{1 + \mu}$$ 
...(C.14)

Using equations (C.1) and (C.2) to express the velocities in equation (C.14) in terms of energies gives

$$E_p^* = \frac{E_{max}}{(1 + \mu)^2}$$ 
...(C.15)

From equation (C.3), the consumed energy measured without a specimen is then given by

$$E_s^* = \frac{E_{max}\mu(\mu + 2)}{(1 + \mu)^2}$$ 
...(C.16)

With $E_c = 0$, equation (C.7) becomes

$$E_s^* = E_{cr,pl} + E_{cr,kin}^*$$ 
...(C.17)

From equations (C.6) and (C.14), it follows that $E_{cr,kin}^*$, the kinetic component of $E_s^*$, is given by

$$E_{cr,kin}^* = \frac{E_{max}\mu}{(1 + \mu)^2}$$ 
...(C.18)

Finally, from equations (C.16) and (C.18) the energy due to the plastic deformation of the crosshead is given by

$$E_{cr,pl}^* = E_{cr,pl} = \frac{E_{max}\mu}{1 + \mu}$$ 
...(C.19)

and the energy correction [see equation (C.10)] can be written as follows:

$$E_q = E_s - E_c = \mu \left[ \frac{E_{max}}{1 + \mu} + (E_{max} - E_s) \right]$$ 
...(C.20)

**C.4** This correction consists of a dominant constant part (representing the energy consumed by the plastic deformation of the crosshead $E_{cr,pl}$) and a smaller part ($E_{max} - E_s$) which decreases from $\mu E_{max}$ to zero with increasing consumed energy (when $E_s \approx E_{max}$). In view of measurement uncertainties, it is sufficient to use a constant correction as an approximation; assuming that

$$E_s = \frac{E_{max}}{2}$$ 
...(C.21)

gives the correction

$$E_q = E_s - E_c = \frac{E_{max}\mu(3 + \mu)}{2(1 + \mu)} \qquad \ldots (C.22)$$

The corrected value of the energy consumed by the impact with the specimen is therefore given by

$$E_c = E_s - E_q = E_s - \frac{E_{max}\mu(3 + \mu)}{2(1 + \mu)} = E_s - \frac{3}{2}\mu E_{max} \qquad \ldots (C.23)$$

―――――

# INTERNATIONAL STANDARD ISO 8256 : 1990
## TECHNICAL CORRIGENDUM 1

Published 1991-05-15

INTERNATIONAL ORGANIZATION FOR STANDARDIZATION· МЕЖДУНАРОДНАЯ ОРГАНИЗАЦИЯ ПО СТАНДАРТИЗАЦИИ· ORGANISATION INTERNATIONALE DE NORMALISATION

# Plastics — Determination of tensile-impact strength

## TECHNICAL CORRIGENDUM 1

*Plastiques — Détermination de la résistance au choc-traction*
*RECTIFICATIF TECHNIQUE 1*

Technical corrigendum 1 to International Standard ISO 8256 : 1990 was prepared by Technical Committee ISO/TC 61, *Plastics*, Sub-Committee SC 2, *Mechanical properties*.

---

*Page 6*

**Table 2**

In the column headed "**Width**, $b$", replace "10 ± 0,5" by "10 ± 0,2"; in the column headed "**Preferred value of dimension**, $x$", delete the comma after the word "**dimension**" and replace "10 ± 0,5" in the third line by "10 ± 0,2".

*Page 7*

In figure 3, replace the drawing of the type 1 specimen by the following:

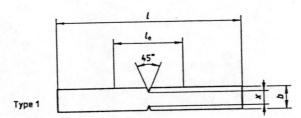

Type 1

UDC 678.5/.8 : 620.178.72

Ref. No. ISO 8256 : 1990/Cor.1 : 1991 (E)

Descriptors : plastics, tests, mechanical tests, tension tests.

# INTERNATIONAL STANDARD

## ISO
## 10350

First edition
1993-12-15

# Plastics — Acquisition and presentation of comparable single-point data

*Plastiques — Acquisition et présentation de caractéristiques intrinsèques comparables*

Reference number
ISO 10350:1993(E)

283

# Foreword

ISO (the International Organization for Standardization) is a worldwide federation of national standards bodies (ISO member bodies). The work of preparing International Standards is normally carried out through ISO technical committees. Each member body interested in a subject for which a technical committee has been established has the right to be represented on that committee. International organizations, governmental and non-governmental, in liaison with ISO, also take part in the work. ISO collaborates closely with the International Electrotechnical Commission (IEC) on all matters of electrotechnical standardization.

Draft International Standards adopted by the technical committees are circulated to the member bodies for voting. Publication as an International Standard requires approval by at least 75 % of the member bodies casting a vote.

International Standard ISO 10350 was prepared by Technical Committee ISO/TC 61, *Plastics*, Sub-Committee SC 1, *Terminology*.

Annex A of this International Standard is for information only.

International Organization for Standardization
Case Postale 56 • CH-1211 Genève 20 • Switzerland
Printed in Switzerland

# Introduction

This International Standard has been prepared because users of plastics find sometimes that available data cannot be used readily to compare the properties of similar materials, especially when the data have been supplied by different sources. Even when the same standard tests have been used, they often allow the adoption of a wide range of alternative test conditions, and the data obtained are not necessarily comparable. The purpose of this International Standard is to identify specific methods and conditions of test to be used for the acquisition and presentation of data in order that valid comparisons between materials can be made.

The present International Standard is concerned with tests employed to present "single-point" data on the limited range of properties commonly included in data sheets and used for the preliminary selection of materials. Such data represent the most basic approach to the specification of properties of materials and this International Standard thus facilitates the first steps towards more efficient selection and use of plastics in the many applications to which they are suited.

A complementary International Standard (to be published as ISO 11403, in several parts) will be concerned with the standardized acquisition and presentation of "multi-point" data, to demonstrate how properties vary with important factors such as time, temperature and the presence of particular natural and chemical environments. In that standard, some additional properties will be included. Its use will provide a more substantial database than one containing only single-point data, and so will enable improved assessment of the fitness of a material for any particular application. In addition, ISO11403-1, which deals with mechanical properties, assists predictions of the performance of components and ISO 11403-2, covering thermal and processing properties, aids predictions of melt-flow behaviour during manufacturing. ISO 11403-3 will be concerned with environmental influences on properties, and other parts may be prepared to cover additional properties.

# Plastics — Acquisition and presentation of comparable single-point data

## 1 Scope

This International Standard identifies specific test procedures for the acquisition and presentation of comparable data for certain basic properties of plastics. In general, each property is specified by a single experimental value although in certain cases properties are represented by two values obtained under different test conditions. The properties included are those presented conventionally in manufacturers' data sheets. The test methods and test conditions apply predominantly to those plastics that may be injection- or compression-moulded or prepared as sheets of specified thickness.

## 2 Normative references

The following standards contain provisions which, through reference in this text, constitute provisions of this International Standard. At the time of publication, the editions indicated were valid. All standards are subject to revision, and parties to agreements based on this International Standard are encouraged to investigate the possibility of applying the most recent editions of the standards indicated below. Members of IEC and ISO maintain registers of currently valid International Standards.

ISO 62:1980, *Plastics — Determination of water absorption*.

ISO 75-1:1993, *Plastics — Determination of temperature of deflection under load — Part 1: General test method*.

ISO 75-2:1993, *Plastics — Determination of temperature of deflection under load — Part 2: Plastics and ebonite*.

ISO 75-3:1993, *Plastics — Determination of temperature of deflection under load — Part 3: High-strength*

thermosetting laminates and long-fibre-reinforced plastics.

ISO 178:1993, *Plastics — Determination of flexural properties*.

ISO 179:1993, *Plastics — Determination of Charpy impact strength*.

ISO 291:1977, *Plastics — Standard atmospheres for conditioning and testing*.

ISO 293:1986, *Plastics — Compression moulding test specimens of thermoplastic materials*.

ISO 294:—[1], *Plastics — Injection moulding of test specimens of thermoplastic materials*.

ISO 295:1991, *Plastics — Compression moulding of test specimens of thermosetting materials*.

ISO 306:1987, *Plastics — Thermoplastic materials — Determination of Vicat softening temperature*.

ISO 527-1:1993, *Plastics — Determination of tensile properties — Part 1: General principles*.

ISO 527-2:1993, *Plastics — Determination of tensile properties — Part 2: Test conditions for moulding and extrusion plastics*.

ISO 899-1:1993, *Plastics — Determination of creep behaviour — Part 1: Tensile creep*.

ISO 1133:1991, *Plastics — Determination of the melt mass-flow rate (MFR) and the melt volume-flow rate (MVR) of thermoplastics*.

ISO 1183:1987, *Plastics — Methods for determining the density and relative density of non-cellular plastics*.

---

1) To be published. (Revision of ISO 294:1975)

ISO 1210:1992, *Plastics — Determination of the burning behaviour of horizontal and vertical specimens in contact with a small-flame ignition source.*

ISO 2577:1984, *Plastics — Thermosetting moulding materials — Determination of shrinkage.*

ISO 2818:1980, *Plastics — Preparation of test specimens by machining.*

ISO 3146:1985, *Plastics — Determination of melting behaviour (melting temperature or melting range) of semi-crystalline polymers.*

ISO 3167:1993, *Plastics — Multipurpose test specimens.*

ISO 4589:1984, *Plastics — Determination of flammability by oxygen index.*

ISO 8256:1990, *Plastics — Determination of tensile-impact strength.*

ISO 10724:—[2], *Plastics — Thermosetting moulding materials — Injection moulding of multipurpose test specimens.*

ISO 11403-1:—[2], *Plastics — Acquisition and presentation of comparable multipoint data — Part 1: Mechanical properties.*

ISO 11403-2:—[2], *Plastics — Acquisition and presentation of comparable multipoint data — Part 2: Thermal and processing properties.*

IEC 93:1980, *Methods of test for volume resistivity and surface resistivity of solid electrical insulating materials.*

IEC 112:1979, *Method for determining the comparative and the proof tracking indices of solid insulating materials under moist conditions.*

IEC 243-1:1988, *Methods of test for electric strength of solid insulating materials — Part 1: Tests at power frequencies.*

IEC 250:1969, *Recommended methods for the determination of the permittivity and dielectric dissipation factor of electrical insulating materials at power, audio and radio frequencies including metre wavelengths.*

IEC 296:1982, *Specification for unused mineral insulating oils for transformers and switchgear.*

IEC 1006:1991, *Methods of test for the determination of the glass transition temperature of electrical insulating materials.*

## 3 Definition

For the purposes of this International Standard, the following definition applies.

**3.1 single-point data:** Data characterizing a plastics material by means of those property tests in which important aspects of performance can be described with a single-value result.

## 4 Specimen preparation and conditioning

In the preparation of specimens by injection- or compression-moulding, the procedures described in ISO 293, ISO 294 or ISO 295 shall be used. The moulding method and the conditions will depend upon the material being moulded. If these conditions are specified in the International Standard appropriate to the material then they shall be adopted for the preparation of every specimen on which data are obtained using this International Standard. For those plastics for which moulding conditions have not yet been standardized, the conditions employed shall be within the range recommended by the polymer manufacturer and shall, for each of the processing methods, be the same for every specimen.

Where moulding conditions are not stipulated in any International Standard, the values used for the parameters in table 1 shall be recorded with the single-point data for that material. Where specimens are prepared by machining from sheet, the machining shall be performed in accordance with ISO 2818 and the dimensions of the specimen shall comply with those for the appropriate specimen in table 2.

Specimen conditioning, including any post-moulding treatment, shall be carried out at 23 °C ± 2 °C and (50 ± 5) % R.H. for a minimum length of time of 88 h, except where special conditioning is required as specified by the appropriate material standard.

---

2) To be published.

## Table 1 — Moulding parameters

| Moulding-material type | Moulding method and standard (where applicable) | Moulding parameters |
|---|---|---|
| Thermoplastic | Injection<br><br>ISO 294 | Melt temperature<br>Mould temperature<br>Average melt velocity<br>Hold pressure[1] |
| | Compression<br><br>ISO 293 | Moulding temperature<br>Moulding time<br>Cooling rate<br>Demoulding temperature |
| Thermosetting | Injection<br><br>ISO 10724 | Temperature at nozzle<br>Mould temperature<br>Average injection velocity<br>Hold pressure[1]<br>Post-cure temperature<br>Post-cure time |
| | Compression<br><br>ISO 295 | Mould temperature<br>Dwell time<br>Post-cure temperature<br>Post-cure time |

1) The hold pressure mainly influences the shrinkage of the specimen, and is recorded principally for this reason.

## 5 Test requirements

The test methods, test conditions and units specified in table 2 shall be used in determining data. All tests shall be conducted at 23 °C ± 2 °C and (50 ± 5) % relative humidity (see ISO 291) unless otherwise stated in table 2 or in the International Standard appropriate to the material.

## 6 Presentation of results

The presentation of data shall be as shown in table 2, and the data shall be preceded by information that identifies the material together with the information required by clause 4 where appropriate.

## Table 2 — Test conditions and format for the presentation of single-point data (see note 1)

| | Property | Standard | Specimen type (dimensions in mm) | Value | Unit | Test conditions and supplementary instructions |
|---|---|---|---|---|---|---|
| **1** | **Rheological properties** | | | | | |
| 1.1 | Melt mass-flow rate | ISO 1133 | Moulding compound | | g/10 min | Use and record test conditions for temperature and load specified in the appropriate material standard |
| 1.2 | Melt volume-flow rate (see note 2) | | | | cm$^3$/10 min | |
| 1.3 | Moulding shrinkage | ISO 2577 (see note 3) | (see note 3) | | % | |
| **2** | **Mechanical properties** | | | | | |
| 2.1 | Tensile modulus | ISO 527-1 and 527-2 | ISO 3167 (see note 4) | | MPa | See note 5 |
| 2.2 | Yield stress | | | | | |
| 2.3 | Yield strain | | | | % | Ductile failure: Test speed 50 mm/min (see notes 5 and 6) |
| 2.4 | Nominal strain at break | | | | | |
| 2.5 | Stress at 50 % strain | | | | MPa | |
| 2.6 | Stress at break | | | | | |
| 2.7 | Strain at break | | | | % | Brittle failure: Test speed 5 mm/min (see notes 5 and 7) |
| 2.8 | Tensile creep modulus | ISO 899-1 | | | MPa | At 1 h / Strain < 0,5 % |
| 2.9 | | | | | | At 1 000 h |
| 2.10 | Flexural modulus | ISO 178 | 80 × 10 × 4 | | MPa | Test speed 2 mm/min (see also note 8) |
| 2.11 | Flexural strength | | | | | |
| 2.12 | Charpy impact strength | ISO 179 | 80 × 10 × 4 | | kJ/m$^2$ | Edgewise impact |
| 2.13 | Charpy notched impact strength | | Machined V-notch, $r = 0,25$ | | | |
| 2.14 | Tensile-impact strength | ISO 8256 | 80 × 10 × 4 Machined double V-notch, $r = 1$ | | | Record if fracture cannot be obtained with notched Charpy test |
| **3** | **Thermal properties** | | | | | |
| 3.1 | Melting temperature | ISO 3146 | Moulding compound | | °C | Method C (DSC or DTA) Use 10 °C/min |
| 3.2 | Glass transition temperature | IEC 1006 | | | | Method A (DSC or DTA) Use 10 °C/min |
| 3.3 | Temperature of deflection under load | ISO 75-1 and 75-2 | 110 × 10 × 4 or 80 × 10 × 4 (see note 9) | | 1,8 | Maximum surface stress (MPa) Use 1,8 MPa and one other value |
| 3.4 | | | | | 0,45 | |
| 3.5 | | | | | 8 | |
| 3.6 | | ISO 75-3 | Variable (see ISO 75-3) | | °C | |
| 3.7 | Vicat softening temperature (see note 10) | ISO 306 | 10 × 10 × 4 (see note 11) | | | Heating rate 50 °C/h Load 50 N |

| | Property | Standard | Specimen type (dimensions in mm) | Value | Unit | Test conditions and supplementary instructions | |
|---|---|---|---|---|---|---|---|
| 3.8 | Coefficient of linear thermal expansion | Thermo-mechanical analysis (see note 12) | Prepared from ISO 3167 (see note 11) | | °C−1 | Parallel | Record the secant value over the temperature range 23 °C to 55 °C |
| 3.9 | | | | | | Normal | |
| 3.10 | Flammability | ISO 1210 | 125 × 13 × 3 | | mm/min | Method A: linear burning rate of horizontal specimens | |
| 3.11 | | | Additional thickness (see note 13) | | | | |
| 3.12 | | | 125 × 13 × 3 | a) | s | Method B: a) after-flame and b) after-glow times of vertical specimens | |
| 3.13 | | | | b) | | | |
| 3.14 | | | Additional thickness (see note 13) | a) | | | |
| 3.15 | | | | b) | | | |
| 3.16 | Ignitability | ISO 4589 | 80 × 10 × 4 | | % | Use procedure A: top surface ignition | |
| **4** | **Electrical properties** | | | | | | |
| 4.1 | Relative permittivity | IEC 250 | ⩾ 80 × ⩾ 80 × 1 (see note 14) | | 100 Hz | Compensate for electrode edge effects | |
| 4.2 | | | | | 1 MHz | | |
| 4.3 | Dissipation factor | | | | 100 Hz | | |
| 4.4 | | | | | 1 MHz | | |
| 4.5 | Volume resistivity | IEC 93 | | | Ω·m | Voltage 100 V | |
| 4.6 | Surface resistivity | | | | Ω | | |
| 4.7 | Electric strength | IEC 243-1 | ⩾ 80 × ⩾ 80 × 1 (see note 15) | | kV/mm | Use 25 mm/75 mm coaxial-cylinder electrode configuration. Immersion in transformer oil in accordance with IEC 296. Use 20 s step-by-step test | |
| 4.8 | | | ⩾ 80 × ⩾ 80 × 3 (see note 15) | | | | |
| 4.9 | Comparative tracking index | IEC 112 | ⩾ 15 × 15 × 4 (see note 16) | | | Use solution A | |
| **5** | **Other properties** | | | | | | |
| 5.1 | Water absorption | ISO 62 | 50 square or diameter × 3 | | % | 24 h immersion in water at 23 °C | |
| 5.2 | | | Thickness ⩽ 1 | | | Saturation value in water at 23 °C | |
| 5.3 | | | | | | Saturation value at 23 °C , 50 % R.H. | |
| 5.4 | Density | ISO 1183 | Use part of the centre of the multipurpose test specimen | | kg/m³ | See note 17 | |

## Notes to table 2

1 Use of the parameters in table 2 is essentially for the comparison of data, and certain of the instructions listed may not be appropriate for all polymers.

2 The ratio of melt mass-flow rate to melt volume-flow rate gives an estimate of the melt density.

3 There is no International Standard for measuring the moulding shrinkage of thermoplastics. ISO 2577 describes how to measure the reaction shrinkage of thermosetting materials.

4 ISO 3167 describes two types of specimen for tensile tests. The type A specimen has a lower value for the radius of the shoulders of 20 mm to 25 mm which thereby enables a central region to be obtained of length at least 80 mm. The standard ISO bar having

dimensions 80 mm × 10 mm × 4 mm can thus be cut from the central region of this type of test specimen which is therefore recommended for directly moulded specimens. The type B specimen has a larger shoulder radius of > 60 mm and is recommended for machined specimens.

5   The data to be recorded for the properties in 2.1 to 2.7 are intended to give a fair impression of the nature of the stress-strain curve to failure.

6   If the polymer shows ductile failure, i.e. yielding or a breaking strain beyond 10 %, the test speed shall be 50 mm/min and the values for yield stress and strain and the nominal strain at break shall be recorded. If rupture occurs above 50 % nominal strain, record either the measured value of the nominal strain at break or simply record > 50. If no yielding is observed up to 50 % strain, record the stress at 50 %.

The determination of the nominal strain is based upon the initial and final grip separations instead of extensometer measurements.

7   If the polymer shows brittle failure, i.e. rupture without yielding and with a strain at break of less than 10 %, the test speed shall be 5 mm/min and the values for stress and strain at break shall be recorded.

8   The flexural test generates a non-uniform stress across the cross-section of the specimen. This will lead to ambiguous results if strains are developed in the specimen which are outside the region of linear deformation. For this reason, the tensile test is favoured. With injection-moulded or reinforced materials, an inhomogeneous structure across the cross-section will also contribute to a difference between results from flexural tests and results from tensile tests. In this case, data from both tests will give information on the extent of inhomogeneity.

9   ISO 75-2 includes the use of the standard 80 mm × 10 mm × 4 mm specimen flatwise-loaded using a support span of 64 mm. (Some comparative tests using this method with a 110-mm-long specimen tested edgewise and an 80-mm-long specimen tested flatwise have indicated no significant difference in the results obtained for the temperature of deflection.)

10   This property is less suitable for thermosets and semicrystalline materials.

11   Use test specimens from the middle of the central region of the multipurpose test specimen where possible.

12   No suitable International Standard exists for determining the linear thermal expansion of plastics but a range of commercial instruments is available for this purpose. For further guidance on sources of error and procedures for achieving high accuracy, reference should be made to scientific publications such as reference [1] in annex A.

Record the thermal expansion coefficient as

$$\frac{L_{55} - L_{23}}{32 \times L_{23}}$$

where $L_{23}$ and $L_{55}$ are the lengths of the test specimen at 23 °C and 55 °C, respectively.

13   A second test specimen of thickness less than 3 mm may be used in addition. Record the thickness with the values measured.

14   A greater thickness may be used for those materials, such as certain thermosetting resins, that cannot be moulded reliably with a thickness of 1 mm. Record the thickness used for the measurement.

15   The specimen shall be sufficiently wide to prevent discharge along the surface.

16   Use test specimens from the shoulder of the multipurpose test specimen.

17   The four methods specified in ISO 1183 are regarded as equivalent for the purposes of this International Standard.

# Annex A
(informative)

# Bibliography

[1] *Fundamentals and applications of thermomechanical methods including dilatometry,* Royal Society of Chemistry Analytical Division, Thermal Methods Group Conference (November 1980), published October 1981.

# INTERNATIONAL STANDARD

ISO
10724

First edition
1994-08-01

# Plastics — Thermosetting moulding materials — Injection moulding of multipurpose test specimens

*Plastiques — Matières à mouler thermodurcissables — Moulage par injection d'éprouvettes à usages multiples*

Reference number
ISO 10724:1994(E)

# Foreword

ISO (the International Organization for Standardization) is a worldwide federation of national standards bodies (ISO member bodies). The work of preparing International Standards is normally carried out through ISO technical committees. Each member body interested in a subject for which a technical committee has been established has the right to be represented on that committee. International organizations, governmental and non-governmental, in liaison with ISO, also take part in the work. ISO collaborates closely with the International Electrotechnical Commission (IEC) on all matters of electrotechnical standardization.

Draft International Standards adopted by the technical committees are circulated to the member bodies for voting. Publication as an International Standard requires approval by at least 75 % of the member bodies casting a vote.

International Standard ISO 10724 was prepared by Technical Committee ISO/TC 61, *Plastics*, Subcommittee SC 12, *Thermosetting materials*.

Annexes A, B, C, D and E of this International Standard are for information only.

International Organization for Standardization
Case Postale 56 • CH-1211 Genève 20 • Switzerland

Printed in Switzerland

# Introduction

In the injection-moulding process, there are many factors that influence the properties of mouldings and the numerical values of test results. The thermal and mechanical properties of test specimens are strongly dependent on the conditions of the moulding process. Precise definition of the essential parameters of the moulding process is the basic requirement for ensuring reproducible operating conditions.

It is important to define the moulding conditions and to take into consideration the correlation between the properties of thermosetting materials and related parameters in the moulding process.

The specimens processed by injection moulding can have differences in orientation of reinforcement fibres and curing. These phenomena can cause scatter in the numerical values of properties. Residual stress as a result of the curing and cooling process can produce the same effect. Due to the crosslinking of thermosets, molecular orientation is of less influence on mechanical properties than it is for thermoplastics.

# Plastics — Thermosetting moulding materials — Injection moulding of multipurpose test specimens

## 1 Scope

This International Standard describes the general principles to be followed for injection moulding test specimens of thermosetting materials. It provides a basis for establishing reproducible moulding conditions. Its purpose is to promote uniformity in describing the various essential parameters of the moulding operation, and also to establish uniform practice in reporting test conditions. The exact conditions required to prepare test specimens in a defined and reproducible way will vary for each material, mould and injection machine. These conditions must be agreed upon between the interested parties, when they do not form a part of the relevant International Standard.

## 2 Normative reference

The following standard contains provisions which, through reference in this text, constitute provisions of this International Standard. At the time of publication, the edition indicated was valid. All standards are subject to revision, and parties to agreements based on this International Standard are encouraged to investigate the possibility of applying the most recent edition of the standard indicated below. Members of IEC and ISO maintain registers of currently valid International Standards.

ISO 3167:1993, *Plastics — Multipurpose test specimens.*

## 3 Definitions

For the purposes of this International Standard, the following definitions apply.

**3.1 mould-surface temperature:** The average temperature of the mould-cavity surface, measured at several points on each half of the mould cavity after the system has attained thermal equilibrium and immediately after opening the mould.

**3.2 temperature of the plasticized material:** The temperature of the plasticized material as measured in a free shot.

**3.3 injection pressure:** The maximum pressure applied to the plastic material in front of the screw during the injection time (see figure 1).

**3.4 hold pressure:** The pressure applied to the plastic material in front of the screw during the hold time.

**3.5 moulding cycle:** The complete sequence of operations in the moulding process required for the production of one set of test specimens (see figure 1).

A moulding cycle consists of the following time intervals:

**3.5.1   injection time:** The time interval from the beginning of screw forward movement until the mould cavity is filled.

**3.5.2   hold time; dwell time:** The time interval from the time the mould cavity is filled until the screw begins to move back.

**3.5.3   cure time:** The time interval from the end of injection until the mould starts to open.

**3.5.4   mould open time:** The time interval beginning at the instant the mould starts to open, including the removal of the moulding, until the mould is closed again.

**3.6   injection speed:** The rate of forward travel, in millimetres per second, of the plasticizing screw during the injection stroke. This parameter may be varied to control the melt velocity (see 3.7).

**3.7   melt velocity:** The velocity of the flow front of the plasticized material in the 4 mm × 10 mm central cross-sectional area of the mould cavity (see also annex E).

$$v_f = \frac{\pi d^2 v_s}{4nS}$$

where

$v_f$   is the melt velocity, in millimetres per second;

$d$   is the screw diameter, in millimetres;

$v_s$   is the injection speed, in millimetres per second (see 3.6);

$n$   is the number of mould cavities;

$S$   is the cross-sectional area, in square millimetres, of the centre of the test specimen.

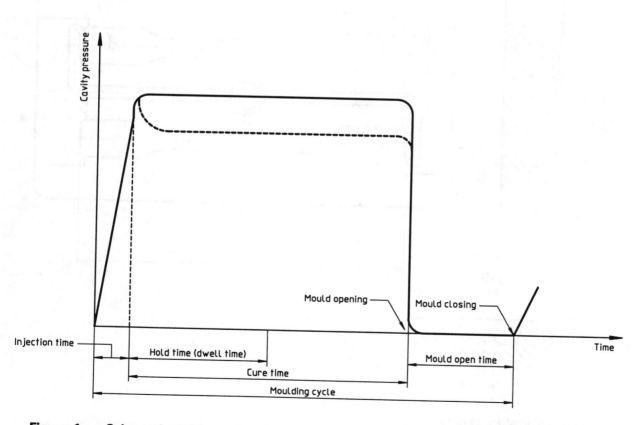

**Figure 1 — Schematic diagram of cavity pressure vs. time during an injection-moulding cycle**

## 4 Apparatus

### 4.1 Injection-mould design

Moulds shall comprise two identical cavities. The flow-path geometry shall be identical for each cavity and the cavities shall be positioned symmetrically in the mould to ensure that the two test specimens from any one shot are equivalent in their properties.

The mould cavities shall be designed to provide moulded specimens in accordance with ISO 3167, type A, with a shoulder radius of 20 mm (see figure 2 and annex A).

The main constructional details of the parts of the ISO injection mould are given below:

a) The sprue diameter on the nozzle side shall be at least 5 mm.

b) The width and height or diameter of the runner system shall each be at least 5 mm.

c) The specimen shall be end-gated (see figure 2).

d) The height and width of the gate shall be at least two-thirds of the height and width of the end of the specimen.

e) The gate shall be as short as possible, and in any case shall not exceed 3 mm.

Dimensions in millimetres

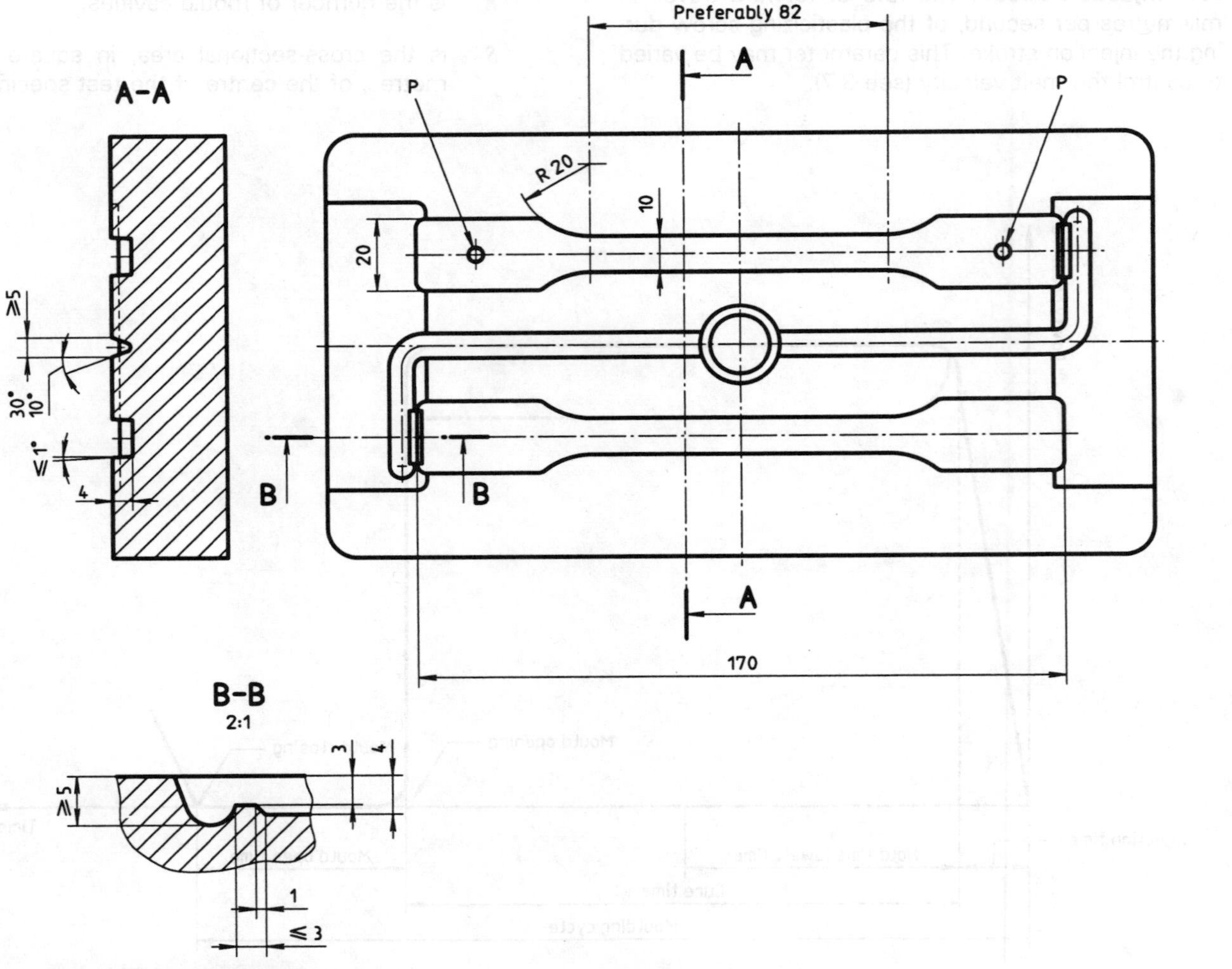


P = pressure transmitters (optional)

**Figure 2 — ISO injection mould for multipurpose test specimens as specified in ISO 3167, type A (preferred type)** (see also annex A)

f) The draft angle of the runner shall be at least 10° and shall not exceed 30° to guarantee ease of release. The test-specimen cavity shall have a maximum draft angle of 1°.

g) Ejector pins, when used, shall be located outside the test area of the test specimen.

   NOTE 1   For low-profile compounds with extremely low mould shrinkage, it is necessary to use flat-shaped ejectors, the size of the cavity itself, instead of ejector pins.

h) The heating system for the mould plates shall be designed so that differences in temperature over the surface of each half of the mould are less than 3 °C.

i) It is recommended that the mould cavities, and hence the multipurpose test specimens, be marked as described in informative annex B.

j) Interchangeable mould plates and gate inserts are recommended to enable the mould to be used for other than the production of multipurpose test specimens in accordance with this International Standard. An example is given in informative annex C.

## 4.2  Injection-moulding machine

Reciprocating-screw injection-moulding machines equipped with the necessary devices for control and maintainance of conditions shall be used.

The shot volume shall not be less than 20 % of the total free-barrel volume.

### 4.2.1  Control system

The control system of the machine shall be set to maintain the operating conditions within the following limits:

| | |
|---|---|
| injection pressure | ± 3 % |
| hold pressure | ± 5 % |
| temperature of the plasticized material | ± 3 °C |
| mould-surface temperature | ± 3 °C |
| hold time | ± 5 % |
| injection time | ± 0,05 s |
| shot mass | ± 1 % |
| injection speed | ± 1 % |

### 4.2.2  Clamp force

The clamp force of the machine shall be sufficient to prevent excessive flashing under all operating conditions.

### 4.2.3  Calibrated pyrometer

For measuring the temperature of the plasticized material, a needle-type pyrometer accurate to ± 1 °C shall be used.

For measuring the surface temperature of the mould cavity, a calibrated surface pyrometer accurate to ± 1 °C shall be used.

## 5  Procedure

### 5.1  Conditioning the material

Prior to moulding of test specimens, the granules or pellets shall be conditioned if required by the relevant International Standard for the thermosetting material or if recommended by the manufacturer (e.g. pre-drying).

### 5.2  Injection moulding

The machine shall be set to the conditions specified in the relevant material International Standard or agreed upon by the interested parties.

By adjustment of moulding-process parameters whose values are not specified, the conditions shall be modified until the mouldings are free of sinkmarks, voids, etc., and have minimum flash.

NOTE 2   For many thermosets, an average injection velocity of 150 mm/s ± 50 mm/s will be found suitable.

The mouldings shall be discarded until the machine has reached equilibrium. At that time, operating conditions shall be recorded accurately.

During the moulding process, steady-state conditions shall be maintained by suitable means. Preferably, the hold pressure shall be maintained until the gate section is cured.

After any change of material, the machine shall be thoroughly cleaned.

### 5.3  Measurement of the temperature of the plasticized material

The temperature of the plasticized material shall be measured after thermal equilibrium has been attained.

299

This is preferably done by injecting a free shot into a non-metallic container of a suitable size. The needle of a rapid-responding preheated pyrometer is inserted immediately into the centre of the plastic shot and gently moved until the reading of the pyrometer has reached its maximum.

The injection conditions for the free shot shall correspond to those used for moulding the specimens, including the allowance of identical cycle time between the ejection of free shots.

The temperature of the plasticized material may be measured by means of any other method using a suitable temperature sensor, provided the values obtained are the same as those obtained using the free-shot method. The sensor used shall have low heat diversion and low thermal inertia. It shall be mounted in contact with the plasticized material at a suitable place, e.g. in the nozzle.

## 5.4 Post-moulding treatment of test specimens

The test specimens removed from the mould shall all be allowed to cool gradually and uniformly to room temperature in order to avoid any differences between test specimens.

NOTE 3   Experience has shown that at least part of this "cooling time" can have a significant influence on the degree of curing.

## 5.5 Recording technical data

The following data shall be recorded:

a) the moulding-cycle time;

b) the mould-surface temperature;

c) the temperature of the plasticized material;

d) the injection time;

e) the injection pressure;

f) the hold pressure;

g) the shot mass;

h) room temperature.

## 6 Test report

The test report shall include the following information:

a) a reference to this International Standard;

b) the date, time and place of moulding;

c) the material used (type, designation and manufacturer);

d) details of the conditioning of the material, if required;

e) the injection-moulding machine used (manufacturer, maximum shot volume, clamp force);

f) the moulding conditions:

1) the temperature of the plasticized material,

2) the mould-surface temperature,

3) the calculated injection velocity in the cavity,

4) the injection pressure,

5) the hold pressure,

6) the shot mass (moulding, sprue, runner),

7) the injection time,

8) the cure time,

9) the total cycle time;

g) any other relevant details (number of mouldings made, number of mouldings discarded before selection of test specimens, etc.), if applicable.

# Annex A
## (informative)

## Mould-cavity size

Because of mould shrinkage, it is recommended that the length of the narrow, parallel-sided portion of the test specimens be 82 mm. This size meets the upper limit of the dimensional tolerance given in ISO 3167 and enables the 80-mm-long test specimen (with a cross-section of 4 mm × 10 mm) to be cut without unnecessary difficulty. For the same reason, the permitted upper tolerance limit of + 0,2 mm should be used for the width and thickness within the narrow, parallel-sided portion of the cavities (see figure A.1).

To prevent the formation of cracks at the sharp edges, particularly within the narrow, parallel-sided portion of the test specimens (e.g. as a consequence of long-term conditioning at elevated temperatures), the edges may be provided with a fillet (see figure A.1). A fillet radius of 0,5 mm has proved suitable.

Dimensions in millimetres

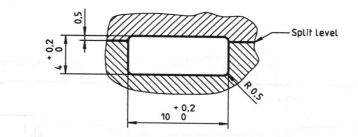

Figure A.1 — Shape and size of the narrow, parallel-sided portion of the cavity

# Annex B
## (informative)

## Marking the test specimens

The purpose of marking is to permit the determination of the original position of the two specimens in the mould, even if the broad tabs at the ends of the specimen are cut off (e.g. to obtain a bar specimen with dimensions of 80 mm × 10 mm × 4 mm).

The numbers used, and their positions in the mould, should be as follows:

— the mirror images of the numbers "1" and "2" should be used;

— the numbers should be legible, upright and aligned with the direction of flow of the material;

— the numbers should be outside the usual test-specimen support spans for flexural-loading tests, but within the length of the 80-mm-long bar specimen;

— the numbers should be just visible (i.e. not very deeply "engraved") to avoid stress concentrations, etc.;

— the numbers should be located at the gate end of the cavity.

Dimensions in millimetres

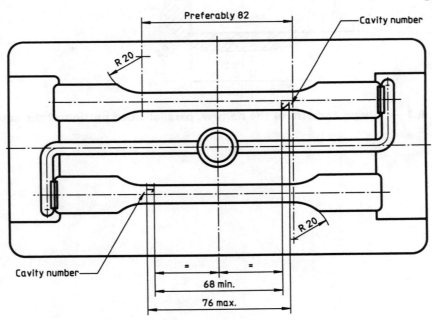

**Figure B.1 — Positions of the cavity numbers**

302

# Annex C

## (informative)

# Examples of possible variations in mould configuration

The layout of the mould may be changed by means of gate inserts (a-a', b-c, b-b') as shown in figure C.1.

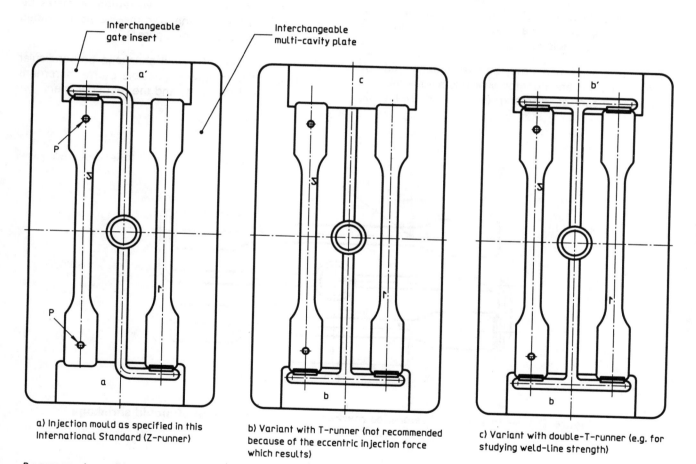

a) Injection mould as specified in this International Standard (Z-runner)

b) Variant with T-runner (not recommended because of the eccentric injection force which results)

c) Variant with double-T-runner (e.g. for studying weld-line strength)

P = pressure transmitter

**Figure C.1 — Variations in the gate position**

303

# Annex D
## (informative)

# Reference marks for the determination of shrinkage

The determination of mould shrinkage parallel to the direction of direct measurement of the specimens is not possible because of the position and size of the gate.

For the contactless measurement of a reference length (e.g. by means of a longitudinal or transverse optical comparator), the test specimens must be marked with suitable reference marks.

These reference marks may be indentations made by means of a Vickers indentor, which is a square-based pyramid made of diamond (see ISO 6507-1:1982,

*Metallic materials — Hardness test — Vickers test — Part 1: HV 5 to HV 100*). The diagonals of the indentations (approximately 0,3 mm long) should be oriented parallel to the long axis of the multipurpose test specimen and perpendicular to it, respectively.

It is recommended that the two reference marks be $\leq 80$ mm apart within the 10-mm-wide parallel-sided portion of each cavity (see figure D.1).

The determination of mould shrinkage perpendicular to the direction of flow is possible by measurement of the width of the cavity and the test specimen.

Dimensions in millimetres

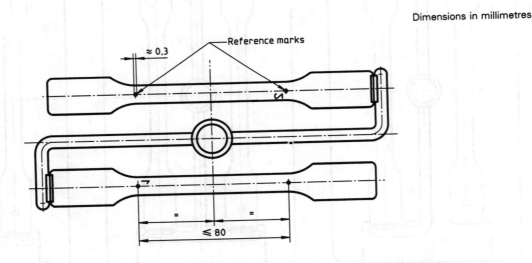

**Figure D.1 — Positions of reference marks for the determination of mould shrinkage**

# Annex E

## (informative)

## Estimation of melt velocity

An estimate of the melt velocity within the 10-mm-wide parallel-sided portion of the cavity is possible if the mould (e.g. one cavity) is equipped with two pressure transducers P1 and P2 (the recommended positions are shown in figure E.1).

This estimate is independent of the injection-moulding machine (unlike the calculation made in 3.7).

Assuming that the cavities are filled at a low enough injection speed to avoid jetting, thus forming an evenly advancing melt front, the average melt velocity $v_F$ is given, in millimetres per second, by the equation

$$v_F = \frac{2s_1 + s_2}{t}$$

where

$s_1$   is the approximate length, in millimetres, of the initial "wide" part of the cavity (cross-section 20 mm × 4 mm), measured from pressure transducer P1 (this distance is multiplied by 2 to allow for the fact that this part of the cavity is twice as wide as the narrow, parallel-sided part);

$s_2$   is the approximate length, in millimetres, of the "narrow" part of the cavity (cross-section 10 mm × 4 mm) (assuming jetting takes place at the far end of the narrow part of the cavity, $s_2$ also includes the length of the subsequent wide part);

$t$   is the flow time, in seconds, of the plasticized material between the pressure transducers.

NOTE 4    For the multipurpose test specimen in accordance with ISO 3167:1993, type A (with $s_1 = 21$ mm and $s_2 = 119$ mm), the estimated average melt velocity is given by $v_F = 161/t$ millimetres per second.

Dimensions in millimetres

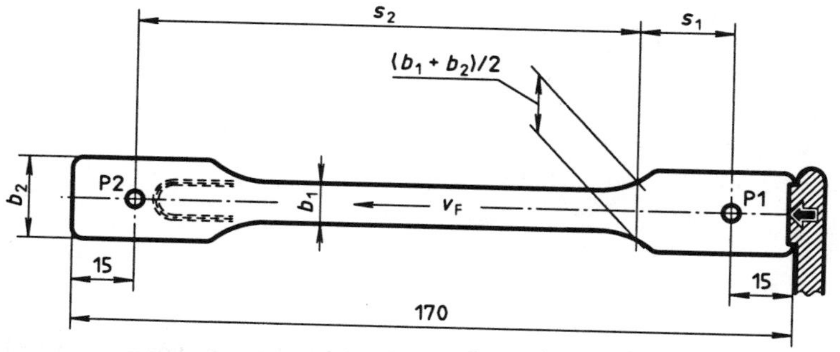

**Figure E.1 — Recommended positions of pressure transducers**

305

# NORME
# INTERNATIONALE

# INTERNATIONAL
# STANDARD

**CEI
IEC
93**

Deuxième édition
Second edition
1980

## Méthodes pour la mesure de la résistivité transversale et de la résistivité superficielle des matériaux isolants électriques solides

## Methods of test for volume resistivity and surface resistivity of solid electrical insulating materials

Bureau Central de la Commission Electrotechnique Internationale  3, rue de Varembé  Genève, Suisse

Commission Electrotechnique Internationale
International Electrotechnical Commission
Международная Электротехническая Комиссия

CODE PRIX
PRICE CODE

R

Pour prix, voir catalogue en vigueur
For price, see current catalogue

# CONTENTS

---

307

# INTERNATIONAL ELECTROTECHNICAL COMMISSION

---

# METHODS OF TEST FOR VOLUME RESISTIVITY
# AND SURFACE RESISTIVITY OF SOLID ELECTRICAL
# INSULATING MATERIALS

---

## FOREWORD

1) The formal decisions or agreements of the IEC on technical matters, prepared by Technical Committees on which all the National Committees having a special interest therein are represented, express, as nearly as possible, an international consensus of opinion on the subjects dealt with.

2) They have the form of recommendations for international use and they are accepted by the National Committees in that sense.

3) In order to promote international unification, the IEC expresses the wish that all National Committees should adopt the text of the IEC recommendation for their national rules in so far as national conditions will permit. Any divergence between the IEC recommendation and the corresponding national rules should, as far as possible, be clearly indicated in the latter.

## PREFACE

This standard has been prepared by Sub-Committee 15A: Short-time Tests, of IEC Technical Committee No. 15: Insulating Materials.

It forms the second edition of IEC Publication 93.

A first draft was discussed at the meeting held in Toronto in 1976. As a result of this meeting, a draft, Document 15A(Central Office)35, was submitted to the National Committees for approval under the Six Months' Rule in November 1977.

Amendments, Document 15A(Central Office)39, were submitted to the National Committees for approval under the Two Months' Procedure in October 1979.

The National Committees of the following countries voted explicitly in favour of publication:

| | |
|---|---|
| Austria | Italy |
| Belgium | Korea (Republic of) |
| Brazil | New Zealand |
| Bulgaria | Norway |
| Canada | Poland |
| China | Spain |
| Czechoslovakia | Sweden |
| Denmark | Switzerland |
| Egypt | United Kingdom |
| France | United States of America |
| Germany | Yugoslavia |
| Ireland | |

*Other IEC publications quoted in this standard:*

Publications Nos. 167: Methods of Test for the Determination of the Insulation Resistance of Solid Insulating Materials.

212: Standard Conditions for Use Prior to and during the Testing of Solid Electrical Insulating Materials.

260: Test Enclosures of Non-injection Type for Constant Relative Humidity.

---

# METHODS OF TEST FOR VOLUME RESISTIVITY AND SURFACE RESISTIVITY OF SOLID ELECTRICAL INSULATING MATERIALS

---

## 1. Scope

These methods of test cover procedures for the determination of volume and surface resistance and calculations for the determination of volume and surface resistivity of solid electrical insulating materials.

Both volume resistance and surface resistance tests are affected by the following factors: the magnitude and time of voltage application, the nature and geometry of the electrodes, and the temperature and humidity of the ambient atmosphere and of the specimens during conditioning and measurement. Recommendations are made for these factors.

## 2. Definitions

### 2.1 *Volume resistance*

The quotient of a direct voltage applied between two electrodes placed on two faces (opposite) of a specimen, and the steady-state current between the electrodes, excluding current along the surface, and neglecting possible polarization phenomena at the electrodes.

*Note.* — Unless otherwise specified, the volume resistance is determined after 1 min of electrification.

### 2.2 *Volume resistivity*

The quotient of a d.c. electric field strength and the steady-state current density within an insulating material. In practice it is taken as the volume resistance reduced to a cubical unit volume.

*Note.* — The SI unit of volume resistivity is the ohm metre. In practice the unit ohm centimetre is also used.

### 2.3 *Surface resistance*

The quotient of a direct voltage applied between two electrodes on a surface of a specimen, and the current between the electrodes at a given time of electrification, neglecting possible polarization phenomena at the electrodes.

*Notes 1.* — Unless otherwise specified, the surface resistance is determined after 1 min of electrification.

*2.* — The current generally passes mainly through a surface layer of the specimen and any associated moisture and surface contaminant, but it also includes a component through the volume of the specimen.

## 2.4 *Surface resistivity*

The quotient of a d.c. electric field strength, and the linear current density in a surface layer of an insulating material. In practice it is taken as the surface resistance reduced to a square area. The size of the square is immaterial.

*Note.* — The SI unit of surface resistivity is the ohm. In practice this is sometimes referred to as "ohms per square".

## 2.5 *Electrodes*

Measuring electrodes are conductors of defined shape, size and configuration in contact with the specimen being measured.

*General note. — Insulation resistance* is the quotient of a direct voltage applied between two electrodes in contact with a specimen and the total current between the electrodes. The insulation resistance depends on both volume and surface resistivity of the specimen (see I E C Publication 167: Methods of Test for the Determination of the Insulation Resistance of Solid Insulating Materials).

## 3. Significance

3.1    Insulating materials are used in general to isolate components of an electrical system from each other and from earth; solid insulating materials may also provide mechanical support. For these purposes it is generally desirable to have the insulation resistance as high as possible, consistent with acceptable mechanical, chemical and heat-resisting properties. Surface resistance changes very rapidly with humidity, while volume resistance changes only slowly, although the final change may be greater.

3.2    Volume resistivity can be used as an aid in the choice of an insulating material for a specific application. The change of resistivity with temperature and humidity may be great and must be known when designing for operating conditions. Volume resistivity measurements are often used in checking the uniformity of an insulating material, either with regard to processing or to detect conductive impurities that affect the quality of the material and that may not be readily detectable by other means.

3.3    When a direct voltage is applied between electrodes in contact with a specimen, the current through it decreases asymptotically towards a steady-state value. The decrease of current with time may be due to dielectric polarization and the sweep of mobile ions to the electrodes. For materials having volume resistivities less than about $10^{10}$ $\Omega \cdot$ m ($10^{12}$ $\Omega \cdot$ cm), the steady-state is in general reached within 1 min, and the resistance is then determined after this time of electrification. For materials of higher volume resistivity the current may continue to decrease for several minutes, hours, days, or even weeks. For such materials, therefore, longer electrification times are used, and, if relevant, the material is characterized by the time dependence of the volume resistivity.

3.4 Surface resistance or surface conductance cannot be measured accurately, only approximated, because more or less volume conductance is nearly always involved in the measurement. The measured value is largely a property of the contamination of the surface of the specimen at the time of measurement. However, the permittivity of the specimen influences the deposition of contaminants, and their conductive capabilities are affected by the surface characteristics of the specimen. Thus surface resistivity is not a material property in the usual sense, but can be considered to be related to material properties when contamination is involved.

Some materials, such as laminates, may have quite different resistivities in a surface layer and in the interior. It may therefore be of interest to measure the intrinsic property of a clean surface. Cleaning procedures aimed at producing consistent results should be fully specified bearing in mind the possible effect of solvents and other factors of the cleaning procedure on the surface characteristics.

The surface resistance, especially when high, often changes in an erratic manner, and in general depends strongly on the time of electrification; for measurements, 1 min of electrification is usually specified.

## 4. Power supply

A source of very steady direct voltage is required. This may be provided either by batteries or by a rectified and stabilized power supply. The degree of stability required is such that the change in current due to any change in voltage is negligible compared with the current to be measured.

Commonly specified test voltages to be applied to the complete specimen are 100 V, 250 V, 500 V, 1000 V, 2500 V, 5000 V, 10000 V and 15000 V. Of these the most frequently used are 100 V, 500 V and 1000 V.

In some cases, the specimen resistance depends upon the polarity of the applied voltage.

If the resistance is polarity dependent, this should be indicated. The geometric (arithmetic mean of the logarithmic exponents) mean of the two resistance values is taken as the result.

Since the specimen resistance may be voltage dependent, the test voltage should be stated.

## 5. Measuring methods and accuracy

### 5.1 Methods

The methods commonly in use for measuring high resistances are either direct methods or comparison methods.

The direct methods depend upon simultaneous measurement of the direct voltage applied to the unknown resistance and the current through it (voltmeter-ammeter method).

The comparison methods establish the ratio of the unknown resistance to the resistance of a known resistor, either in a bridge circuit, or by comparison of currents through the resistances at fixed voltage.

Examples illustrating the principles are described in Appendix A.

The voltmeter-ammeter method requires a reasonably accurate voltmeter, but the sensitivity and accuracy of the method depend mainly on the properties of the current measuring device, which may be a galvanometer, an electronic amplifier instrument, or an electrometer.

The bridge method requires only a sensitive current detector as null indicator, and the accuracy is mainly determined by the known bridge arm resistors, which are obtainable with high precision and stability over a wide range of resistance values.

The accuracy of the current comparison method depends on the accuracy of the known resistor, and on the stability and linearity of the current measuring device, including associated measuring resistors, etc., whereas the exact values of current are insignificant, as long as the voltage is constant.

Determination of volume resistivity in accordance with Sub-clause 10.1 using a galvanometer in the voltmeter-ammeter method is feasible for resistances up to about $10^{11}$ $\Omega$. For higher values, the use of a d.c. amplifier or electrometer is recommended.

In the bridge method, it is not possible to measure the current directly in the short-circuited specimen (see Sub-clause 10.1).

The methods utilizing current measuring devices permit automatic recording of the current to facilitate determination of the steady state (Sub-clause 10.1).

Special circuits and instruments for measuring high resistance are available. These may be used, provided that they are sufficiently accurate and stable, and that, where needed, they enable the specimen to be properly short-circuited, and the current measured before electrification.

## 5.2 *Accuracy*

The measuring device should be capable of determining the unknown resistance with an overall accuracy of at least $\pm 10\%$ for resistances below $10^{10}$ $\Omega$, and $\pm 20\%$ for higher values. See also Appendix A.

## 5.3 *Guarding*

The insulation of the measuring circuit is composed of materials which, at best, have properties comparable with those of the material under test. Errors in the measurement of the specimen may arise from:

a) stray current from spurious external voltages which are usually unknown in magnitude and often sporadic in character;
b) undue shunting of the specimen resistance, reference resistors, or the current measuring device by insulation, having resistance of unknown, and possibly variable magnitude.

An approximate correction of these difficulties may be obtained by making the insulation resistance of all parts of the circuit as high as possible under the conditions of use. This may lead to unwieldy apparatus which is still inadequate for measurement of insulation resistances higher than a few hundred megohms. A more satisfactory correction is obtained by using the technique of guarding.

Guarding depends on interposing, in all critical insulated parts, guard conductors which intercept all stray currents that might otherwise cause errors. The guard conductors are connected together, constituting the guard system and forming with the measuring terminals a three terminal network. When suitable connections are made, stray currents from spurious external voltages are shunted away from the measuring circuit by the guard system, the insulation resistance from either measuring terminal to the guard system shunts a circuit element which should be of very much lower resistance, and the specimen resistance constitutes the only direct path between the measuring terminals. By this technique the probability of error is considerably reduced. Figure 1, page 36, shows the basic connections for guarded electrodes used for volume resistance and surface resistance measurements.

Proper use of the guard system for the method involving current measurement is illustrated in Figures 5 and 7, pages 39 and 40, where the guard system is shown connected to the junction of the voltage source and current-measuring device. In Figure 6, page 40, for the Wheatstone bridge method, the guard system is shown connected to the junction of the two lower-valued resistance arms. In all cases, to be effective, guarding shall be complete, and shall include any control operated by the observer in making the measurement.

Electrolytic, contact, or thermal e.m.f.'s. existing between guard and guarded terminals can be compensated if they are small. Care must be taken that such e.m.f.'s do not introduce appreciable errors in the measurements.

Errors in current measurements may result from the fact that the current-measuring device is shunted by the resistance between the guarded terminal and the guard system. This resistance should be at least 10 and preferably 100 times that of the current-measuring device. In some bridge techniques, the guard and measuring terminal are brought to nearly the same potential but a standard resistor in the bridge is shunted by the resistance between the unguarded terminal and the guard system. This resistance should be at least 10 and preferably 100 times that of the reference resistor.

To ensure satisfactory operation of the equipment, a measurement should be made with the lead from the voltage source to the specimen disconnected. Under this condition, the equipment should indicate infinite resistance within its sensitivity. If suitable standards of known values are available, they may be used to test the operation of the equipment.

## 6. Test specimens

### 6.1 Volume resistivity

For the determination of volume resistivity the test specimen may have any practicable form that allows the use of a third electrode to guard against error from surface effect. For specimens that have negligible surface leakage, the guard may be omitted when measuring volume resistance, provided that it has been shown that its omission has negligible effect on the result.

The gap on the surface of the specimen between the guarded and guard electrodes should be of uniform width and as narrow as possible provided that the surface leakage does not cause error in the measurement. A gap of 1 mm is usually the smallest practicable.

Examples of electrode arrangements with three electrodes are shown in Figures 2 and 3, pages 37 and 38. In the measurement of volume resistance, electrode No. 1 is the guarded electrode, No. 2 is the guard electrode, and No. 3 is the unguarded electrode. The diameter $d_1$ (Figure 2), or length $l_1$ (Figure 3) of the guarded electrode should be at least ten times the specimen thickness $h$ and for practical reasons usually at least 25 mm. The diameter $d_4$ (or length $l_4$) of the unguarded electrode, and the outer diameter $d_3$ of the guard electrode (or length $l_3$ between the outer edges of the guard electrodes) should be equal to the inner diameter $d_2$ of the guard electrode (or length $l_2$ between the inner edges of the guard electrodes) plus at least twice the specimen thickness.

## 6.2  *Surface resistivity*

For the determination of surface resistivity the test specimen may have any practicable form that allows the use of a third electrode to guard against error from volume effects. The three electrode arrangements of Figures 2 and 3 are recommended. The resistance of the surface gap between electrodes Nos. 1 and 2 is measured directly by using electrode No. 1 as the guarded electrode, electrode No. 3 as the guard electrode and electrode No. 2 as the unguarded electrode. The resistance so measured includes the surface resistance between electrodes Nos. 1 and 2 and the volume resistance between the same two electrodes. With suitable dimensioning of the electrodes, however, the effect of the volume resistance can be made negligible for wide ranges of ambient conditions and material properties. This condition may be achieved for the arrangement of Figures 2 and 3 when the electrodes are dimensioned so that the surface gap width $g$ is at least twice the specimen thickness; 1 mm is normally the smallest practicable. The diameter $d_1$ (or length $l_1$) of the guarded electrode should be at least ten times the specimen thickness $h$, and for practical reasons usually at least 25 mm.

Alternatively, straight electrodes or other arrangements with suitable dimensions may be used.

*Note.* — Due to the influence of current through the interior of the test specimen the calculated value of surface resistivity may depend strongly on the specimen and electrode dimensions. For comparative determinations it is therefore recommended to use specimens of identical form with the electrode arrangement of Figure 2 with $d_1 = 50$ mm, $d_2 = 60$ mm, and $d_3 = 80$ mm.

## 7.  Electrode material

## 7.1  *General*

The electrodes for insulating materials should be of a material that is readily applied, allows intimate contact with the specimen surface and introduces no appreciable error because of electrode resistance or contamination of the specimen. The electrode material should be corrosion resistant under the conditions of the test. The following are typical electrode materials that may be used. The electrodes shall be used with suitable backing plates of the given form and dimensions.

It may be advantageous to use two different electrode materials or two methods of application to see if appreciable error is introduced.

## 7.2 Conductive silver paint

Certain types of commercially available, high-conductivity silver paints, either air-drying or low-temperature-baking varieties are sufficiently porous to permit diffusion of moisture through them and thereby allow the test specimens to be conditioned after application of the electrodes. This is a particularly useful feature in studying resistance-humidity effects as well as changes with temperature. However, before conductive paint is used as an electrode material, it should be established that the solvent in the paint does not affect the electrical properties of the specimen. Reasonably smooth edges of guard electrodes may be obtained with a fine-bristle brush. However, for circular electrodes, sharper edges may be obtained by the use of a compass for drawing the outline circles of the electrodes and filling in the enclosed areas by brush. Clamp-on masks may be used if the electrode paint is sprayed on.

## 7.3 Sprayed metal

Sprayed metal may be used if satisfactory adhesion to the test specimen can be obtained. Thin sprayed electrodes may have certain advantages in that they are ready for use as soon as applied. They may be sufficiently porous to allow the specimen to be conditioned, but this should be verified. Clamp-on masks may be used to produce a gap between the guarded electrode and the guard electrode.

## 7.4 Evaporated or sputtered metal

Evaporated or sputtered metal may be used under the same conditions as given in Sub-clause 7.3 where it can be shown that the material is not affected by ion bombardment or vacuum treatment.

## 7.5 Liquid electrodes

Liquid electrodes may be used and give satisfactory results. The liquid forming the upper electrode should be confined, for example, by stainless steel rings, each of which should have its lower rim reduced to a sharp edge by bevelling on the side away from the liquid. Figure 4, page 39, shows the electrode arrangement. Mercury is not recommended for continuous use or at elevated temperatures due to toxic effects.

## 7.6 Colloidal graphite

Colloidal graphite dispersed in water or other suitable medium, may be used under the same conditions as given in Sub-clause 7.2.

## 7.7 Conducting rubber

Conducting rubber may be used as an electrode material. It has the advantage that it can be applied and removed from the specimen quickly and easily. As the electrodes are applied only during the time of measurement they do not interfere with the conditioning of the specimen. The conducting rubber material shall be soft enough to ensure that effective contact to the specimen is obtained when a reasonable pressure, for example 2 kPa (0.2 N/cm$^2$), is applied.

### 7.8   *Metal foil*

Metal foil may be applied to specimen surfaces as electrodes for volume resistance measurement, but it is not suitable for surface resistance measurement. Lead, antimonial lead, aluminium, and tin foil are in common use. They are usually attached to the specimen by a minimum quantity of petrolatum, silicone grease, oil or other suitable material, as an adhesive. A pharmaceutically obtainable jelly of the following composition is suitable as a conductive adhesive:

| | |
|---|---|
| Anhydrous polyethylene glycol of molecular mass 600 | 800 parts by mass |
| Water | 200 parts by mass |
| Soft soap (pharmaceutical quality) | 1 part  by mass |
| Potassium chloride | 10 parts by mass |

The electrodes shall be applied under a smoothing pressure sufficient to eliminate all wrinkles and to work excess adhesive towards the edge of the foil where it can be wiped off with a cleansing tissue. Rubbing with a soft material such as the finger, has been used successfully. This technique can be used satisfactorily only on specimens that have very smooth surfaces. With care, the adhesive film can be reduced to 0.0025 mm or less.

## 8.   Specimen handling and mounting

It is important that stray currents between the electrodes or between the measuring electrodes and earth do not have a significant effect on the reading of the measuring instrument. Great care shall be used in applying the electrodes, in handling the specimens, and in mounting the specimens for measurement to avoid the possibility of creating stray paths that may adversely affect the result of the measurement.

When surface resistance is to be measured, the surface shall not be cleaned unless agreed or specified. That part of the surface which is to be measured shall not be touched by anything other than an untouched surface of another specimen of the same material.

## 9.   Conditioning

The conditioning that a specimen should receive depends upon the material being tested and should be specified in the material specification.

Recommended conditions are given in I EC Publication 212: Standard Conditions for Use Prior to and during the Testing of Solid Electrical Insulating Materials, and the relative humidities associated with various salt solutions are given in I EC Publication 260: Test Enclosures of Non-injection Type for Constant Relative Humidity. Mechanical vaporization systems may be used.

Both volume resistivity and surface resistivity are particularly sensitive to temperature changes. The change is exponential. It is therefore necessary to measure the volume resistance and surface resistance of the specimen while under specified conditions. Extended periods of conditioning are required to determine the effect of humidity on volume resistivity since the absorption of water into the body of the dielectric is a relatively slow process. Water absorption usually decreases volume resistance. Some specimens may require months to reach equilibrium.

## 10.   Test procedure

A number of specimens as prescribed in the relevant specification are prepared in accordance with Clauses 6, 7, 8 and 9.

The specimen and electrode dimensions, and the width of the surface gap $g$ are measured with an accuracy of $\pm 1\%$. For thin specimens, however, a different accuracy may be stated in the relevant specification, when appropriate.

For the determination of volume resistivity, the average thickness of each specimen is determined in accordance with the relevant specification, the measuring points being distributed uniformly over the area to be covered by the guarded measuring electrode.

*Note.* — For thin specimens, at least, the thickness should be measured before applying the electrodes.

In general, the resistance measurements should be made at the same humidity (except for conditioning by immersion in a liquid) and temperature as used during conditioning. In some cases however it may be sufficient to make the measurements within a specified time after stopping the conditioning.

### 10.1 *Volume resistance*

Before measurement the specimen shall be brought into a dielectrically stable condition. To obtain this, short-circuit the measuring electrodes Nos. 1 and 3 of the specimen, (Figure 1a) through the measuring device and observe the changing short-circuit current, while increasing the sensitivity of the current-measuring device as required. Continue until the short-circuit current attains a fairly constant value, small compared with the expected steady-state value of the current under electrification, or if relevant, the current at 100 min of electrification. As there is a possibility of a change in the direction of the short-circuit current, the short circuit should be maintained even if the current passes zero. The magnitude and direction of the short-circuit current $I_0$ are noted when it becomes essentially constant, which may require several hours.

Then apply the specified direct voltage and start a timing device simultaneously. Unless otherwise specified, make a measurement after each of the following times of electrification: 1 min, 2 min, 5 min, 10 min, 50 min, 100 min. If two successive measurements give the same results, the test may be terminated, and the value thus found used to calculate the volume resistance. The electrification time until the first of the identical measurements is recorded. If the steady state is not reached within 100 min, the volume resistance is reported as a function of electrification time.

For acceptance tests, the value after a fixed time of electrification, for example 1 min, is used, as specified in the relevant specification.

### 10.2 *Surface resistance*

Apply the specified direct voltage, and determine the resistance between the measuring electrodes on the specimen surface (Nos. 1 and 2, Figure 1b). The resistance shall be determined after 1 min of electrification, even though the current has not necessarily reached a steady-state value within this time.

## 11. Calculation

### 11.1 *Volume resistivity*

The volume resistivity shall be calculated from the following formula:

$$\rho = R_x \cdot A/h$$

where:

$p$ is the volume resistivity in ohms metres (ohms centimetres)

$R_x$ is the volume resistance in ohms measured as specified in Sub-clause 10.1

$A$ is the effective area of the guarded electrode in square metres (square centimetres)

$h$ is the average thickness of the specimen in metres (centimetres)

Formulae for calculating the effective area $A$ for some particular electrode arrangements are given in Appendix B.

For some materials with high resistivity, the short-circuit current $I_0$ prior to electrification (see Sub-clause 10.1) may not be negligible compared with the steady-state current $I_s$ during electrification. In such cases the volume resistance is determined as

$$R_x = U_x/(I_s \pm I_0)$$

where:

$R_x$ is the volume resistance in ohms

$U_x$ is the applied voltage in volts

$I_s$ is the steady-state current in amperes during electrification, or the values of current in amperes after 1 min, 10 min and 100 min if the current changes during electrification

$I_0$ is the short-circuit current in amperes prior to electrification

The minus sign is used when $I_0$ is in the same direction as $I_s$, otherwise the plus sign is used.

### 11.2 *Surface resistivity*

The surface resistivity shall be calculated from the following formula:

$$\sigma = R_x \cdot p/g$$

where:

$\sigma$ is the surface resistivity in ohms

$R_x$ is the surface resistance in ohms measured as specified in Sub-clause 10.2

$p$ is the effective perimeter in metres (centimetres) of the guarded electrode for the particular electrode arrangement employed

$g$ is the distance in metres (centimetres) between the electrodes

### 11.3 *Reproducibility*

Because of the variability of the resistance of a given specimen with test conditions, and because of non-uniformity of the material from specimen to specimen, determinations are usually not reproducible to closer than $\pm 10\%$ and are often even more widely divergent (a range of values of 10 to 1 may be obtained under apparently identical conditions).

In order that measurements on similar specimens are to be comparable, they must be made with approximately equal voltage gradients.

## 12. Report

The report shall include at least the following information:

*a)* description and identification of the material (name, grade, colour, manufacturer, etc.);

*b)* shape and dimensions of the specimen;

*c)* type, material and dimensions of the electrodes and guards;

*d)* conditioning of the specimen (cleaning, pre-drying, conditioning time, humidity and temperature, etc.);

*e)* test condition (specimen temperature, relative humidity);

*f)* method of measurement;

*g)* applied voltage;

*h)* volume resistivity (when relevant);

Notes 1. — When a fixed electrification time is specified, state this time, give the individual results, and report the central value as the volume resistivity.

2. — When measurements have been made after different electrification times report as follows:
Where specimens reach a steady state in the same electrification time, give the individual results, and report the central value as the volume resistivity. Where some specimens do not reach the steady state in this electrification time, report the number failing to do so and give the results on them separately. Where results are dependent on electrification time, report this relationship, e.g., in the form of a graph, or as the central value of the volume resistivity after 1 min, 10 min and 100 min.

*i)* surface resistivity (when relevant):

give the individual values after 1 min of electrification, and report the central value as the surface resistivity.

# APPENDIX A

## EXAMPLES OF MEASURING METHODS AND THEIR ACCURACY

### A1. Voltmeter-ammeter method

This direct method employs the circuit shown in Figure 5, page 39. The applied voltage is measured by the d.c. voltmeter. The current is measured by a current-measuring device, which may be a galvanometer (now seldom used), an electronic amplifier instrument, or an electrometer.

In general, while the specimen is being charged, the measuring device should be short-circuited to avoid damage to it during this period.

The galvanometer should have high current sensitivity and be provided with a universal shunt (also known as Ayrton shunt). The unknown resistance in ohms is calculated as

$$R_x = U/k\alpha$$

where:

$U$ is the applied voltage in volts

$k$ is the sensitivity of the shunted galvanometer in ampere per scale division

$\alpha$ is the deflection in scale divisions

Resistances up to about $10^{10}$ to $10^{11}$ $\Omega$ can be measured at 100 V with the required accuracy by means of a galvanometer.

An electronic amplifier instrument or an electrometer with high input resistance shunted by a resistor of known, high resistance $R_s$, may be used as the current-measuring device. The current is measured in terms of a voltage drop $U_s$ across $R_s$. The unknown resistance $R_x$, is calculated as

$$R_x = U \cdot R_s/U_s$$

where:

$U$ is the applied voltage (provided $R_s \ll R_x$)

A number of different resistors $R_s$ may be included in the instrument case, and the instrument is then often graduated directly in amperes or submultiples thereof.

Here also the maximum resistance that can be measured with the required accuracy depends on the properties of the current-measuring device. The error in $U_s$ is determined by the indicator error, the amplifier zero drift and gain stability. In adequately designed amplifiers and electrometers the instability in gain is negligible, and the zero drift can be held so low that it is of no concern in relation to the times involved in these measurements. The indicator error for high gain electronic voltmeters is typically $\pm 2\%$ to 5% of full-scale deflection, and resistors up to $10^{12}$ $\Omega$ with about the same degree of accuracy are feasible. If the voltage-measuring device has an input resistance greater than $10^{14}$ $\Omega$, and full-scale deflection at an input voltage of 10 mV, a current of $10^{-14}$ A can be measured with an accuracy of about $\pm 10\%$.

A resistance of $10^{16}$ $\Omega$ can thus be measured at 100 V with the required accuracy by means of a precision resistor with very high resistance and an electronic amplifier voltmeter or electrometer.

## A2.  Comparison methods

### A2.1  *Wheatstone bridge method*

The test specimen is connected in one arm of a Wheatstone bridge as shown in Figure 6, page 40. The three known arms shall be of as high a resistance as praticable, limited by the errors inherent in such resistors. Usually, the resistance $R_B$ is changed in decade steps and the resistance $R_A$ is used for fine balance adjustment, and $R_N$ is fixed for the duration of measurement. The detector shall be a d.c. amplifier, with an input resistance high compared with any of these arms. The unknown resistance $R_x$ is calculated as follows:

$$R_x = R_N \cdot R_B / R_A$$

where $R_A$, $R_B$ and $R_N$ are as shown in Figure 6.

The maximum percentage error in the computed resistance is the sum of the percentage errors in $R_A$, $R_B$ and $R_N$, when the null detector has adequate sensitivity. If $R_A$ and $R_B$ are wire-wound resistors with values below, for example, 1 M$\Omega$, their errors can be made negligible, and for measuring very high resistances, $R_N$ could be, for instance, $10^9$ $\Omega$, which may be known with an accuracy of $\pm 2\%$. The accuracy with which the ratio $R_B/R_A$ can be determined depends essentially on the sensitivity of the null detector. If the unknown resistance $R_x \gg R_N$, the uncertainty $\Delta r$ in the determination of the ratio $r = R_B/R_A$ is determined by $\Delta r/r = I_g R_x / U$, where $I_g$ is the minimum perceivable null detector current and $U$ the voltage applied to the bridge. If, for example, an electronic amplifier instrument with input resistance 1 M$\Omega$ and full-scale deflection for an input voltage of $10^{-5}$ is used, the lowest perceivable current will be about $2 \cdot 10^{-13}$ A, corresponding to 2% of full-scale deflection. With this value of $I_g$, $U = 100$ V, and $R_x = 10^{13}$ $\Omega$, $\Delta r/r = 0.02$ or 2% is obtained.

Resistances up to $10^{13}$ to $10^{14}$ $\Omega$ can thus be measured at 100 V with the required accuracy by the Wheatstone bridge method.

### A2.2  *Ammeter method*

This method employs the circuit shown in Figure 7, page 40, and the components are the same as those described in Clause A1 with the addition of a resistor $R_N$ of known value, and a switch to short-circuit the unknown resistance. It is very important that the resistance of this switch in the open position be much higher than the unknown resistance $R_x$ in order not to affect the measurement of the latter. This is most easily obtained by short-circuiting $R_x$ with a copper wire, which is removed when measuring $R_x$. In general, it is preferable to leave $R_N$ in the circuit at all times in order to limit the current in case of failure of the specimen and thus protect the current-measuring device.

With the switch open, the current through $R_x$ and $R_N$ is determined as specified in Clause 10 by noting the instrument deflection $\alpha_x$ and the shunt ratio $F_x$, the shunt being adjusted to give as near as possible maximum scale deflection. Thereafter, $R_x$ is short-circuited and the current through $R_N$ determined by noting the instrument deflection $\alpha_N$ and the shunt ratio $F_N$, the shunt again being adjusted to give as near as possible maximum scale deflection, starting from the least sensitivity. Provided the applied voltage $U$ does not change during the measurement, $R_x$ can be calculated from

$$R_x = R_N [(\alpha_N F_N / \alpha_x F_x) - 1]$$

if $\alpha_N F_N/\alpha_x F_x > 100$, the approximated formula may be used

$$R_x = R_N(\alpha_N F_N/\alpha_x F_x)$$

This method allows $R_x$ to be determined with about the same accuracy as by the direct method described in Clause A1, but has the advantage that the current-measuring device is checked *in situ* by the measurement of $R_N$, the error of which can be made negligible by using a wirewound resistor, which is readily obtainable with an accuracy of 0.1% or better. The measurement of the current through $R_x$ may thus be more reliable.

# APPENDIX B

## FORMULAE FOR CALCULATING $A$ AND $p$

For most purposes, the following approximate formulae are sufficiently accurate for calculating the effective area $A$ and the effective perimeter $p$ of the guarded electrode.

### B1.  The effective area $A$

a) Circular electrodes (Figure 2, page 37) . . . . . . . . . . . . . . $A = \pi(d_1 + g)^2/4$

b) Rectangular electrodes . . . . . . . . . . . . . . . . . . . $A = (a + g)(b + g)$

c) Square electrodes . . . . . . . . . . . . . . . . . . . . . $A = (a + g)^2$

d) Tubular electrodes (Figure 3, page 38) . . . . . . . . . . . . $A = \pi(d_0 - h)(l_1 + g)$

where $d_0$, $d_1$, $g$, $h$ and $l_1$ are the dimensions indicated in Figures 2 and 3, and $a$ and $b$ are the length and width, respectively, of the guarded electrode when rectangular or square. The dimensions are expressed in metres (centimetres).

### B2.  The effective perimeter

a) Circular electrodes (Figure 2) . . . . . . . . . . . . . . . . $p = \pi(d_1 + g)$

b) Rectangular electrodes . . . . . . . . . . . . . . . . . . . $p = 2(a + b + 2g)$

c) Square electrodes . . . . . . . . . . . . . . . . . . . . . $p = 4(a + g)$

d) Tubular electrodes . . . . . . . . . . . . . . . . . . . . . $p = 2\pi d_0$

where the meaning of the symbols is the same as in Clause B1.

---

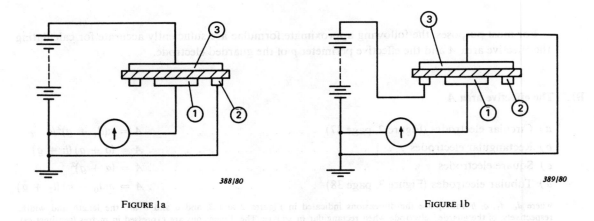

388/80

389/80

**FIGURE 1a**                    **FIGURE 1b**

FIG. 1. — Branchements types des électrodes gardées pour la mesure de la :
Basic connections for guarded electrodes used for :

a) résistivité transversale          b) résistivité superficielle

volume resistivity                   surface resistivity

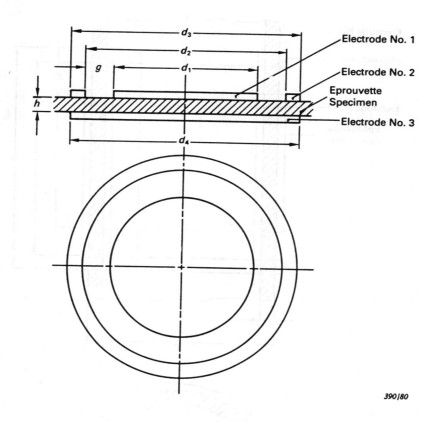

FIG. 2. — Exemple de disposition des électrodes sur des éprouvettes planes.
Example of electrode arrangement on flat specimen.

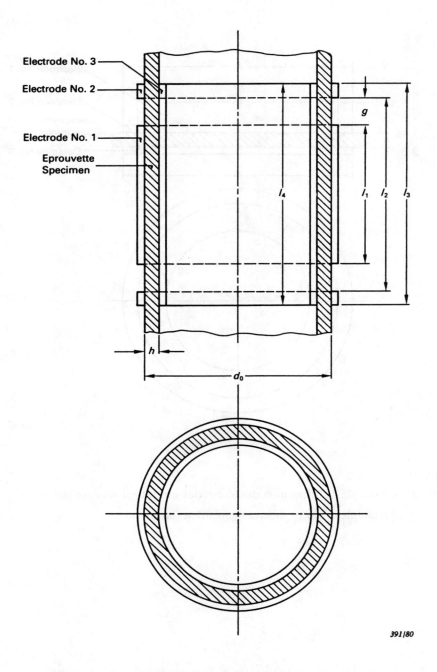

Electrode No. 3

Electrode No. 2

Electrode No. 1

Eprouvette
Specimen

FIG. 3. — Exemple de disposition d'électrodes sur éprouvette tubulaire.

Example of electrode arrangement on tubular specimen.

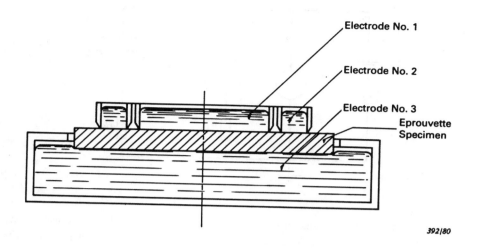

Electrode No. 1
Electrode No. 2
Electrode No. 3
Eprouvette
Specimen

*392/80*

FIG. 4. — Disposition pour électrodes liquides.

Arrangement of liquid electrodes.

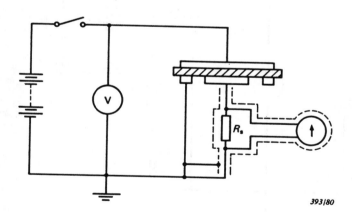

*393/80*

FIG. 5. — Méthode voltmètre-ampèremètre utilisée pour la mesure de la résistance transversale. Pour les mesures de résistance superficielle, les connexions à l'éprouvette sont réalisées comme représenté à la figure 1b.

Voltmeter-ammeter method used for measuring volume resistance. For surface resistance measurements, the connections to the specimen are made as shown in Figure 1b.

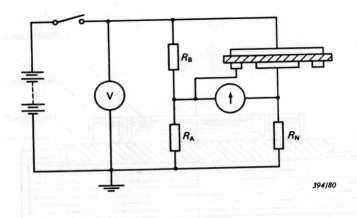

394/80

FIG. 6. — Méthode du pont de Wheatstone utilisée pour la mesure de la résistance transversale. Pour les mesures de résistance superficielle, les connexions à l'éprouvette sont réalisées comme représenté à la figure 1b.

Wheatstone bridge method used for measuring volume resistance. For surface resistance measurements, the connections to the specimen are made as shown in Figure 1b.

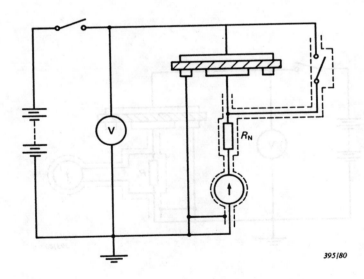

395/80

FIG. 7. — Méthode de l'ampèremètre utilisée pour la mesure de la résistance transversale. Pour les mesures de résistance superficielle, les connexions à l'éprouvette sont réalisées comme représenté à la figure 1b.

Ammeter method used for measuring volume resistance. For surface resistance measurements, the connections to the specimen are made as shown in Figure 1b.

COMMISSION ÉLECTROTECHNIQUE INTERNATIONALE

NORME DE LA CEI

INTERNATIONAL ELECTROTECHNICAL COMMISSION

IEC STANDARD

Publication 112

Troisième édition — Third edition

1979

DEUXIÈME IMPRESSION 1986                                    SECOND IMPRESSION 1986

# Méthode pour déterminer les indices de résistance et de tenue au cheminement des matériaux isolants solides dans des conditions humides

# Method for determining the comparative and the proof tracking indices of solid insulating materials under moist conditions

**Descripteurs :** Matériaux isolants solides, mesure de la résistance au cheminement, exigences, définitions, essais dans les conditions humides.

**Descriptors :** solid insulating materials, measurement of resistance to tracking, requirements, definitions, testing under moist conditions.

Bureau Central de la Commission Electrotechnique Internationale

1, rue de Varembé

Genève, Suisse

Code prix  **10**
Price code

*Pour prix, voir catalogue en vigueur*
*For price, see current catalogue*

# CONTENTS

# INTERNATIONAL ELECTROTECHNICAL COMMISSION

## METHOD FOR DETERMINING THE COMPARATIVE AND THE PROOF TRACKING INDICES OF SOLID INSULATING MATERIALS UNDER MOIST CONDITIONS

### FOREWORD

1) The formal decisions or agreements of the I E C on technical matters, prepared by Technical Committees on which all the National Committees having a special interest therein are represented, express, as nearly as possible, an international consensus of opinion on the subjects dealt with.

2) They have the form of recommendations for international use and they are accepted by the National Committees in that sense.

3) In order to promote international unification, the I E C expresses the wish that all National Committees should adopt the text of the I E C recommendation for their national rules in so far as national conditions will permit. Any divergence between the I E C recommendation and the corresponding national rules should, as far as possible, be clearly indicated in the latter.

### PREFACE

This standard has been prepared by Sub-Committee 15A, Short-time Tests, of I E C Technical Committee No. 15, Insulating Materials.

It constitutes the third edition of I E C Publication 112.

Drafts were discussed at the meetings held in Zurich in 1973 and in Toronto in 1976. As a result of this latter meeting, a draft, Document 15A(Central Office)32, was submitted to the National Committees for approval under the Six Months' Rule in June 1976.

The following countries voted explicitly in favour of publication:

| | |
|---|---|
| Austria | Norway |
| Belgium | Poland |
| Canada | Romania |
| China | South Africa (Republic of) |
| Czechoslovakia | Sweden |
| Denmark | Switzerland |
| Egypt | Turkey |
| France | Union of Soviet Socialist Republics |
| Germany | United Kingdom |
| Japan | Yugoslavia |

The United States of America voted against the publication of this revision objecting that the use of two test solutions would lead to two different and confusing sets of ratings.

*Other I E C publication quoted in this standard:*

Publication No. 587: Test Method for Evaluating Resistance to Tracking and Erosion of Electrical Insulating Materials Used under Severe Ambient Conditions.

# METHOD FOR DETERMINING THE COMPARATIVE AND THE PROOF TRACKING INDICES OF SOLID INSULATING MATERIALS UNDER MOIST CONDITIONS

## 1. Scope

This method of test indicates the relative resistance of solid electrical insulating materials to tracking for voltages up to 600 V when the surface is exposed under electric stress to water with the addition of contaminants.

Tracking may occur during this test when voltage is applied between a defined electrode arrangement on the surface of a material and drops of electrolyte are applied between them at defined intervals of time. The number of drops needed to cause failure increases with the reduction of the applied voltage and, below a critical value, tracking ceases to occur.

Materials which do not track at the highest test voltage may erode differently. The depth of erosion can be measured. Some materials may ignite during the test.

*Notes 1.* — The grading of materials reached by this method may possibly differ from that obtained by other testing methods for assessing tracking resistance, for example tests based on the use of high voltage, low-current discharges. This test method provides good resolution between materials with relatively poor tracking resistance. It lacks resolution for materials normally required for outdoor use for which the test method in IEC Publication 587: Test Method for Evaluating Resistance to Tracking and Erosion of Electrical Insulating Materials Used under Severe Ambient Conditions, should be used.

*2.* — The test results as such cannot be used directly for the evaluation of safe creepage distances when designing electrical apparatus.

## 2. Definitions

### 2.1 *Tracking*

The progressive formation of conducting paths, which are produced on the surface of a solid insulating material, due to the combined effects of electric stress and electrolytic contamination on this surface.

### 2.2 *Electrical erosion*

The wearing away of insulating material by action of electrical discharges.

### 2.3 *Comparative tracking index (CTI)*

The numerical value of the maximum voltage in volts at which a material withstands 50 drops without tracking.

*Note.* — The value of each test voltage and the CTI should be divisible by 25.

### 2.4 *Proof tracking index (PTI)*

The numerical value of the proof voltage in volts at which a material withstands 50 drops without tracking.

## 3. Test specimen

Any flat surface may be used, provided that the area is sufficient to ensure that during the test no liquid flows over the edges of the specimen. Flat surfaces of not less than 15 mm × 15 mm are recommended. The thickness of the specimen should be 3 mm or more and should be reported.

*Notes 1.* — In special cases, in order to obtain a flat surface, grinding can be applied; this fact should, however, be mentioned in the test report.

*2.* — The values of the CTI obtained on specimens of thicknesses below about 3 mm may not be comparable; for example, if thin specimens are mounted on a metal or glass supporting plate, this may remove heat quickly and so alter the CTI. Therefore, if the thickness of the specimen is less than 3 mm, two, or if necessary more, specimens should be stacked.

*3.* — Where the direction of the electrodes relative to any feature of the material is significant, the direction shall be reported. The direction giving the lowest CTI should be used.

*4.* — The test should be carried out on areas which are free from scratches. If this is impossible, the results obtained on the scratched area should be reported together with a statement describing the surface of the specimen.

Scratches on the surface of the test specimen will add to the dispersion of the test results. If the tracking current follows the scratches, failure may occur at a lower voltage (or a lower number of drops) than when the tracking current passes across the scratches.

## 4. Conditioning

The surface of the test specimen shall be clean and free of dust, dirt, fingerprints, grease, oil, mould release or other contaminants which can influence the test results. Care should be used in cleaning to avoid swelling, softening, substantial abrasion or other damage to the material. The conditioning and cleaning procedure should be stated in the test report.

## 5. Test apparatus

### 5.1 *Electrodes*

The two platinum electrodes shall have a rectangular cross-section of 5 mm × 2 mm, with one end chisel-edged with an angle of 30° (Figure 1, page 16). The chisel-edge shall be slightly rounded.

The electrodes shall be symmetrically arranged in a vertical plane, the total angle between them being 60°, and with opposing electrode faces vertical and 4.0 ± 0.1 mm apart on a flat horizontal surface of the specimen (Figure 2, page 16). The force exerted by each electrode on the surface shall be 1 ± 0.05 N. An arrangement for applying the electrodes to the specimen is shown in Figure 3, page 16.

*Note.* — Where any metal other than platinum is used to simulate practical conditions, the metal should be stated in the test report. The results shall not be designated as CTI or PTI.

### 5.2 *Test circuit*

The electrodes shall be supplied with a substantially sinusoidal voltage, variable between 100 V and 600 V at a frequency of 48 Hz to 60 Hz. The power of the source should not be less than 0.5 kVA. The basic circuit is shown in Figure 4, page 17.

The variable resistor shall be capable of adjusting the current between the short-circuited electrodes to $1.0 \pm 0.1$ A and the voltage indicated by the voltmeter shall not fall by more than $10\%$ when this current flows.

An over-current relay in the test circuit shall trip when 0.5 A or more has persisted for 2 s.

## 5.3 *Dropping device*

The surface between the electrodes shall be wetted with drops of the test solution at intervals of $30 \pm 5$ s. The drops shall fall centrally between the electrodes from a height of 30 mm to 40 mm. The size of the drops shall be $20 ^{+3}_{-0}$ mm³. Prior to each test, the needle or other outlet for the drops shall be cleaned and sufficient drops let out to ensure that the correct concentration of the test solution is used.

*Notes 1.* — When the test solution is left in the needle between tests evaporation will increase its concentration. Letting out about 5 to 20 drops depending upon the delay between tests, will normally remove any liquid with too high a concentration.

*2.* — To determine the drop size check that 1 cm³ of liquid is delivered by not less than 44 and not more than 50 drops. The drop size shall be checked periodically.

*3.* — A hypodermic needle, having an outer diameter of 0.9 mm to 1.1 mm with the tip cut off square, is suitable for a dropping device.

*4.* — In some special cases, the $\pm 5$ s deviation of the intervals may be too broad and may influence the results. Then a deviation of $\pm 1$ s should be arranged.

## 5.4 *Test solutions*

*Solution A:* $0.1 \pm 0.002\%$ by mass ammonium chloride ($NH_4Cl$) in distilled or deionized water. The resistivity at $23 \pm 1$ °C is $395 \pm 5$ $\Omega \cdot$cm.

*Solution B:* $0.1 \pm 0.002\%$ by mass ammonium chloride and $0.5 \pm 0.002\%$ by mass sodium-alkylnaphthalene-sulphonate in distilled or deionized water. The resistivity at $23 \pm 1$ °C is $170 \pm 5$ $\Omega \cdot$cm.

Solution A is the preferred solution.

When a more aggressive contaminant is required, solution B is to be used. To indicate the use of solution B the CTI or PTI value shall be followed by the letter "M" (e.g. CTI 250 M).

Where a solution other than solution A or B is used, this shall be stated in the test report. The results shall not be designated CTI or PTI.

*Note.* — Tracking will be accelerated by decreasing the solution resistivity and will be influenced by the chemical nature of the test solution.

## 6. Procedure

## 6.1 *General*

The test shall be carried out with the specimen screened from draughts and at an ambient temperature of $23 \pm 5$ °C. Contamination of the electrodes may have an influence on the test results. The electrodes shall be cleaned before every test.

The specimen to be tested shall be placed, with the test surface horizontal, on a metal or glass support so that the chisel edges of both electrodes are pressed against the specimen with the specified force.

The distance between the electrodes shall be checked and they shall make good contact with the specimen. If the electrode edges have become eroded, the edges shall be reshaped. The voltage is set at some suitable value, divisible by 25, and the circuit resistance is adjusted so that the short-circuit current is within the given tolerance. The drops of electrolyte are then allowed to fall on the test surface until failure by tracking occurs or 50 drops fall.

A failure has occurred if a current of 0.5 A or more flows for at least 2 s in a conducting path between the electrodes on the surface of the specimen, thus operating the overcurrent relay; or if the specimen burns without releasing the overcurrent relay.

Notes 1. — If several tests are carried out on the same test specimen, care must be taken that the testing points are sufficiently far from each other so that splashes from the testing point will not contaminate the other areas to be tested.

2. — If the support is of metal, it can be connected into the test circuit to indicate when erosion of a specimen causes puncture.

3. — Since the test may produce poisonous or noxious gases, it is advisable to provide means for their safe removal or containment.

## 6.2 Determination of the CTI

Set the voltage at a selected level and carry out the test for 50 drops or until previous failure. Repeat the test on other test sites at lower or higher voltages until the maximum voltage is established at which no failure occurs at 50 drops in tests on five sites. The numerical value of this voltage is the CTI (e.g. CTI 425) provided no failure occurs below 100 drops when the voltage is reduced by 25 V for a further set of five test sites. A few materials may not meet this last requirement. For these materials, the maximum voltage at which five sites will each withstand 100 drops or more is established and the numerical value of this voltage indicated in addition to the CTI, e.g. CTI 425 (375).

Notes 1. — If the behaviour of the material is unknown, the initial test voltage should be in the middle of the test range e.g. 300 V. This voltage should be increased if a specimen withstands 50 drops or decreased if it fails before 50 drops. The voltage change should be 25 V or a multiple of 25 V. The process is continued until the highest voltage at which five specimens withstand 50 drops is found.

2. — For most materials, the 50 drop voltage (for which they withstand 50 drops without tracking) is acceptably close to the asymptotic value. The test at 25 V below the 50 drop voltage is intended to confirm this. The more the 100 drop voltage has to be reduced below the 50 drop voltage, the further this voltage is from the asymptote.

3. — Warning: At higher voltages and more than 50 drops, a failure (indicated by the trip of the overcurrent relay) may occur by accumulation of solution and contaminents in grooves and holes on the surface and not by conductive paths as defined in Sub-clause 2.1. In this case, new tests are necessary or this should be reported, if no test result as defined is possible.

## 6.3 Proof tracking test

Where, in I E C standards for material specification or for electrical equipment or in other standards, a proof test only is required, tests shall be made in accordance with Sub-clause 6.1 but at the single specified voltage. The required number of specimens shall withstand 50 drops without failure.

The recommended number of specimens is five; less may be specified in special circumstances.

Preferred test voltages are 175 V, 250 V, 300 V, 375 V or 500 V. The recommended abbreviation is PTI (proof tracking index).

### 6.4 *Determination of erosion*

The specimens which have not tracked shall be cleaned of any debris or loosely attached degradation products and placed on a plate of a depth gauge. The maximum depth of erosion of each specimen shall be measured to an accuracy of 0.1 mm using a 1 mm diameter probe having a hemispherical end. The maximum of the five values measured shall be stated in the test report.

6.4.1   In the case of tests according to Sub-clause 6.2, the erosion shall be measured on the five specimens tested at the voltage corresponding to the CTI.

6.4.2   In the case of tests according to Sub-clause 6.3, the erosion shall be measured on the specimens which withstand 50 drops at the specified voltage.

## 7. Report

The report shall include the following information:

1. Identification of material tested.

2. Thickness of specimen.

3. Nature of surface.
   3.1 Whether or not the original surface of the specimen was tested.
   3.2 Whether the surface tested was ground.
   3.3 Whether or not the surface tested was varnished.
   3.4 State of surface with regard to scratches.

4. The conditioning and cleaning procedure.

5. Electrode metal if platinum was not used.

6. The contaminating solution if solution A or B was not used.

7. Comparative tracking index.
   7.1 The CTI, e.g. "CTI 400", "CTI 400 M" or "CTI 400 (350)".
   7.2 The depth of erosion, e.g. "CTI 275 — 1.2", "CTI 275 M — 1.2" or "CTI 275 (200) — 1.2".

8. Proof tracking index.
   8.1 Pass or fail at specified voltage, e.g. "Pass at PTI 175" or "Fail at PTI 175 M".

   8.2 Pass or fail at the specified depth of erosion together with the specified voltage, e.g. "Pass at PTI 250 — 0.8" or "Fail at PTI 250 M — 0.8".

9. Where Items 7 or 8 cannot be reported because the specimen ignited, this shall be reported.

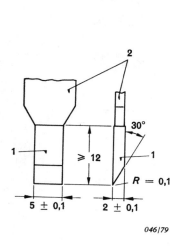

FIG. 1. — Electrode.

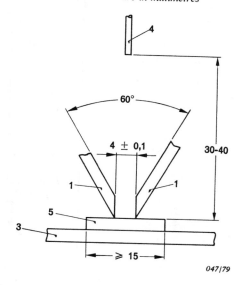

FIG. 2. — Disposition des électrodes.
Electrode arrangement.

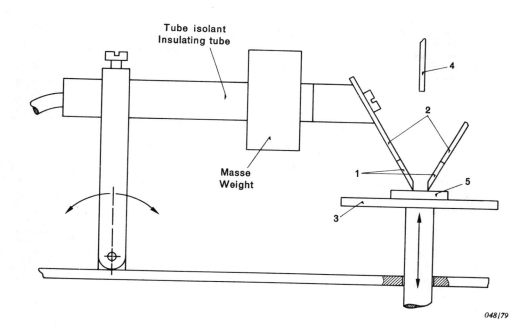

FIG. 3. — Exemple d'appareil d'essai.
Example of test apparatus.

| | |
|---|---|
| 1 = électrode de platine | 1 = platinum electrode |
| 2 = rallonge laiton | 2 = brass extension |
| 3 = support | 3 = support |
| 4 = pointe du dispositif de production de gouttes | 4 = tip of dropping device |
| 5 = éprouvette | 5 = specimen |

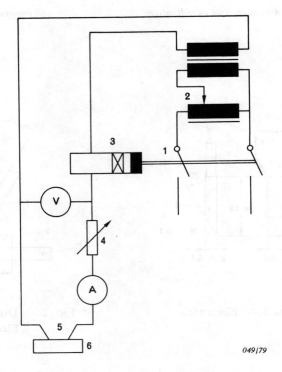

049/79

| | |
|---|---|
| 1 = contacteur | 1 = switch |
| 2 = source de courant alternatif 100 V à 600 V | 2 = a.c. source 100 V to 600 V |
| 3 = relais de surintensité temporisé | 3 = delay overcurrent relay |
| 4 = résistance variable | 4 = variable resistor |
| 5 = électrodes | 5 = electrodes |
| 6 = éprouvette | 6 = specimen |

FIG. 4. — Exemple de circuit d'essai.
Example of test circuit.

# NORME INTERNATIONALE
# INTERNATIONAL STANDARD

**Commission Electrotechnique Internationale**

**International Electrotechnical Commission**

**Международная Электротехническая Комиссия**

CEI
IEC
243-1

Première édition
First edition
1988

## Méthodes d'essai pour la détermination de la rigidité diélectrique des matériaux isolants solides

Première partie: Mesure aux fréquences industrielles

## Methods of test for electric strength of solid insulating materials

Part 1: Tests at power frequencies

Bureau Central de la Commission Electrotechnique Internationale    3, rue de Varembé    Genève, Suisse

Code prix   **12**
Price code

Pour prix, voir catalogue en vigueur
For price, see current catalogue

# CONTENTS

FOREWORD

PREFACE

INTRODUCTION

IEC Publication 243–1
(First edition – 1988)

Methods of test for electric strength of solid insulating materials

Part 1:
Tests at power frequencies

# CORRIGENDUM

## Page 5

*In the preface:*

*a) Insert the following text between the first and second paragraphs:*

It replaces IEC Publication 243 (1967).

*b) In the publications quoted, delete the reference to Publication 243 and its title.*

## Page 31

*Clause 11, in the last note in this clause, instead of:*

Note.– A separate part of IEC Publication 243 is being prepared...

*read:*

Note.– IEC Publication 243–4 is being prepared...

March 1989

# INTERNATIONAL ELECTROTECHNICAL COMMISSION

---

## METHODS OF TEST FOR ELECTRIC STRENGTH
## OF SOLID INSULATING MATERIALS

### Part 1: Tests at power frequencies

---

**FOREWORD**

1) The formal decisions or agreements of the IEC on technical matters, prepared by Technical Committees on which all the National Committees having a special interest therein are represented, express, as nearly as possible, an international consensus of opinion on the subjects dealt with.

2) They have the form of recommendations for international use and they are accepted by the National Committees in that sense.

3) In order to promote international unification, the IEC expresses the wish that all National Committees should adopt the text of the IEC recommendation for their national rules in so far as national conditions will permit. Any divergence between the IEC recommendation and the corresponding national rules should, as far as possible, be clearly indicated in the latter.

## PREFACE

This standard has been prepared by Sub-Committee 15A: Short-time tests, of IEC Technical Committee No. 15: Insulating materials.

The text of this standard is based on the following documents:

| Six Months' Rule | Report on Voting |
|---|---|
| 15A(CO)52 | 15A(CO)55 |

Full information on the voting for the approval of this standard can be found in the Voting Report indicated in the above table.

*The following IEC publications are quoted in this standard:*

Publications Nos. 212 (1971): Standard conditions for use prior to and during the testing of solid electrical insulating materials.

243 (1967): Recommended methods of test for electric strength of solid insulating materials at power frequencies.

296 (1982): Specification for unused mineral insulating oils for transformers and switchgear.

455: Specification for solventless polymerisable resinous compounds used for electrical insulation.

464-2 (1974): Specification for insulating varnishes containing solvent, Part 2: Test methods.

674: Specification for plastic films for electrical purposes.

684: Specification for flexible insulating sleeving.

---

# METHODS OF TEST FOR ELECTRIC STRENGTH
## OF SOLID INSULATING MATERIALS
### Part 1: Tests at power frequencies

---

## Introduction

This standard is one of a series which deals with tests for electric strength of solid insulating materials. The series will consist of four parts:

Part 1: Tests at power frequencies (IEC Publication 243-1);

Part 2: Additional requirements for tests using direct voltage (IEC Publication 243-2);

Part 3: Additional requirements for impulse tests (IEC Publication 243-3);

Part 4: Statistical treatment and interpretation (IEC Publication 243-4).

## 1. Scope

This standard gives methods of test for the determination of the short-time electric strength of solid insulating materials at power frequencies, that is, those between 48 Hz and 62 Hz. It does not consider the testing of liquids and gases although these are specified and used as impregnants or surrounding media for the solid insulating materials being tested.

*Note.-* Methods for making tests on sheet materials with reduced levels of surface discharges with spherical electrodes are also being prepared.

## 2. Definitions

For the purpose of this standard, the following definitions apply:

### 2.1 *Electric breakdown (electric failure)*

The annihilation, at least temporary, of the insulating properties of an insulating medium under electrical stress.

### 2.2 *Breakdown voltage*

The voltage at which breakdown occurs, under prescribed test conditions or in use.

## 2.3  *Electric strength*

The quotient of the breakdown voltage and the distance between the conducting parts between which the voltage is applied under prescribed test conditions. In the case of step-by-step tests, the breakdown voltage is taken as the highest voltage maintained for the entire duration of a step. The term is also used to describe the corresponding property of a material.

## 2.4  *Flashover*

A breakdown between electrodes in a gas, a liquid or in vacuum, at least partly along the surface of solid insulation.

## 3.  Significance of the test

3.1    Electric strength test results obtained in accordance with this standard can be used for detecting changes or deviations from normal characteristics resulting from processing variables, ageing conditions or other manufacturing or environmental situations but can seldom be used directly to determine the behaviour of insulating materials in an actual application.

3.2    Measured values of the electric strength of a material may be affected by many factors, including:

*a)*  The frequency, waveform and rate of rise or time of application of the voltage.

*b)*  The thickness and homogeneity of the specimen and the presence of mechanical strain.

*c)*  Previous conditioning of the specimens, in particular drying and impregnation procedures.

*d)*  The ambient temperature, pressure and humidity.

*e)*  The presence of gaseous inclusions, moisture or other contamination.

*f)*  The system, the dimensions and thermal conductivity of the test electrodes.

*g)*  The intensity of surface discharges prior to breakdown.

*h)*  The area or volume between electrodes under maximum voltage stress.

*i)*  The electrical and thermal characteristics of the surrounding medium.

3.3    The effects of all these factors should be considered when investigating materials for which no experience exists. This standard defines particular conditions which give rapid discrimination between materials and which can be used for quality control and similar purposes.

The results given by different methods are not directly comparable but may provide information on relative electric strengths of materials. It should be noted that the electric strength of most materials decreases as the thickness of the specimen between the electrodes increases and as the time of voltage application increases.

3.4 The electric strength of most materials is different, depending on the intensity and the duration of surface discharges prior to break-down. For designs which are free from partial discharges up to the test voltage, it is very important to know the electric strength without discharges prior to breakdown but the methods in this standard are generally not suitable for providing this information.

3.5 Materials with high electric strength will not necessarily resist long-term degradation processes such as heat, erosion or chemical deterioration by partial discharges, or electrochemical deterioration in the presence of moisture, all of which may cause eventual failure in service at much lower stress.

## 4. Electrodes and specimens

The metal electrodes shall be maintained smooth, clean and free from defects at all times.

*Note.-* This maintenance becomes more important when thinner specimens are being tested. Stainless steel electrodes are preferred to minimize electrode damage at breakdown.

The leads to the electrodes shall not tilt or otherwise move the electrodes or affect the pressure on the specimen, nor appreciably affect the electric field configuration in the neighbourhood of the specimen.

4.1 *Test normal to the surface of non-laminated materials and normal to laminae of laminated materials*

4.1.1 *Boards and sheet materials, including press boards, papers, fabrics fabrics and films*

4.1.1.1 *Unequal electrodes*

The electrodes shall consist of two metal cylinders with the edges rounded to give a radius of 3 mm. One electrode shall be 25 ± 1 mm in diameter and approximately 25 mm high. The other electrode shall be 75 ± 1 mm in diameter and approximately 15 mm high. These electrodes shall be arranged coaxially as in Figure 1a.

When specified, boards and sheets over 3 mm thick shall be reduced to 3 ± 0.2 mm and then tested with the 25 mm diameter electrode on an original non-machined surface.

### 4.1.1.2 Equal diameter electrodes

If a fixture is employed, which accurately aligns upper and lower electrodes, the lower electrode may be reduced to a 25 mm diameter but the results obtained will not necessarily be the same as those obtained with 25/75 mm diameter electrodes (see Figure 1b).

### 4.1.2 Tapes, films and narrow strips

The electrodes shall be two metal rods, each 6 ± 0.2 mm in diameter, mounted vertically one above the other in a jig so that the specimen is held between the faces of the squared ends of the rods.

The upper and lower electrodes shall be coaxial. The edges of the squared ends shall be rounded to a radius of approximately 1 mm. The upper electrode shall have a mass of 50 ± 2 g and shall be moveable in the jig with a minimum of friction.

Figure 2 shows a convenient arrangement. If specimens are to be tested while extended, they shall be clamped in a frame holding them in the required position relative to the assembly shown in Figure 2. Wrapping one end of the specimen round a rotatable rod is one convenient way of achieving the required extension.

To prevent flashover around the edges of narrow tapes, the test specimen may be clamped using strips of varnished cloth overlapping the edges of the tape. Alternatively, gaskets that surround the electrodes may be used, provided that there is an annular space between electrode and gasket of about 1.5 mm.

Note.- For testing films see IEC Publication 674.

### 4.1.3 Flexible tubing and sleeving

To be tested according to IEC Publication 684.

### 4.1.4 Rigid tubes (having an internal diameter up to and including 100 mm)

The outer electrode shall be a band of metal foil 25 ± 1 mm wide. The inner electrode is a closely fitting internal conductor, e.g. rod, tube, metal foil or a packing of spheres 0.75 mm to 2.0 mm in diameter, making good contact with the inner surface. In each case, the ends of the inner electrode shall extend for at least 25 mm beyond the ends of the outer electrode.

Note.- Where no adverse effect will result, petroleum jelly may be used for attaching the foil to the inner and outer surfaces.

### 4.1.5 *Tubes and cylinders (having an internal diameter greater than 100 mm)*

The outer electrode shall be a band of metal foil 75 ± 1 mm wide and the inner electrode, a disk of metal foil 25 ± 1 mm in diameter, flexible enough to conform with the curvature of the cylinder. The arrangement is shown in Figure 3.

### 4.1.6 *Moulded and cast materials*

Make a test piece, at least 100 mm in diameter and 3 ± 0.2 mm thick and test according to IEC Publication 455.

### 4.1.7 *Varnishes*

To be tested according to IEC Publication 464-2.

### 4.1.8 *Filling compounds*

The electrodes shall be two metal spheres, each 12.5 to 13.0 mm in diameter, arranged horizontally along the same axis either 0.75 ± 0.05 mm, 1.0 ± 0.10 mm, or 1.25 ± 0.10 mm apart and embedded in the compound. Care shall be taken to avoid cavities, particularly between the electrodes. As values obtained with the different electrode spacings are not directly comparable, the gap length shall be detailed in the specification for the compound and mentioned in the test report.

## 4.2 *Tests parallel to the surface of non-laminated materials and parallel to the laminae of laminated materials*

If it is not necessary to differentiate between failure by puncture of the specimen and failure across its surface, the electrodes of Sub-clause 4.2.1 or Sub-clause 4.2.2 may be used, those of Sub-clause 4.2.1 being preferred.

When the prevention of surface failure is required, the electrodes of Sub-clause 4.2.3 shall be used.

### 4.2.1 *Parallel plate electrodes*

### 4.2.1.1 *Boards and sheets*

For tests on boards and sheets the test specimen shall be of the thickness of the material to be tested and rectangular, 100 ± 2 mm long and 25 ± 0.2 mm wide. The long edges shall be cut as parallel planes at right angles to the surface of the material. The test specimen is placed with the 25 mm width between parallel metal plates, not less than 10 mm thick, forming the electrodes between which the voltage shall be applied. For thin materials, two or three test specimens are used suitably placed (i.e. with their long edges at a convenient angle) to support the upper electrode. The electrodes shall be of sufficient size to overlap the edges of the test specimens by not less than 15 mm and care shall be taken to ensure good contact over the whole area of those edges. The edges of the electrodes shall be suitably radiused (3 to 5 mm) to avoid flashover from edge to edge of the electrodes (see Figure 4).

This type of electrode is suitable only for tests on rigid materials at least 1.5 mm thick.

### 4.2.1.2 *Tubes and cylinders*

For tests on tubes and cylinders, the test specimen shall be a complete ring or a 100 mm portion of a ring of 25 ± 0.2 mm axial length. Both edges of the specimen shall be finished as parallel planes at right angles to the axis of the tube or cylinder. The specimen is tested between parallel plates as described in Sub-clause 4.2.1.1 for boards and sheets. Where necessary to support the upper electrode, two or three specimens are used. The electrodes shall be of sufficient size to overlap the edges of the specimens by not less than 15 mm and care shall be taken to ensure good contact over the whole area of the edges of the specimens.

### 4.2.2 *Taper pin electrodes*

Two parallel holes are drilled normal to the surface, with centres 25 ± 1 mm apart and of such a diameter that, after reaming with a reamer having a taper of approximately 2%, the diameter of each hole at the larger end is not less than 4.5 mm and not greater than 5.5 mm.

The holes shall be drilled completely through the specimen or, in the case of large tubes, through one wall only, and shall be reamed throughout their full length.

When the specimens are drilled and reamed, the material adjacent to the holes shall not be damaged, e.g. split, broken or charred, in any way.

The taper pins used as electrodes shall be pressed, not hammered into the holes so that they fit tightly and extend on each side of the test specimen by not less than 2 mm (see Figures 5a and 5b).

This type of electrode is suitable only for tests on rigid materials at least 1.5 mm thick.

### 4.2.3 *Parallel cylindrical electrodes*

For tests on specimens of high electric strength and which are more than 15 mm thick, specimens 100 mm x 50 mm shall be cut and two holes drilled as shown in Figure 6 so that each is not more than 0.1 mm greater in diameter than each actual cylindrical electrode which shall be 6 ± 0.1 mm in diameter and have hemispherical ends. The base of each hole is hemispherical to mate with the end of the electrode. If not otherwise specified in the material specification, the holes shall be 10 ± 1 mm apart throughout their length and extend to within 2.25 ± 0.25 mm of the surface opposite that through which they are drilled. Alternative forms of vented electrodes are shown in Figure 6. When electrodes with slots are used, these slots shall be diametrically opposed to the gap between the electrodes.

## 4.3 Test specimens

In addition to the information concerning specimens given in the preceding sub-clauses, the following general points shall be noted.

In the preparation of test specimens from solid materials, care shall be taken that the surfaces in contact with the electrodes are parallel and as flat and smooth as the material allows.

For tests made normal to the surface of the material, test specimens need only be of sufficient area to prevent flashover under the conditions of test.

In tests made normal to the surface of the material, the results on specimens of different thicknesses are not directly comparable (see Clause 3).

## 4.4 Distance between electrodes

The value to be used in calculating the electric strength shall be one of the following, as specified for the material under test:

a) nominal thickness or distance between electrodes (use this value unless otherwise specified);

b) average thickness of the test specimen or distance between electrodes for tests parallel to the surface;

c) thickness or distance between electrodes measured immediately adjacent to the breakdown on each test specimen.

## 5. Conditioning before tests

The electric strength of insulating materials varies with temperature and moisture content. Where a specification is available for the material to be tested, this shall be followed. Otherwise, specimens shall be conditioned for not less than 24 h at 23 ± 2 °C, 50 ± 5% relative humidity, i.e. standard ambient atmosphere of IEC Publication 212, unless other conditions from that publication are agreed upon.

## 6. Surrounding medium

Materials shall be tested:

a) in air;

b) in oil (IEC Publication 296, unless otherwise specified); or

c) in the medium in which they are to be used.

If this presents difficulties on account of flashover or excessive discharges, the tests shall be made in a medium of higher electric strength or permittivity. Any medium used shall have no deleterious effect on the material under test.

The effect of the ambient medium on the results may be great, particularly in the case of absorbent materials such as pressboard, and it is essential that procedures for specimen preparation define fully all the necessary steps (e.g. drying and impregnation).

Sufficient time shall be allowed for the specimens to attain the required temperature but some materials may be affected by prolonged exposure to high temperatures.

## 6.1 Tests in air

Tests in air at elevated temperature may be made in any well-designed oven of sufficient size to accommodate the test specimen and the electrodes without flashover occurring during the tests. It should be provided with some means of circulating the air so that a substantially uniform temperature is maintained around the test specimen, and with a thermometer, thermocouple, or other means for measuring the temperature as near the point of test as practicable.

## 6.2 Tests in liquids

When tests are conducted in an insulating liquid, the liquid used shall be transformer oil according to IEC Publication 296, unless otherwise specified. It is necessary to ensure adequate electric strength of the liquid to avoid flashover. Reduction of the resistivity of the liquid may result in a spurious increase in apparent electric strength of the specimen.

Tests at elevated temperature may be made either in a container of liquid in an oven (see Sub-clause 6.1) or in a thermostatically controlled bath using the insulating liquid for heat transfer. In this case, suitable means for circulating the liquid, so that the temperature is substantially uniform around the test specimen, shall be provided.

## 7. Electrical apparatus

### 7.1 Voltage source

The test voltage shall be obtained from a step-up transformer supplied from a variable sinusoidal low-voltage source. The transformer, its voltage source, and the associated controls shall have the following properties:

7.1.1 The ratio of crest to root-mean-square (r.m.s.) test voltage shall be equal to $\sqrt{2} \pm 5\%$ (1.34...1.48), with the test specimen in the circuit, at all voltages greater than 50% of the breakdown voltage.

**7.1.2** The power rating of the source shall be sufficient to maintain the test voltage until electric breakdown occurs. For most materials, using electrodes as recommended, an output current capacity of 40 mA is usually adequate. The power rating for most tests will vary from 0.5 kVA, for testing low-capacitance specimens at voltages up to 10 kV, to 5 kVA for voltages up to 100 kV.

**7.1.3** The controls on the variable low-voltage source shall be capable of varying the test voltage smoothly, uniformly, and without overshoots or transients. When applying voltage in accordance with Clause 9, the incremental increase produced, e.g. by a variable auto-transformer, shall not exceed 2% of the expected breakdown voltage.

Under no circumstances shall the peak of any voltage transient exceed 1.48 times the indicated rms test voltage. Motor-driven controls are preferable for making short-time or rapid-rise tests.

**7.1.4** To protect the voltage source from damage, it shall be equipped with a device which disconnects the power supply within three cycles on breakdown of the specimen. It may consist of a current-sensitive element in the HV supply to the electrodes.

**7.1.5** To restrict damage by current or voltage surges at breakdown, it is desirable to include a resistor with a suitable value in series with the electrodes. The value of the resistor will depend on the damage which can be tolerated on the electrodes.

**7.2** *Voltage measurement*

Provision shall be made for measuring the r.m.s. test voltage. It may be preferable to use a peak-reading voltmeter, in which case the reading is divided by $\sqrt{2}$ to get the r.m.s. value (see Sub-clause 7.1.1). The overall error of the voltage-measuring circuit shall not exceed 5% of the measured value including the error due to the response time of the voltmeter which shall be such that its time lag will not be greater than 1% of full scale at any rate-of-rise used.

**7.2.1** A voltmeter complying with the requirements of the previous sub-clause shall be used to measure the voltage applied to the electrodes. It should preferably be connected directly to them, or via a potential divider or a potential transformer. If a potential winding on the step-up transformer is used for measurement, the accuracy of indica-tion of the voltage applied to the electrodes shall be unaffected by the loading of the step-up transformer and the series resistor.

**7.2.2** It is desirable for the reading of the maximum applied test voltage to be retained on the voltmeter after breakdown so that the breakdown voltage can be accurately read and recorded but the indicator shall not be sensitive to transients which can occur at breakdown.

## 8. Procedure

Electrodes complying with Clause 4 shall be applied to the specimen in such a manner that damage to the specimen is avoided. Using apparatus providing a voltage complying with Clause 7, a voltage is applied between the electrodes and increased in accordance with one of the Sub-clauses 9.1 to 9.5. Observe whether breakdown was by puncture or flashover (see Clause 10).

The document calling for the test shall state the following:

- specimen to be tested;
- method for measurement of specimen thickness (if not nominal);

- any treatment or conditioning prior to test;
- number of specimens, if other than five;
- temperature of test;
- surrounding medium;
- electrodes to be used;
- mode of increase of voltage;
- whether result is to be reported as electric strength or breakdown voltage.

## 9. Mode of increase of voltage

### 9.1 Short-time (rapid-rise) test

9.1.1 The voltage shall be raised from zero at a uniform rate until break-down occurs.

9.1.2 A rate-of-rise shall be selected for the material under test which will cause breakdown to occur most commonly between 10 s and 20 s. For materials which differ considerably in their breakdown voltage, some samples may fail outside these limits. This is satisfactory if the majority of breakdowns occur between 10 s and 20 s.

9.1.3 A commonly used rate-of-rise for a broad spectrum of materials is 500 V/s. It is recommended that wherever possible a rate shall be chosen from the following:

100, 200, 500, 1 000, 2 000, 5 000 V/s.

Note.- For short times to breakdown, the response time requirement of the voltmeter, as specified in Sub-clause 7.2, is more difficult to achieve.

### 9.2 20 s step-by-step test

9.2.1 The nearest voltage to 40% of the probable short-time breakdown voltage shall be selected from Table I and applied to the specimen. If the probable rapidly applied value is not known, it shall be obtained in accordance with the method in Sub-clause 9.1.

9.2.2 If the test specimen withstands this voltage for 20 s without failure, the next highest and subsequent voltages in Table I shall each be immediately and successively applied for 20 s until failure occurs.

## TABLE I

### Successive voltages to be applied
### (kilovolts, peak/√2)

| 0.50 | 0.55 | 0.60 | 0.65 | 0.70 | 0.75 | 0.80 | 0.85 | 0.90 | 0.95 | 4.0 | 4.2 | 4.4 | 4.6 | 4.8 |
|------|------|------|------|------|------|------|------|------|------|-----|-----|-----|-----|-----|
| 1.0 | 1.1 | 1.2 | 1.3 | 1.4 | 1.5 | 1.6 | 1.7 | 1.8 | 1.9 | | | | | |
| 2.0 | 2.2 | 2.4 | 2.6 | 2.8 | 3.0 | 3.2 | 3.4 | 3.6 | 3.8 | | | | | |
| 5.0 | 5.5 | 6.0 | 6.5 | 7.0 | 7.5 | 8.0 | 8.5 | 9.0 | 9.5 | | | | | |
| 10 | 11 | 12 | 13 | 14 | 15 | 16 | 17 | 18 | 19 | | | | | |
| 20 | 22 | 24 | 26 | 28 | 30 | 32 | 34 | 36 | 38 | | | | | |
| 50 | 55 | 60 | 65 | 70 | 75 | 80 | 85 | 90 | 95 | 40 | 42 | 44 | 46 | 48 |
| 100 | 110 | 120 | 130 | 140 | 150 | 160 | 170 | 180 | 190 | 200 | | | | |

9.2.3 The increases of voltage shall be made as quickly as possible and the time spent in raising the voltage shall be included in the period of 20 s at the higher voltage.

If breakdown occurs in less than six voltage steps from the start of the test, a further five specimens shall be tested, using a lower starting voltage.

The electric strength and/or the breakdown voltage shall be based on the highest voltage which is withstood for 20 s without breakdown.

9.3 *Slow rate-of-rise test (120...240 s)*

The voltage shall be raised from 40% of the probable short-time breakdown voltage at a uniform rate such that breakdown occurs between 120 s and 240 s. For materials which differ considerably in their breakdown voltage, some samples may fail outside this limit. This is satisfactory if the majority of breakdowns occur between 120 s and 240 s. The rate-of-rise of voltage shall be initially selected from the following:

2, 5, 10, 20, 50, 100, 200, 500, 1 000 V/s.

9.4 *60 s step-by-step test*

Unless otherwise specified, the test shall be carried out in accordance with Sub-clause 9.2 but with a step duration of 60 s.

9.5 *Very slow rate-of-rise test (300...600 s)*

Unless otherwise specified, this test is carried out in accordance with Sub-clause 9.3 but with breakdowns occurring between 300 s and 600 s with a rate-of-rise of voltage selected from the following:

1, 2, 5, 10, 20, 50, 100, 200 V/s.

*Note.-* The slow rate-of-rise tests of 120...240 s in Sub-clause 9.3 and 300...600 s in Sub-clause 9.5 produce approximately the same results as the 20 s (Sub-clause 9.2) or 60 s (Sub-clause 9.4) step-by-step tests. They are more convenient when using modern automated equipment and they are introduced to enable such equipment to be used.

## 9.6 *Proof tests*

When it is required to apply a pre-determined proof voltage for the purpose of a proof or withstand test, the voltage shall be raised to the required value as rapidly as possible, consistent with its accurate attainment without any transient overvoltage. This voltage is then maintained at the required value for the duration of the specified time.

## 10. Criterion of breakdown

Electric breakdown is accompanied by an increase of current flowing in the circuit and by a decrease of voltage across the specimen. The increased current may trip a circuit-breaker or blow a fuse. However, tripping of a circuit-breaker may sometimes be influenced by flash-over, specimen charging, leakage or partial discharge currents, equipment magnetizing current or malfunctioning. It is therefore essential that the circuit-breaker is well co-ordinated with the charac-teristics of the test equipment and the material under test, otherwise the circuit-breaker may operate without breakdown of the specimen, or fail to operate when breakdown has occurred and thus not provide a positive criterion of breakdown. Even under the best conditions, premature breakdowns in the ambient medium may occur, and observa-tions shall be made to detect them during tests. If breakdowns in the medium are observed, they shall be reported.

Where tests are made perpendicular to the surface of a material, there is usually no doubt when breakage has occurred and subsequent visual inspection readily shows the actual breakdown channel whether this is filled with carbon or not.

In tests parallel to the surface it is required by Sub-clause 4.2 that failures by puncture and failures by flashover are differentiated. This can be done by examination of the specimen or in some cases by re-applying a voltage less than that of the first apparent breakdown. A convenient practice that has been found is the re-application of half the breakdown voltage, followed by increasing into failure by the same procedure as in the first test.

## 11. Number of tests

Unless otherwise specified, five tests shall be conducted and the electric strength or breakdown voltage determined from the central value of the test results. If any test result deviates by more than 15% from the central value, five additional tests shall be made. The electric strength or breakdown voltage shall then be determined from the central value of the ten results.

Note.- *Central value:* Middle result of an odd number of tests or the mean of the two middle results of an even number of tests when arranged in order of magnitude.

When tests are being made for purposes other than routine quality control, larger numbers of specimens will be necessary depending on the variability of the material and the statistical analysis to be applied.

Note.- A separate part of IEC Publication 243 is being prepared dealing with the statistical treatment and interpretation of results.

## 12. Report

Unless otherwise specified, the report shall include the following:

a) A complete identification of the material tested, a description of the specimens and their method of preparation.

b) The central value of the electric strengths in kV/mm and/or breakdown voltages in kV.

c) The thickness of each test specimen (see Sub-clause 4.4).

d) The surrounding medium during the test and its properties.

e) The electrode system.

f) The individual values of electric strengths in kV/mm and/or breakdown voltages in kV.

g) The temperature, pressure and humidity during tests in air or other gas; or the temperature of the surrounding medium when this is a liquid.

h) The conditioning treatment before test.

i) The mode of application of the voltage and the frequency.

j) An indication of the type and position of breakdown.

When the shortest statement of results is required, the first four items shall be included.

---

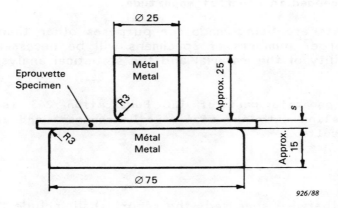

**Fig. 1a. - Electrodes de dimensions inégales.**
**Unequal electrodes.**

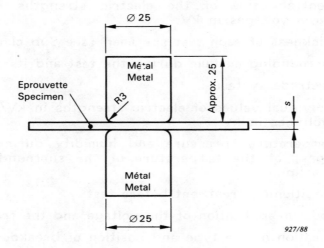

**Fig. 1b. - Electrodes de même diamètre.**
**Equal diameter electrodes.**

**Dimensions en millimètres**                    **Dimensions in millimetres**

**Fig. 1. - Disposition des électrodes pour les essais des planches et feuilles perpendiculairement à la surface.**

**Electrode arrangements for tests on boards and sheets normal to the surface.**

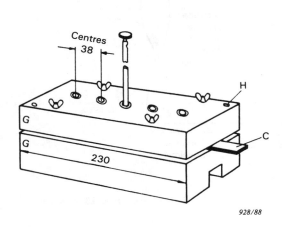

Centres
38

H

G

G

C

230

*928/88*

Dimensions en millimètres

Fig. 2a. - Disposition générale de l'appareil.

General arrangement of apparatus.

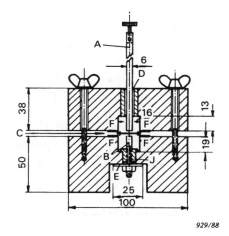

A

6

D

38

16

13

C

F

F

F

F

19

50

B

J

E

25

100

*929/88*

Dimensions in millimetres

Fig. 2b. - Coupe de l'appareil passant par les électrodes (partie supérieure légèrement surélevée).

Section of apparatus through electrodes with top slightly raised.

A = électrode supérieure ajustée dans la douille D
upper electrode to be an easy fit in bush D

B = électrode inférieure
lower electrode

C = éprouvette en essai
specimen under test

D = douille de laiton avec diamètre intérieur juste suffisant pour laisser passer une tige de 6 mm
brass bush with inside diameter just sufficient to clear 6 mm rod

E = bande de laiton de 25 mm de large reliant toutes les électrodes inférieures
brass strip 25 mm wide connecting all lower electrodes

F = bandes de tissu verni à cheval sur les bords de l'éprouvette
pieces of varnished cloth strip overlapping edges of specimen

G = blocs de matériau isolant approprié, par exemple en papier stratifié imprégné
blocks of suitable insulating material, for example a paper-filled laminate

H = trou du goujon
dowel hole

J = douille en laiton avec filetage intérieur
brass bushing with internal thread

Fig. 2. - Exemple type de dispositif d'électrodes pour essais sur bandes perpendiculairement à la surface.

Typical example of electrode arrangement for tests on tapes normal to the surface.

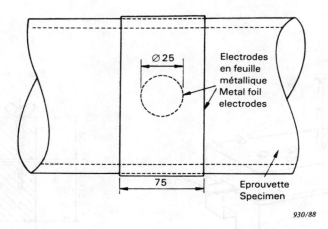

Ø 25

Electrodes
en feuille
métallique
Metal foil
electrodes

75

Eprouvette
Specimen

*930/88*

**Dimensions en millimètres**                    **Dimensions in millimetres**

Fig. 3. - Disposition des électrodes pour essais perpendiculairement
         à la surface de tubes et de cylindres d'un diamètre intérieur
         supérieur à 100 mm.

         Electrode arrangement for tests normal to the surface on tubes
         and cylinders with internal diameter greater than 100 mm.

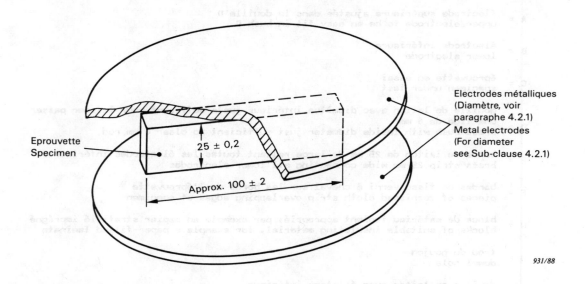

Electrodes métalliques
(Diamètre, voir
paragraphe 4.2.1)
Metal electrodes
(For diameter
see Sub-clause 4.2.1)

Eprouvette
Specimen

25 ± 0,2

Approx. 100 ± 2

*931/88*

**Dimensions en millimètres**                    **Dimensions in millimetres**

Fig. 4. - Disposition des électrodes pour essais parallèlement à la surface
         des matériaux (et parallèlement aux strates des matériaux
         stratifiés, s'il y a lieu).

         Electrode arrangement for tests parallel to the surface
         (and along the laminae, if present).

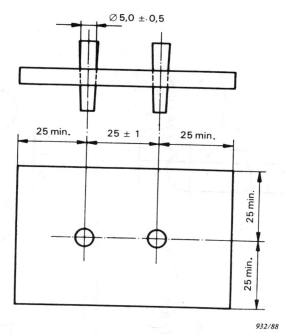

Fig. 5a. - Eprouvette plate avec électrodes à broches coniques.
Plate specimen with taper pin electrodes.

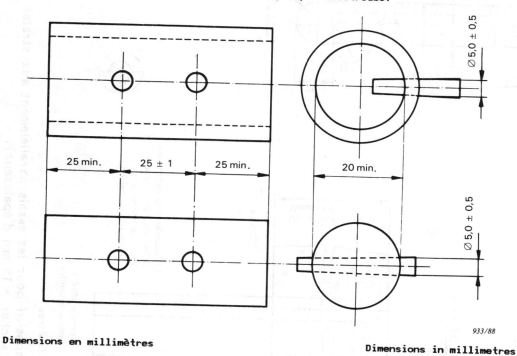

Dimensions en millimètres

Dimensions in millimetres

Fig. 5b. - Eprouvettes tubulaires et tiges cylindriques avec électrodes à broches coniques.
Tube or rod specimens with taper pin electrodes.

Fig. 5. - Disposition des électrodes à broches coniques pour essais
parallèlement à la surface des matériaux (et parallèlement aux strates s'il y a lieu).

Electrode arrangement for tests parallel to the surface
(and along the laminae if present).

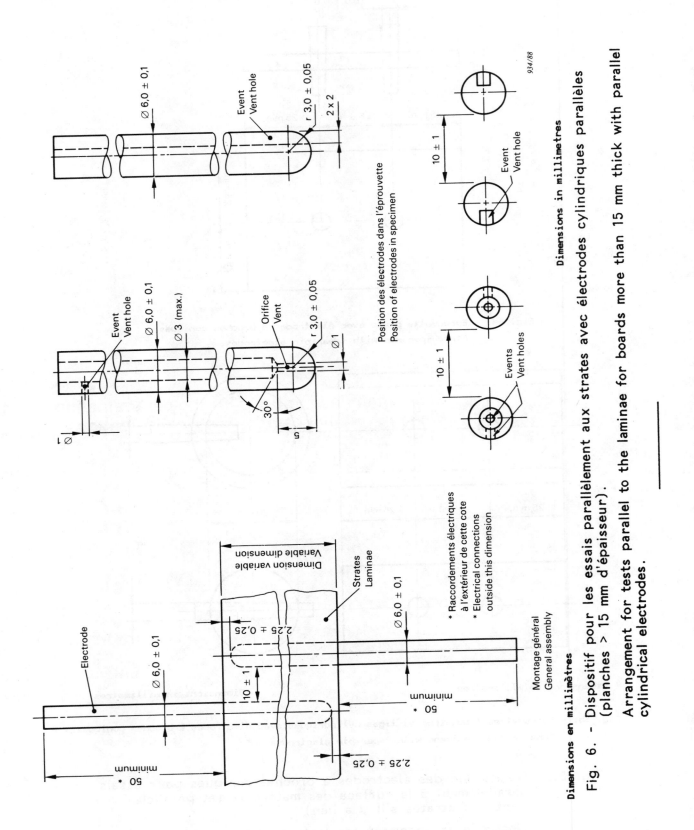

934/88

Position des électrodes dans l'éprouvette
Position of electrodes in specimen

Dimensions in millimetres

Ø 6,0 ± 0,1

Event
Vent hole

r 3,0 ± 0,05

2 x 2

Event
Vent hole

Ø 6,0 ± 0,1

Ø 3 (max.)

Orifice
Vent

r 3,0 ± 0,05

Ø 1

30°

5

Ø 1

10 ± 1

Event
Vent hole

10 ± 1

Events
Vent holes

Electrode

Ø 6,0 ± 0,1

Dimension variable
Variable dimension

Strates
Laminae

2,25 ± 0,25

10 ± 1

Ø 6,0 ± 0,1

2,25 ± 0,25

50 *
minimum

50 *
minimum

* Raccordements électriques
à l'extérieur de cette cote
* Electrical connections
outside this dimension

Montage général
General assembly

Dimensions en millimètres

Fig. 6. - Dispositif pour les essais parallèlement aux strates avec électrodes cylindriques parallèles
(planches > 15 mm d'épaisseur).

Arrangement for tests parallel to the laminae for boards more than 15 mm thick with parallel
cylindrical electrodes.

# NORME INTERNATIONALE

# INTERNATIONAL STANDARD

**CEI
IEC
250**

Première édition
First edition
1969

Méthodes recommandées pour la détermination
de la permittivité et du facteur de dissipation des
isolants électriques aux fréquences industrielles,
audibles et radioélectriques (ondes métriques
comprises)

Recommended methods for the determination
of the permittivity and dielectric dissipation
factor of electrical insulating materials at power,
audio and radio frequencies including metre
wavelengths

Bureau Central de la Commission Electrotechnique Internationale  3, rue de Varembé  Genève, Suisse

Commission Electrotechnique Internationale
International Electrotechnical Commission
Международная Электротехническая Комиссия

CODE PRIX
PRICE CODE          **U**

*Pour prix, voir catalogue en vigueur*
*For price, see current catalogue*

# CONTENTS

# INTERNATIONAL ELECTROTECHNICAL COMMISSION

---

# RECOMMENDED METHODS FOR THE DETERMINATION OF THE PERMITTIVITY AND DIELECTRIC DISSIPATION FACTOR OF ELECTRICAL INSULATING MATERIALS AT POWER, AUDIO AND RADIO FREQUENCIES INCLUDING METRE WAVELENGTHS

---

## FOREWORD

1) The formal decisions or agreements of the I E C on technical matters, prepared by Technical Committees on which all the National Committees having a special interest therein are represented, express, as nearly as possible, an international consensus of opinion on the subjects dealt with.

2) They have the form of recommendations for international use and they are accepted by the National Committees in that sense.

3) In order to promote this international unification, the I E C expresses the wish that all National Committees having as yet no national rules, when preparing such rules, should use the I E C recommendations as the fundamental basis for these rules in so far as national conditions will permit.

4) The desirability is recognized of extending international agreement on these matters through an endeavour to harmonize national standardization rules with these recommendations in so far as national conditions will permit. The National Committees pledge their influence towards that end.

## PREFACE

This Recommendation has been prepared by I E C Technical Committee No. 15, Insulating Materials.

A first draft was discussed at the meeting held in Venice in 1963, as a result of which a final draft was submitted to the National Committees for approval under the Six Months' Rule in November 1964.

The following countries voted explicitly in favour of publication:

| | |
|---|---|
| Australia | Korea (Republic of) |
| Austria | Netherlands |
| Belgium | Norway |
| Canada | Romania |
| Czechoslovakia | South Africa |
| Denmark | Sweden |
| Finland | Switzerland |
| France | Turkey |
| Germany | Union of Soviet Socialist Republics |
| Israel | United Kingdom |
| Italy | United States of America |
| Japan | |

---

# RECOMMENDED METHODS FOR THE DETERMINATION OF THE PERMITTIVITY AND DIELECTRIC DISSIPATION FACTOR OF ELECTRICAL INSULATING MATERIALS AT POWER, AUDIO AND RADIO FREQUENCIES INCLUDING METRE WAVELENGTHS

1. **Object and scope**

This Recommendation applies to the procedures for determination of permittivity and dissipation factor and of quantities calculated from them, such as loss index, within the frequency range 15 Hz to 300 MHz approximately. Some of the methods described in this Recommendation can, with special precautions, be used for measurements at frequencies considerably lower or higher than the given limits.

Liquids and fusible materials, as well as solid materials, can be measured by the methods described. The measured values are dependent on physical conditions such as frequency, temperature and moisture content, and in special cases, on field strength as well.

In some cases, tests at voltages exceeding 1 000 V may introduce effects not related to permittivity and dielectric dissipation factor, and are not described here.

2. **Definitions**

2.1    *The relative permittivity* $\varepsilon_r$ of an insulating material is the ratio of capacitance $C_x$ of a capacitor, in which the space between and around the electrodes is entirely and exclusively filled with the insulating material in question, to the capacitance $C_o$ of the same configuration of electrodes in vacuum:

$$\varepsilon_r = \frac{C_x}{C_o} \tag{1}$$

The relative permittivity $\varepsilon_r$ of dry air free from carbon dioxide, at normal atmospheric pressure, equals 1.000 53 so that in practice, the capacitance $C_a$ of the configuration of electrodes in air can normally be used instead of $C_o$ to determine the relative permittivity $\varepsilon_r$ with sufficient accuracy.

The *permittivity* of an insulating material is, in a measurement system, the product of its relative permittivity $\varepsilon_r$ and the electric constant (or permittivity of vacuum) $\varepsilon_o$ in that measurement system.

In the SI system, the absolute permittivity is expressed in farad per metre (F/m); furthermore, in SI units, the electric constant $\varepsilon_o$ has the following value:

$$\varepsilon_o = 8.854 \times 10^{-12} \text{ F/m} \approx \frac{1}{36\,\pi} \times 10^{-9} \text{ F/m} \tag{2}$$

For the purpose of this Recommendation, where picofarads and centimetres are used in calculating capacitance, the electric constant is:

$$\varepsilon_o = 0.088\,54 \text{ pF/cm}$$

**2.2**  *The dielectric loss angle* $\delta$ of an insulating material is the angle by which the phase difference between applied voltage and resulting current deviates from $\pi/2$ rad, when the dielectric of the capacitor consists exclusively of the dielectric material.

**2.3**  *The dielectric dissipation factor* [1]) *tan* $\delta$ of an insulating material is the tangent of the loss angle $\delta$.

**2.4**  *The loss index* $\varepsilon_r''$ of an insulating material is equal to the product of its dissipation factor tan $\delta$ and its relative permittivity $\varepsilon_r$.

**2.5**  *The relative complex permittivity* $\varepsilon_r^*$ is derived by combining the relative permittivity and the loss index:

$$\varepsilon_r^* = \varepsilon_r' - j\underline{\varepsilon}_r'' \tag{3}$$

$$\varepsilon_r' = \varepsilon_r \tag{4}$$

$\varepsilon_r$ being the relative permittivity defined in Sub-clause 2.1:

$$\varepsilon_r'' = \varepsilon_r \tan \delta \tag{5}$$

$$\tan \delta = \frac{\varepsilon_r''}{\varepsilon_r'} \tag{6}$$

*Note.* — A capacitor with losses can be represented at any given frequency either by capacitance $C_s$ and resistance $R_s$ in series, or by capacitance $C_p$ and resistance $R_p$ (or conductance $G_p$) in parallel.

Equivalent parallel circuit:                                                Equivalent series circuit:

$$\tan \delta = \frac{1}{\omega C_p R_p} = \frac{G_p}{\omega C_p} \tag{7} \qquad\qquad\qquad \tan \delta = \omega C_s R_s \tag{8}$$

While the parallel representation of an insulating material having a dielectric loss is usually the more proper representation, it is always possible and occasionally desirable to represent a capacitor at a single frequency by a capacitance $C_s$ in series with a resistance $R_s$.

---

[1]) Certain countries refer to "loss tangent" in preference to "dielectric dissipation factor" because the result of the measurement of the loss is reported as the tangent of the loss angle.

Between the series components and the parallel components, the following relations hold:

$$C_p = \frac{C_s}{1 + \tan^2 \delta} \tag{9}$$

$$R_p = \frac{1 + \tan^2 \delta}{\tan^2 \delta} R_s \tag{10}$$

$$\omega C_s R_s = \frac{1}{\omega C_p R_p} \tag{11}$$

The dielectric dissipation factor tan $\delta$ is the same for the series and parallel representations.

If a measuring circuit yields results in terms of series components, and if $\tan^2 \delta$ is too large to be ignored in equation (9), then the parallel capacitance must be evaluated before the permittivity is calculated.

The calculations and measurements in this Recommendation are based on a sinusoidal wave-form of the current with $\omega = 2\pi f$.

## 3. Uses and properties of electrical insulating materials

### 3.1 Object of dielectrics

Dielectric materials are used in general in two distinct ways:

— to support components of an electrical network and insulate them from each other and from ground, and

— to function as the dielectric of a capacitor.

### 3.2 Factors influencing dielectric properties

In the following, the influences of frequency, temperature, moisture and field strength on dielectric properties are separately treated.

#### 3.2.1 Frequency

As only a few materials, such as fused silica, polystyrene, or polyethylene, have $\varepsilon_r$ and tan $\delta$ practically constant over the wide frequency range through which dielectric materials are used for technical purposes, it is necessary to measure the dissipation factor and the permittivity at those frequencies at which the dielectric material will be used.

Changes in permittivity and in dissipation factor are produced by the dielectric polarization and conductivity. The most important changes are caused by dipole polarization due to polar molecules and interfacial polarization caused by inhomogeneities in the material.

#### 3.2.2 Temperature

The loss index may show a maximum at a frequency which depends upon the temperature of the dielectric material. The temperature coefficients of dissipation factor and permittivity can be positive or negative depending on the position of the loss index maximum with respect to the measuring temperature.

366

### 3.2.3 *Moisture*

The degree of polarization is increased by absorption of water or by the formation of a water film on the surface of the dielectric material, thus raising the permittivity, the dissipation factor and the d.c. conductivity. Conditioning of test specimens is therefore of decisive importance and control of the moisture content, both before and during testing, is imperative if test results are to be interpreted correctly.

*Note.* — The gross effects of humidity usually occur at frequencies below approximately 1 MHz and in the micro-wave frequency region.

### 3.2.4 *Field strength*

When interfacial polarization exists, the number of free ions increases with the field strength, and the magnitude and the position of the loss index maximum is altered.

At higher frequencies, permittivity and dissipation factor are independent of the field strength, so long as no partial discharge occurs in the dielectric.

## 4. Form of specimen and electrode arrangement

### 4.1 *Solid insulating materials*

#### 4.1.1 *Geometry of solid specimens*

For determining the permittivity and dissipation factor of a material, sheet specimens are preferable; but material may be available only in tubular form.

When high accuracy is needed in measuring permittivity, the source of the greatest uncertainty is the dimensions of the specimen, and particularly its thickness, which should therefore be large enough to allow its measurement with the required accuracy. The choice of thickness depends on the method of producing the specimen and the likely variation in thickness from point to point. For 1% accuracy, a thickness such as 1.5 mm is usually enough, although for greater accuracy it may be desirable to use a thicker specimen, for example 6 mm – 12 mm. The thickness must be determined by measurements distributed systematically over the area of the specimen which is used in the electrical measurement, and should be uniform to within ± 1% of the average thickness. When the electrodes extend to the edge of the specimen, the thickness can be determined by weighing, if the density of the material is known. The *area* chosen for the specimen should be such as to provide a specimen capacitance which can be measured to the desired accuracy. With well guarded and screened apparatus, there need be no difficulty in measuring capacitances of 10 pF. Much existing apparatus, however, is limited in resolution to about 1 pF and then the specimen should be thin and of a diameter of 10 cm or more.

When small values of dissipation factor are being measured, it is essential that the loss introduced by the series resistance of the leads be as small as possible, that is, the product of the resistance and the capacitance being measured should be as small as possible. Also, the ratio of the measured capacitance to the total capacitance should be as large as possible. The first point indicates a need for keeping the lead resistances as low as possible and the desirability of having a small specimen capacitance. The second point indicates the need for low total capacitance in the arm of the bridge to which the specimen is connected and the desirability of having a large specimen capacitance. Frequently the best compromise is a specimen having a capacitance of about 20 pF, used with a measuring circuit which does not connect more than about 5 pF in parallel with the specimen.

### 4.1.2 *Electrode systems*

#### 4.1.2.1 *Electrodes applied to the specimen*

Electrodes may be applied to the surface of the specimen using one of the materials listed in Sub-clause 4.1.3. When a guard-ring is not used and when there is difficulty in locating electrodes accurately opposite one another on the two faces of the specimen, one electrode should be larger than the other. The specimen with its own electrodes should then be mounted between metal backing electrodes, these being slightly smaller than the specimen electrodes. The equations for computing the capacitance of different arrangements of disk-shaped or cylindrical electrodes as well as empirical equations for computing the approximate edge capacitances for this condition are given in Table I, page 45. These equations hold only for a restricted range of shapes of specimens.

For measurement of dissipation factor, electrodes of this type are unsatisfactory at high frequencies, unless the surfaces of the specimen and the metal plates are very flat. The electrode system of Figure 1, page 50, requires specimens also to be of uniform thickness.

#### 4.1.2.2 *No electrodes applied to the specimen*

Specimens of sufficiently low surface conductivity can be tested without applied electrodes by inserting them in an electrode system in which there is an intentional gap, occupied by air or liquid, on one or both sides of the specimen.

The equations for computing the capacitances of arrangements of plane or cylindrical electrodes are given in Table III.

Two forms of apparatus are particularly convenient:

#### 4.1.2.2.1 *In air,* with micrometer-controlled parallel electrodes.

The capacitance can be adjusted to the same value with and without the specimen inserted, and the permittivity determined without reference to the electrical calibration of the measuring system.

A guard electrode can be included in the electrode system.

#### 4.1.2.2.2 *Fluid displacement method*

In a liquid, whose permittivity is nearly equal to that of the specimen and whose dissipation factor is negligible, the measurement depends less critically than usual on exact knowledge of the thickness of the specimen. By using two fluids in turn, the thickness of the specimen and the dimensions of the electrode system can be eliminated from the equations.

The test specimen should be a disk having the same diameter as the cell electrodes, or for micrometer electrodes, the specimen may be sufficiently small to render edge effects negligible. To make the edge effects negligible in the micrometer electrodes, the specimen diameter should be smaller than that of the micrometer electrodes by at least twice the thickness of the specimen.

#### 4.1.2.3 *Edge effects*

To avoid errors in permittivity caused by edge effects, the electrode system may include a guard electrode. If so, its width should be at least twice the thickness of the specimen, and the gap between it and the main electrode should be small compared with the thickness of the specimen. If guard rings cannot be used, a correction must usually be made for edge capacitance; approximate equations are given in Table I. These equations are empirical and hold only for a restricted range of shapes of specimen.

Alternatively, the edge capacitance may be found from measurements both with and without a guard ring at a convenient frequency and temperature; the edge capacitance so found will be sufficiently accurate for use as a correction at other frequencies and temperatures.

### 4.1.3 *Materials for the constitution of electrodes*

#### 4.1.3.1 *Metal foil electrodes*

Metal foil electrodes can be applied to the specimen by using the smallest possible quantity of silicone grease or of any other suitable low-loss adhesive. The metal foil may be of pure tin or lead or an alloy of these metals, of thickness up to 100 μm, or of aluminium of thickness less than 10 μm. Aluminium foils, however, are liable to form an electrically insulating oxide film which may influence the measuring results. Gold foil can also be used.

#### 4.1.3.2 *Fired-on metal electrodes*

Metal films can be fired on glass, mica and ceramics, and are suitable for tests on these materials. Silver is commonly used, but migrates under a potential difference at high temperatures or high humidities. Gold is better.

#### 4.1.3.3 *Electrodes produced by spraying metal*

Zinc or copper electrodes may be sprayed on to the specimen. They conform readily to a rough surface. They may be applied even to cloth since they do not penetrate very small holes.

#### 4.1.3.4 *Metal electrodes applied by cathodic evaporation or evaporation in high vacuum*

These procedures may be used if the resultant stresses neither change nor damage the insulating material, and if the material does not emit excessive gas when subjected to vacuum. The edges of any such electrodes shall be sharply defined.

#### 4.1.3.5 *Mercury and other liquid metal electrodes*

These can be used by clamping the specimen between suitable hollow blocks and filling with liquid metal which must be clean. Mercury should not be used at high temperatures, and precautions should be taken even when using it at room temperature, as its vapour is toxic.

Wood's metal or other low-melting alloy can be used instead of mercury. These alloys frequently contain cadmium which, like mercury, is a toxic element. These alloys should be used above 100 °C only in a well-ventilated room or preferably in a hood and the staff told of the possible health hazards.

#### 4.1.3.6 *Conducting paint*

Certain types of high-conductivity silver paints, either air-drying or low-temperature-baking varieties, are commercially available for use as electrode material. They may be sufficiently porous to permit diffusion of moisture through them and thereby allow the test specimen to condition after application of the electrodes. This is particularly useful in studying humidity effects. The paint has the disadvantage of not being ready for use immediately after application. It usually requires overnight air-drying or low-temperature baking to remove all traces of solvent, which otherwise may increase both permittivity and dissipation factor.

The edges of any such electrodes shall be sharply defined; this may be difficult when the paint is brushed on, but this limitation can usually be overcome by spraying the paint and employing either clamp-on or pressure-sensitive masks. The conductivity of silver-paint electrodes may be low enough to give trouble at the highest frequencies.

It is essential that the solvent of the paint has no permanent effect on the specimen.

### 4.1.3.7 *Graphite*

Graphite is not recommended but may sometimes be used, especially at lower frequencies. Its resistance may cause an appreciable increase in loss angle and if it is applied from a suspension in a liquid it may penetrate the specimen.

### 4.1.4 *Choice of electrodes*

### 4.1.4.1 *Sheet specimens*

Two considerations are important:

*a*) Working without applied electrodes is quick and convenient, and avoids uncertainty about the effectiveness of the contact between electrodes and specimen.

*b*) The proportional error in relative permittivity:

$$\frac{\Delta \varepsilon_r}{\varepsilon_r}$$

resulting from an error:

$$\frac{\Delta h}{h}$$

in the measurement of the specimen thickness $h$, is given by:

$$\frac{\Delta \varepsilon_r}{\varepsilon_r} = \frac{\Delta h}{h} \qquad (12)$$

if electrodes are applied to the specimen, but if the specimen is placed between electrodes of a fixed spacing $s > h$, it is given by:

$$\frac{\Delta \varepsilon_r}{\varepsilon_r} = \left(1 - \frac{\varepsilon_r}{\varepsilon_t}\right) \cdot \frac{\Delta h}{h} \qquad (13)$$

when $\varepsilon_t$ is the relative permittivity of the fluid in which the specimen is immersed, being unity for the measurement in air.

For non-porous materials having relative permittivities above about 10, deposited metal electrodes should be used. For such materials, the electrodes should cover the whole surfaces of the specimen, and no guard electrode is necessary. For materials having relative permittivities between about 3 and 10, the electrodes allowing the best accuracy are foils, mercury or deposited metal, and they must be chosen to suit the properties of the material. But, if sufficient accuracy can be obtained in the measurement of thickness, the method with no electrodes applied to the specimen may be preferable on the grounds of convenience. The liquid-immersion method is excellent if suitable liquids exist and their relative permittivities are known or can be determined with sufficient accuracy.

#### 4.1.4.2 *Tube specimens*

The most appropriate electrode system for a tube specimen will depend on its permittivity, wall thickness, diameter, and the accuracy of measurement required. In general, the electrode system should consist of an inner electrode and a somewhat narrower outer electrode, with a guard electrode at each end. The gap between the outer and guard electrodes should be small compared with the thickness of the tube wall. For tube specimens of small and medium diameters, three bands of foil or deposited metal can be applied to the outside of the tube, the centre band serving as the working outer electrode with the two bands of foil or deposited metal, one on each side, serving as guard electrodes. Inner electrodes of mercury, deposited metal film, or a tightly fitting mandrel may be used.

For tube specimens of high permittivity, the inner and outer electrodes may extend the complete length of the tube and the guard electrodes may be dispensed with.

For tubes or cylinders of large diameter, the electrode system can be either a circular or rectangular patch, a portion only of the tube periphery being tested. Such specimens can be treated as sheet specimens. Inner electrodes of metal foil, deposited metal film, or a tightly fitting mandrel are employed with outer and guard electrodes of metal foil, or deposited metal. A flexible, expanding clamp may be necessary inside the tube to ensure satisfactory contact between the inner electrode and the specimen if a foil electrode is used.

For very accurate measurements, a system with no electrodes applied to the specimen may be used provided sufficient accuracy can be obtained in the measurement of thickness. For tube specimens having relative permittivities $\varepsilon_r$ up to about 10, the most convenient electrodes are foils, mercury or deposited metal. For tube specimens having relative permittivities above about 10, deposited metal electrodes should be employed; fired-on electrodes should be used for ceramic tubes. The electrodes may be applied to the complete circumference of the tube as bands or to only a portion of the circumference.

### 4.2 *Liquid insulating materials*

#### 4.2.1 *Design of cells*

The essential features of an electrode system for testing liquids having low dielectric dissipation factor are: that it can easily be cleaned, reassembled if necessary and filled without disturbing the relative positions of the electrodes. Other desirable features are: that it should need only a small amount of liquid, that the electrode materials do not affect the liquid and vice versa, that its temperature should be easily controllable, that the terminals and connections should be adequately screened and that the insulating supports for the electrodes should not be immersed in the liquid. Furthermore, the cell should not contain too short creepage distances and sharp edges which otherwise could influence the measuring accuracy.

Details of cells meeting the above requirements are given in Figures 2 to 4, pages 51 to 53. The electrodes are of stainless steel and the insulation of borosilicate glass or fused quartz. Cells of Figures 2 and 3, which are also usable for resistivity measurements, are described in detail in I E C Publication 247, Recommended Test Cells for Measuring the Resistivity of Insulating Liquids and Methods of Cleaning the Cells.

As some liquids, as for example chlorides, reveal some significant dependence of the dielectric dissipation factor on the electrode material, electrodes made of stainless steel are not always appropriate. Much more stable results sometimes have been obtained working with electrodes made of aluminium and duraluminium.

### 4.2.2 *Preparation of cells*

The cell should be cleaned with one or more appropriate solvents or a succession of solvents which have previously been checked to ensure that they do not contain unstable compounds, either by chemical tests for purity or by ascertaining that they lead to correct results on a sample of liquid of known low permittivity and dielectric dissipation factor. When cells are used for testing some types of insulating fluids, it may be necessary to clean the electrode surfaces of the cell with a mildly abrasive detergent and water as the use of solvents alone does not always result in the removal of contamination products. If a series of solvents is used, it should end with the use of analytical grade petroleum ether with a maximum boiling point less than 100 °C, or, alternatively, with any solvent which is known to give the correct values for a liquid of known low permittivity and dielectric dissipation factor and chemically similar to the liquid to be tested. The technique described below is recommended.

The cell should be dismantled completely and all parts thoroughly cleaned with the chosen solvents, either by a reflux procedure or by repeated washings with agitation in a fresh solvent. All parts should be shaken free of solvent and placed in an uncontaminated oven at approximately 110 °C for 30 min.

The parts should be allowed to cool to a few degrees above room temperature and then reassembled. The cell should then be filled with some of the liquid to be measured, allowed to stand for a few minutes, emptied and refilled. The supporting insulation should not be wetted by the liquid.

At all stages, the parts should be manipulated with clean hooks or tongs so that no effective internal surfaces are touched with the hands.

Notes 1. — During routine testing of oils of the same quality, the cleaning procedure described above can be foregone by merely rubbing the cell after each test with a dry paper which leaves no waste.

2. — Appropriate precautions against fire and toxic effects on personnel must be observed when using solvents: some, notably benzene, carbon tetrachloride, toluene, and xylene, are particularly toxic. Further, chlorinated solvents are subject to decomposition by light.

### 4.2.3 *Calibration of cells*

When high accuracy in determining the relative permittivity of liquid dielectrics is needed, the "electrode constant" should be determined preliminarily by means of a calibration liquid of known relative permittivity, e.g. benzene.

The "electrode constant" $C_c$ is determined by the formula:

$$C_c = \frac{C_n - C_o}{\varepsilon_n - 1} \tag{14}$$

where:

$C_o$ = the capacitance of electrode arrangement in air

$C_n$ = the capacitance of electrode arrangement filled with the calibration liquid

$\varepsilon_n$ = the relative permittivity of the calibration liquid.

The difference of values $C_o$ and $C_c$ gives the correction capacitance:

$$C_g = C_o - C_c \qquad (15)$$

which is taken into account while calculating the relative permittivity $\varepsilon_x$ of the unknown liquid according to:

$$\varepsilon_x = \frac{C_x - C_g}{C_c} \qquad (16)$$

where:

$C_x$ = the capacitance of electrode arrangement filled with the liquid to be tested.

Maximum accuracy in $\varepsilon_x$ is obtained if values $C_o$, $C_n$, and $C_x$ are determined at that temperature for which the value $\varepsilon_n$ is known.

The application of the described method ensures that sufficiently accurate results are obtained in determining the relative permittivity of liquid dielectrics, as it eliminates errors made either because of parasitic capacitances or through inaccurate measuring of the value of the gap between the electrodes.

## 5. Choice of measuring methods

Methods for measuring permittivity and dissipation factor can be divided into two groups: null methods and resonance methods.

5.1 *Null methods* are used at frequencies up to about 50 MHz. For measurements of permittivity and dissipation factor, substitution techniques can be used; that is, the bridge is balanced, by adjustment mainly in one arm of the network, both with and without the specimen connected. The networks normally used are the Schering bridge, the transformer bridge (i.e. a bridge with ratio arms coupled by mutual inductance) and the parallel-T. The transformer bridge has the advantage of allowing the use of a guard electrode without any additional components or operations; it has no disadvantages in comparison with the other networks.

5.2 *Resonance methods* can be used in the range 10 kHz to several hundred megahertz. They are invariably substitution methods. The method commonly used is that of reactance-variation. These methods cannot easily be adapted for use with guard electrodes.

*Note.* — Examples of typical bridges and circuits are to be found in the Appendix. This enumeration can by no means be complete. Further information describing the bridges and the methods for making the measurements may be found in the literature and also in pamphlets of the firms producing such apparatus.

## 6. Testing procedure

6.1 *Preparation of specimens*

.The specimen shall be cut from the solid material or prepared by an appropriately standardized technique in order to obtain a determined initial condition.

The subsequent measurement of thickness shall be made accurately with a tolerance of ± (0.2% + 0.005 mm). The measuring points shall be distributed uniformly over the surface of the specimen. The effective area is also determined if necessary.

## 6.2 *Conditioning*

Conditioning should be made in accordance with the relevant specifications.

## 6.3 *Measurement*

Electrical measurements are made appropriate to the method employed, following this Recommendation and the recommendations of the makers of the equipment used.

At frequencies of the order of 1 MHz or more, care must be taken that the inductance of the connecting leads does not influence the results. A coaxial-lead system is useful. (The system in Figure 1, page 50, also provides a built-in vernier capacitor for use in the reactance-variation method of measurement.)

## 7. Results

### 7.1 *Relative permittivity $\varepsilon_r$*

The relative permittivity $\varepsilon_r$ of a specimen provided with its own electrodes is calculated according to equation (1). As the measured capacitance $C'_x$ of a specimen without guard-rings contains a small amount of edge capacitance $C_e$, the relative permittivity is:

$$\varepsilon_r = \frac{C'_x - C_e}{C_0} \tag{17}$$

where:

$C_0$ and $C_e$ can be calculated from Table I, page 45.

If necessary, corrections should be made similarly for capacitance of the specimen to earth, capacitance between switch contacts and the difference between equivalent series and parallel capacitances.

Relative permittivity of specimens measured between micrometer electrodes or between non-contacting electrodes is calculated according to the appropriate equations of Tables II and III, pages 47 and 49.

### 7.2 *The dielectric dissipation factor tan δ*

The dielectric dissipation factor tan δ shall be calculated from the measured values in accordance with the equations given for the particular measuring arrangement used.

### 7.3 *Accuracy to be expected*

The methods outlined in Clause 5 and in the Appendix envisage a degree of accuracy in the determination of permittivity of ± 1% and of dissipation factor of ± (5% + 0.0005). These accuracies depend upon at least three factors: the accuracy of the observations for capacitance and dissipation factor, the accuracy of the corrections to these quantities caused by the electrode arrangement used and the accuracy of the calculation of the direct interelectrode vacuum capacitance (Table I).

Under favourable conditions and at the lower frequencies, capacitance can be measured with an accuracy of ± (0.1% + 0.02 pF) and dissipation factor of ± (2% + 0.000 05). At the higher frequencies, these limits may increase for capacitance to ± (0.5% + 0.1 pF) and for dissipation factor to ± (2% + 0.0002).

Dielectric specimens provided with a guard electrode are subject only to an error in the calculation of the direct interelectrode vacuum capacitance. The error caused by too wide a gap between the guarded and the guard electrodes will generally amount to several tenths per cent, and the correction can be calculated to a few per cent of itself. The error in measuring the thickness of the specimen can amount to a few tenths per cent for an average thickness of 1.6 mm, on the assumption that it can be measured to $\pm$ 0.005 mm. The diameter of a circular specimen can be measured to an accuracy of $\pm$ 0.1%, but enters as the square. Combining these errors, the direct interelectrode vacuum capacitance can be determined to an accuracy of $\pm$ 0.5%.

The capacitance of specimens with electrodes applied to the surface, measured with micrometer electrodes, needs no corrections other than that for direct interelectrode capacitance, provided the specimens are sufficiently smaller in diameter than the micrometer electrodes. When two-terminal specimens are measured in any other manner, the calculation of edge and ground capacitance will involve considerable error, since each may be from 2% to 40% of the specimen capacitance. With the present knowledge of these capacitances, there may be a 10% error in calculating the edge capacitance and a 25% error in evaluating the ground capacitance. Hence the total error involved may be from several tenths to a few per cent. However, when neither electrode is grounded, the ground capacitance error is greatly reduced.

With micrometer electrodes, it is possible to measure dissipation factor of the order of 0.03 to within $\pm$ 0.0003 of the true value and dissipation factor of the order of 0.0002 to within $\pm$ 0.000 05 of it. The range of dissipation factor is normally 0.0001 to 0.1 but may be extended above 0.1. Between 10 MHz and 20 MHz it is possible to detect a dissipation factor of 0.000 02. Relative permittivities from 1 to 5 may be measured to $\pm$ 2% of the true value. The accuracy is limited by the accuracy of the measurements required in the calculation of direct interelectrode vacuum capacitance and by errors in the micrometer electrode system.

8.  **Test report**

In the test report, the following information shall be given when relevant:

– Type and designation of the insulating material as well as the form in which it is delivered. Method of sampling, shape, dimensions of the test specimen and date of sampling (statements on the specimen thickness and, if necessary, exact information on the treatment of the specimens at the contact areas of the electrodes are important).

– Method and dur___ of conditioning the specimens.
– Electrode arrange___nt and type of electrode, applied to specimen, if any.
– Measuring apparatus.
– Temperature and relative humidity during the test and temperature of specimen.
– Applied voltage.
– Applied frequency.
– Relative permittivity $\varepsilon_r$ (average value).
– Dielectric dissipation factor tan $\delta$ (average value).
– Date of test.

Values of relative permittivity and dielectric dissipation factor and the values calculated from them as loss index and loss angle shall be given, if necessary, in relation to temperature and frequency. Not all are necessary or even appropriate in all cases.

# APPENDIX

## APPARATUS

1. **Schering bridge**

1.1 *General*

The Schering bridge represents the most classical device for the measurement of permittivity and dielectric losses. It is used in the frequency range from below power frequencies (50 Hz and 60 Hz) up to the order of 100 kHz and for capacitances of 50 pF to 1 000 pF (usual capacitances of specimens or equipment to be tested).

It is a four-arm network (Figure 5, page 53), two of these arms being primarily capacitive (the unknown capacitance $C_X$ and a capacitance $C_N$ without losses) and the two others (often called measuring arms) consisting of non-reactive resistances $R_1$ and $R_2$, the resistance ($R_1$) opposite to the unknown capacitance $C_X$, at least, being shunted by a capacitance ($C_1$). This last capacitance, and generally one at least of the two resistances $R_1$ and $R_2$ are adjustable.

If one chooses for the capacitance $C_X$ the equivalent representation of a resistance $R_S$ and a (pure) capacitance $C_S$ in series, the balance of the bridge illustrated in Figure 5 leads to:

$$C_S = C_N \frac{R_1}{R_2} \tag{18}$$

and:

$$\tan \delta_X = \omega C_S R_S = \omega C_1 R_1 \tag{19}$$

If the resistance $R_2$ is shunted by a capacitance $C_2$, the expression for $\tan \delta$ becomes:

$$\tan \delta_X = \omega C_1 R_1 - \omega C_2 R_2 \tag{20}$$

The practical realizations of the bridge will differ appreciably according to the frequency range. This is a result of the fact that a capacitance of 50 pF to 1 000 pF represents an impedance of about 60 MΩ to 3 MΩ at 50 Hz and only about 30 000 Ω to 1 500 Ω at 100 kHz.

In the latter case, the four arms of the bridge may easily have impedances of the same order of magnitude, while this will never be the case at frequencies of 50 Hz or 60 Hz. We are thus led to distinguish two different types of use, at low and at (relatively) high frequency.

1.2 *Low-frequency bridge*

This will normally be a high-voltage bridge, not only for reasons of sensitivity but also because, at low frequencies, it is high-voltage technology which is nearly exclusively interested in the problem of dielectric losses. The very great difference in the order of magnitude of the impedances between the so-called capacitive and measuring arms, consequently, results in a similar inequality in the distribution of the voltage, by far the largest part of it being found across the capacitances $C_X$ and $C_N$. The balance conditions given above are only valid if the low-voltage elements are screened from the high-voltage elements; this screening must be earthed in order to secure the stability of the balance. As Figure 6, page 54, shows, the screening is compatible with the use of guarded capacitances $C_X$ and $C_N$, the guard for $C_N$ being practically indispensable.

The choice for the earthing procedure substantially leads to two types of bridges.

### 1.2.1 Simple Schering bridge, with screening

Point B of the bridge (supply terminal of the bridge on the side of the measuring arms) is connected to the screening and to earth.

The screening acts well as a protection towards the high-voltage side, but it increases the capacitance between it and the various conductors leading to the terminals M and N of the detector arm, this capacitance being subjected to the voltage across the measuring arms. This may introduce an error which normally limits the accuracy on $\tan \delta$ to the order of 0.1%, especially when capacitances $C_X$ and $C_N$ are unequal.

### 1.2.2 Schering bridge with Wagner earth circuit

This is the solution which is represented in Figure 6, page 54, and by which the detector arm and the screening are brought to the same potential. This is secured by using additional external arms $Z_A$ and $Z_B$ (Wagner earth circuit) and the intermediate point P being connected to the screening and to earth. The additional arms (practically $Z_B$) are adjusted in order that the voltages across $Z_A$ and $Z_B$ are identical to those across the capacitive and measuring arms of the bridge respectively. It is obvious that the solution consists in the simultaneous balance of two bridges, the main bridge AMNB and the auxiliary bridge AMPB (or ANPB). The double balance is reached by successive approximations, the detector being displaced from one bridge to the other. With this method, one order of magnitude may be gained on the accuracy, which is now in practice only limited by the precision of the elements used for the bridge.

It will be noted that the particular solution which has been considered requires that both terminals of the supply may be insulated from earth. If this is not possible, the use of a more complex arrangement may be required (bridge with double screen).

### 1.3 Schering bridge for high frequencies

This bridge will normally be operated with moderate voltages and will thus be more flexible; the capacitance $C_N$ will very often be adjustable (while for high voltages, this capacitance is ordinarily of fixed value) and substitution methods can be more easily adopted.

Screening and Wagner earth circuit can still be used with advantage as undesirable capacitive effects increase with the frequency.

### 1.4 Note on detectors

When point B of the Schering bridge is earthed, detectors with asymmetrical input (which are most common with electronic devices) must be avoided.

Such detectors, however, are readily used with bridges with a Wagner earth circuit provided that the earthed input terminal is always connected to point P.

## 2. Transformer bridge (inductive ratio-arm bridges)

### 2.1 General

This bridge is based on a simpler principle than the Schering bridge. Its fundamental arrangement is given in Figure 7, page 54.

When balanced, the ratio between the (complex) impedances $Z_X$ and $Z_M$ is identical to the (vector) ratio of voltages $U_1$ and $U_2$. If this last ratio be known, then $Z_X$ is readily derived from the knowledge of $Z_M$. In the ideal bridge, this ratio $U_1 / U_2$ is a pure number $k$ and thus $Z_X = kZ_M$; in particular, the argument of $Z_M$ yields directly $\delta_X$.

The great advantage of the transformer bridge over the Schering bridge is to allow a direct and rational earthing of the screens and guard electrodes without any need of additional auxiliary arms.

The bridge is used from power frequencies up to radio frequencies of several tens of megahertz; this range is thus wider than the previous one for the Schering bridge but, again, the realization will differ appreciably according to the frequency range involved.

This will again normally be a high-voltage bridge (more precisely, the voltage $U_1$ is high while the voltage $U_2$ remains moderate) and its technique is related to that of voltage transformers.

Two methods of supply are used:

1) the supply voltage is directly applied to one of the windings, the other working as the secondary of a transformer;

2) (which is illustrated in Figure 7, page 54), the supply voltage is applied to a separate primary winding while both windings of the bridge itself constitute either two separate secondary circuits or a single secondary winding with an intermediate tapping allowing both voltages $U_1$ and $U_2$ to be obtained.

As with any measuring transformer, the bridge will show errors (that is, the vector ratio $U_1 / U_2$ will differ from its theoretical value) and these errors will be load dependant. The most important is actually the phase error between voltages $U_1$ and $U_2$ which influences directly the measurement of tan $\delta$.

It is thus necessary to calibrate the bridge. This will be done by replacing $Z_X$ by a capacitance $C_N$ without losses (similar to that used in the Schering bridge). If $C_N$ has the same value as $C_X$, the method is practically a substitution and the calibration is immediate. But, as $C_N$ will seldom be adjustable, such a calibration would not remain valid for $C_X$ on account of the load variation. It is then possible to operate under constant load, as indicated in Figure 8, page 55: a switch connects $C_X$ to earth when $C_N$ is measured and vice-versa; the constant load for the high-voltage winding is then the sum of both. (Strictly speaking, a similar disposition should be used on the low-voltage side but, because the loads are much smaller, such an arrangement, although easy to use, is less essential.)

In addition, when the calibration is made with a pure capacitance $C_N$ connected across the voltage $U_1$, the measuring impedance $Z_M$, submitted to the voltage $U_2$, will consist of either:

1) a *pure* capacitance $C_M$ if $U_2$ and $U_1$ have the *same phase* (ideal case); or

2) a capacitance $C_M$ and a resistance $R_M$, if the voltage $U_2$ is *leading* on the voltage $U_1$; or

3) if the voltage $U_2$ is *lagging* with respect to $U_1$, this resistance $R_M$ should become negative. This means that, in order to restore balance, a resistive component of the current has to be introduced on the side of $U_1$. But, as very high and adjustable resistors for high voltage practically do not exist, this resistive component is usually obtained through an auxiliary winding supplying a voltage $U_3$ (of low value) in phase with $U_1$ (Figure 9, page 55).

Note. — The addition of a resistance in series with $C_N$ is impossible for the following reasons: if it were inserted below the capacitor, the potentials of the measuring electrode and of the guard of $C_N$ would no longer be the same; and if inserted before $C_N$ on the high-voltage conductor, the current in the resistance would also include the guard circuit current and no calibration would be possible.

These remarks may also be applied to the resistance $R_M$ mentioned above for the second case. But on the low-voltage side, a resistance of high and adjustable value shunting the capacitance may easily be obtained by using the star connection of three resistances $R_1$, $R_2$ and $R'$ illustrated in dotted lines at bottom of Figure 9. One has then:

$$R_M = R_1 + R_2 + \frac{R_1 R_2}{R'} \tag{21}$$

The adjustable capacitance $C_M$ for the measurement, however, must be a pure one or of low and well-known losses (while the measuring capacitance $C_1$ of the Schering bridge need not comply with such severe conditions).

## 2.3 *High-frequency bridge*

Several of the previous considerations will apply again, but as the bridge will normally no longer be a high-voltage one, the arm across voltage $U_1$ may easily admit adjustable elements; *substitution methods*, which are highly recommendable whenever applicable, may then readily be used.

It should moreover be noted that a bridge, with separate primary winding, allows the source and detector to be interchanged. The balance corresponds to the compensation of the opposite ampere-turns in the secondary windings.

## 2.4 *Note on detectors*

Since one terminal of the detector arm is always earthed, the use of detectors with a symmetrical input is not required.

## 3. Parallel-T networks

Parallel-T networks are bridge circuits in which the currents flowing from the oscillator to the detector through two T networks are equal and opposite at the detector input. In such a circuit, the oscillator and detector can each have one terminal connected to earth; and in some of the possible circuits, the specimen and each of the variable components used for balancing also have a terminal connected to earth.

Figure 10, page 56, shows the simplest parallel-T circuit, using resistors and capacitors only. The circuit more commonly used for measurements on dielectrics is in principle that of Figure 11, page 56, for which the balance conditions are (with terminals X, X open):

$$\frac{1}{C_A} + \frac{1}{C_N} + \frac{1}{C_B} = \frac{1}{\omega^2 \, C_A \, C_N \, L} \tag{22}$$

$$R_T \left(1 + \frac{C_H}{C_B}\right) = \frac{1}{\omega^2 \, C_A \, C_N \, R_F} \tag{23}$$

In practice, a variable capacitor is connected to the terminals X, X, and its capacitance $C_V$ and conductance modify the apparent values of $L$ and $R_F$. The circuit is balanced in this condition; the specimen is then also connected to the terminals X, X and balance is regained by varying the capacitances $C_V$ and $C_H$.

Then

1) the capacitance of the specimen equals the decrease $\Delta C_V$ of $C_V$;

2) the conductance $G$ of the specimen is:

$$G = \frac{\omega^2 C_A \, C_N \, R_T}{C_B} \cdot \Delta C_H \quad \text{and} \tag{24}$$

3) the dissipation factor $\tan \delta$ of the specimen:

$$\tan \delta = \frac{\omega C_A \, C_N \, R_T}{C_B} \cdot \frac{\Delta C_H}{\Delta C_V} \tag{25}$$

where:

$\Delta C_H$ is the increase of $C_H$.

These networks can conveniently be constructed for the frequency range 50 kHz to 50 MHz and are easy to screen satisfactorily. A serious disadvantage is that the balance is strongly sensitive to frequency, so that harmonics of the supply frequency are badly unbalanced. To cover a wide range of frequencies, components must be changed or switched. At the highest frequencies, the impedances of connecting wires and of switches (if used) may introduce significant errors.

4.     **Resonant-rise ("Q-meter") method**

The resonant-rise or Q-meter method is used in the frequency range 10 kHz to 260 MHz. It is based on measurement of the voltage appearing across a resonant circuit when a small known voltage is induced in it. Figure 12, page 56, shows the usual form of the circuit, in which the resonant circuit is coupled to the oscillator by means of a common resistance $R$. Other methods of coupling are equally acceptable.

The procedure is to adjust the input voltage or current, at the required frequency, to a known value; to tune the resonant circuit to its maximum response, and to observe the voltage $U_0$ then appearing. The specimen is then connected to the appropriate terminals, the circuit brought again to its maximum response by adjusting the variable capacitor, and the new value of voltage, $U_1$, is observed.

When the specimen is connected and the circuit re-tuned, the total capacitance remains nearly constant, provided that $R_L G \ll 1$ (see Figure 12). The capacitance of the specimen is therefore approximately $\Delta C$, the change of capacitance of the variable capacitor.

The dissipation factor of the specimen is approximately:

$$\tan \delta \approx \frac{C_t}{\Delta C} \left( \frac{1}{Q_1} - \frac{1}{Q_0} \right) \tag{26}$$

where $C_t$ is the total capacitance in the circuit including that of the voltmeter and the self-capacitance of the inductor, and $Q_1$, $Q_0$ are the values of $Q$ respectively with and without the specimen connected.

The main sources of error in the method are in the calibrations of the two indicating instruments and in the unwanted impedances introduced in the wiring, especially between the variable capacitor and the specimen. For high values of dissipation factor, the condition $R_L G \ll 1$ may not hold, and the approximate equation quoted above then fails.

5.     **Susceptance variation method (reactance-variation method)**

The *micrometer electrode system*, as shown in Figure 1, page 50, was developed by Hartshorn to eliminate the errors caused at high frequencies by the series inductance and resistance of the connecting leads and of the measuring capacitor. It accomplishes this by using a coaxial go-and-return path to the specimen, and by maintaining these inductances and resistances relatively constant, regardless of whether the test specimen is in or out of the circuit. The specimen, which is either the same size as, or smaller than, the electrodes, is clamped between the electrodes. Unless the surfaces of the specimen and electrodes are lapped or ground very flat, metal foil or its equivalent must be applied to the specimen before it is placed in the electrode system. Upon removal of the specimen, the electrode system can be made to have the same capacitance by moving the micrometer electrodes closer together.

When the micrometer electrode system is carefully calibrated for capacitance changes, its use eliminates the corrections for edge capacitance, ground capacitance, and connection capacitance. A disadvantage is that the capacitance calibration is not as accurate as that of a conventional variable multiple plate capacitor, and also it is not direct reading.

At frequencies below 1 MHz, where the effect of series inductance and resistance in the leads is negligible, the capacitance calibration of the micrometer electrodes can be replaced by that of a standard capacitor in parallel with the micrometer electrode system.

The change in capacitance with the specimen in and out is measured in terms of this capacitor.

A source of minor error in a micrometer electrode system is that the edge capacitance of the electrodes, which is included in their calibration, is slightly changed by the presence of a dielectric having the same diameter as the electrodes. This error can be practically eliminated by making the diameter of the specimen less than that of the electrodes by twice its thickness.

The specimen is first clamped between the micrometer electrodes and the network used for measurement is tuned. The specimen is then removed, and the total capacitance in the circuit is restored to its original value either by moving the micrometer electrodes closer together or by readjusting the standard capacitor.

Capacitance $C_p$ of the specimen is calculated according to Table II.

The dissipation factor is then:

$$\tan \delta_1 = (\varDelta C_1 - \varDelta C_0) / 2\, C_p \tag{27}$$

where:

$\varDelta C_1$ is the difference between the two capacitance readings on either side of the resonance setting of the variable capacitor $M_2$ (Figure 1, page 50) which give detector input voltages equal to $\sqrt{2}/2$ of the resonance voltage, with the specimen in place

and:

$\varDelta C_0$ is the same difference with the specimen removed.

Care is to be taken that the test frequency remains unchanged during this procedure.

*Note.* — The resistance of the electrodes applied to the specimen becomes appreciable at high frequencies and causes a spurious increase in dissipation factor if the specimen is not flat or of uniform thickness. The frequency at which this becomes noticeable depends on the surface flatness of the specimen, but it can be as low as 10 MHz. Additional measurements of capacitance and dissipation factor must therefore be made at frequencies of 10 MHz and upwards on the specimen with no electrodes. If $C_w$ and $\tan \delta_w$ are the capacitance and dissipation factor of the specimen with no electrodes, the true dissipation factor is:

$$\tan \delta = \frac{C_p}{C_w} \tan \delta_w \tag{28}$$

where:

$C_p$ is the capacitance of the specimen with electrodes.

## 6. Screening of apparatus

An earthed screen between two points in the apparatus removes any capacitance between them and substitutes capacitance to earth from both of them. Screening of wires and components can therefore be used freely in circuits in which capacitance to earth from any point is unimportant; the Schering bridge with Wagner earth, and the transformer bridge, are circuits of this kind.

On the other hand, screening is unnecessary in a substitution bridge in parts of the circuit which remain unaltered whether the specimen is in or out.

In practice, these two considerations imply that leads to the specimen, to the oscillator, and to the detector, should be screened. As much as possible of the apparatus should be enclosed within a metal screen, to prevent variation in capacitance between the observer's body (which may not be at earth potential, and may not remain stationary) and the components of the bridge circuit.

For frequencies of the order of 100 kHz or more, "go" and "return" leads should be kept close together, so as to minimize self and mutual inductances; and if several wires are intended to be connected together, they should meet as nearly as possible at a single point, since the impedance of even a short length of wire may be appreciable at these frequencies.

If a switch is used to disconnect the specimen, it must be such that the capacitance between its contacts when they are open does not cause an error in the measurement. In three-terminal measuring systems, this can be achieved either by inserting an earthed screen between the contacts; or by using two switches in series and, when they are opened, earthing the connection between them; or by earthing the electrode which would otherwise be left disconnected.

## 7. Oscillators and detectors for bridges

### 7.1 A.C. voltage sources

The generator can be any source of power capable of supplying the necessary voltage and current with a total harmonic content of less than 1%.

### 7.2 Detectors

The following forms of detector are available. All may be used with an amplifier to increase the sensitivity:
1) Telephone (with frequency changer if necessary).
2) Electronic voltmeter or wave analyser.
3) Cathode-ray oscillograph.
4) "Magic eye" tuning indicator.
5) Vibration galvanometer (for low frequencies only).

A transformer may be needed between bridge network and the detector, either to match their impedances or because either side of the bridge output is grounded.

Harmonics may obscure or change the balance-point. This trouble can be avoided either by tuning the amplifier or by introducing a low-pass filter. A discrimination of 40 dB against the second harmonic of the measuring frequency is adequate.

## 8. Frequency range

| Method | Recommended range of frequency | Form of specimen | Remarks |
|---|---|---|---|
| 1. Schering bridge | Up to        0.10 MHz | Sheet or tube | |
| 2. Transformer bridge | 15 Hz  –  50 MHz | | |
| 3. Parallel-T networks | 50 kHz  –  30 MHz | Sheet or tube | |
| 4. Resonant-rise method | 10 kHz – 260 MHz | | |
| 5. Susceptance-variation method | 10 kHz – 100 MHz | Sheet or tube | |

TABLE I

*Calculation of vacuum capacitance and edge corrections*

| (1) | Direct interelectrode capacitance (units: pF and cm) (2) | Correction for edge capacitance (units: pF and cm) (3) |
|---|---|---|
| **1. Disk electrodes with guard-ring** Eprouvette Specimen | $C_0 = \varepsilon_0 \cdot \dfrac{A}{h} = 0.088\ 54\ \dfrac{A}{h}$ <br><br> $A = \dfrac{\pi}{4}\ (d_1 + g)^2$ | $C_e = 0$ |
| **2. Disk electrodes without guard-ring** | | |
| *a)* Diameter of the electrodes = diameter of the specimen Eprouvette Specimen | | When $a \ll h$ <br> $\dfrac{C_e}{P} = 0.029 - 0.058 \log h$ <br><br> $P = \pi d_1$ |
| *b)* Equal electrodes smaller than the specimen Eprouvette Specimen | $C_0 = \varepsilon_0 \cdot \dfrac{\pi}{4} \cdot \dfrac{d_1^2}{h}$ <br><br> $= 0.069\ 54\ \dfrac{d_1^2}{h}$ | $\dfrac{C_e}{P} = 0.019\,\varepsilon_1 - 0.058 \log h + 0.010$ <br><br> $P = \pi d_1$ <br><br> Where: $\varepsilon_1$ is an approximate value of the specimen relative permittivity, and $a \ll h$ |
| *c)* Unequal electrodes Eprouvette Specimen | | $\dfrac{C_e}{P} = 0.041\,\varepsilon_1 - 0.077 \log h + 0.045$ <br><br> $P = \pi d_1$ <br><br> Where: $\varepsilon_1$ is an approximate value of the specimen relative permittivity, and $a \ll h$ |
| **3. Cylindrical electrodes with guard-ring** Eprouvette Specimen | $C_0 = \varepsilon_0 \cdot \dfrac{2\,\pi\,(l_1 + g)}{\ln d_2/d_1}$ <br><br> $= 0.2416\ \dfrac{(l_1 + g)}{\log d_2/d_1}$ | $C_e = 0$ |
| **4. Cylindrical electrodes without guard-ring** Eprouvette Specimen | $C_0 = \varepsilon_0 \cdot \dfrac{2\,\pi\,l_1}{\ln d_2/d_1}$ <br><br> $= 0.2416\ \dfrac{l_1}{\log d_2/d_1}$ | If $\dfrac{h}{h + d_1} < \dfrac{1}{10}$, <br> $\dfrac{C_e}{2\,P} = 0.019\,\varepsilon_1 - 0.058 \log h + 0.010$ <br> $P = \pi\,(d_1 + h)$ <br> Where: $\varepsilon_1$ is an approximate value of the specimen relative permittivity |

Relative permittivity of specimen: $\varepsilon_r = \dfrac{C_x' - C_e}{C_0}$

Where:

$C_x'$    is the measured capacitance between electrodes

ln    = natural logarithm

log    = common logarithm

TABLE II

*Calculation of specimen capacitance – Contacting micrometer electrodes*

| Specimen capacitance | Remarks | Definition of symbols | |
|---|---|---|---|
| 1. Substitution of the specimen capacitance by a standard capacitor in parallel | | $C_p$ | = parallel capacitance of the specimen |
| $C_p = \Delta C + C_{or}$ | Specimen diameter is less than the micrometer electrode diameter by at least 2 $r$. The true thickness $h$ and area $A$ of the specimen must be used in calculating the permittivity | $\Delta C$ | = increase of the capacitance of the standard capacitors as to restore balance after removal of the specimen |
| | | $C_r$ | = calibration capacitance of the micrometer electrodes at spacing $r$ |
| 2. Substitution of the specimen capacitance by lowering the spacing of the micrometer electrodes after removal of the specimen | | $C_s$ | = calibration capacitance of the micrometer electrodes at spacing $s$, restoring balance after removal of the specimen |
| $C_p = C_s - C_r + C_{or}$ | Specimen diameter is less than the micrometer electrodes diameter by at least 2 $r$. The true thickness $h$ and area $A$ of the specimen must be used in calculating the permittivity | | |
| | | $C_{or}, C_{oh}$ | = air capacitance for the area between the micrometer electrodes which was occupied by the specimen, at spacing $r$ or $h$ respectively, calculated using equation 1 of Table I |
| The double calculation of the air capacitance can be avoided with only a small error (0.2% to 0.5% due to fringing at the electrode edge), when the specimen has the same diameter as the electrodes, by | | $r$ | = thickness of specimen and applied electrodes |
| 3. Substitution of the specimen capacitance by a standard capacitor in parallel | | | |
| $C_p = \Delta C + C_{oh}$ | Specimen diameter equals micrometer electrode diameter | $h$ | = thickness of specimen |
| | Electrodes applied to the specimen are of zero thickness | Relative permittivity = $\varepsilon_r = \dfrac{C_p}{C_{oh}}$ | |

TABLE III

*Calculation of relative permittivity and dissipation factor — Non-contacting electrodes*

| Relative permittivity (1) | Dissipation factor (2) | Definition of symbols (3) |
|---|---|---|
| **1. Micrometer electrodes in air** | | $\Delta C$ = capacitance change when specimen is inserted (+ when capacitance increases) |
| $$\varepsilon_r = \cfrac{1}{1 - \cfrac{\Delta C}{C_1} \cdot \cfrac{h_o}{h}}$$ or, if $h_o$ is adjusted to a new value $h_o'$, such that $\Delta C = 0$ $$\varepsilon_r = \frac{h}{h - (h_o - h_o')}$$ | $$\tan \delta_x = \tan \delta_c + M \cdot \varepsilon_r \cdot \Delta \tan \delta$$ | $C_1$ = capacitance with specimen in place <br><br> $C_f$ = $\varepsilon_f \cdot C_o$ = capacitance with fluid alone <br><br> $C_o$ = vacuum capacitance for area considered (= $\varepsilon_o \cdot A/h_o$) <br><br> $A$ = area of one face of specimen in square centimetres (or area of electrodes when specimen is exactly the same size or larger than the electrodes) |
| **2. Plane electrodes – fluid displacement** | | $\varepsilon_f$ = relative permittivity of fluid at test temperature (= 1.00 for air) |
| $$\varepsilon_r = \frac{\varepsilon_f}{1 + \tan^2 \delta_x} \cdot$$ $$\cdot \left\{ \frac{(C_f + \Delta C)(1 + \tan^2 \delta_c)}{C_f + M[C_f - (C_f + \Delta C)(1 + \tan^2 \delta_c)]} \right\}$$ | $$\tan \delta_x = \tan \delta_c + M \cdot \Delta \tan \delta \cdot$$ $$\cdot \left\{ \frac{(C_f + \Delta C)(1 + \tan^2 \delta_c)}{C_f + M[C_f - (C_f + \Delta C)(1 + \tan^2 \delta_c)]} \right\}$$ | $\varepsilon_o$ = electric constant rated in pF/cm <br><br> $\Delta \tan \delta$ = increase in dissipation factor when specimen is inserted |
| **When the dissipation factor of the specimen is less than about 0.1, the following equations can be used:** | | $\tan \delta_c$ = dissipation factor with specimen in place |
| $$\varepsilon_r = \cfrac{\varepsilon_f}{1 - \cfrac{\Delta C}{\varepsilon_f \cdot C_o + \Delta C} \cdot \cfrac{h_o}{h}}$$ | $$\tan \delta_x = \tan \delta_c + M \frac{\varepsilon_r}{\varepsilon_f} \cdot \Delta \tan \delta$$ | $\tan \delta_x$ = calculated dissipation factor of specimen <br><br> $d_o$ = outer diameter of inner electrode <br><br> $d_1$ = inner diameter of specimen <br><br> $d_2$ = outer diameter of specimen |
| **3. Cylindrical electrodes – fluid displacement (for tan $\delta_x$ less than about 0.1)** | | $d_3$ = inner diameter of outer electrode <br><br> $h_o$ = parallel-plate spacing |
| $$\varepsilon_r = \cfrac{\varepsilon_f}{1 - \cfrac{\Delta C}{C_1} \cdot \cfrac{\log d_3/d_o}{\log d_2/d_1}}$$ | $$\tan \delta_x = \tan \delta_c + \Delta \tan \delta \cdot \frac{\varepsilon_r}{\varepsilon_f} \left[ \frac{\log d_3/d_o}{\log d_2/d_1} - 1 \right]$$ | $h$ = average thickness of specimen <br><br> $M$ = $h_o/h - 1$ <br><br> $\log$ = common logarithm |
| **4. Two-fluid method – plane electrodes (for tan $\delta_x$ less than about 0.1)** | | *Note.*—In the equation for the two-fluid method, subscripts 1 and 2 refer to the first and second fluids respectively. |
| $$\varepsilon_r = \varepsilon_{f_1} + \frac{\Delta C_1 \cdot C_2 (\varepsilon_{f_2} - \varepsilon_{f_1})}{\Delta C_1 \cdot C_2 - \Delta C_2 \cdot C_1}$$ | $$\tan \delta_x = \tan \delta_{c_1} + \frac{\varepsilon_r C_o - C_1}{\Delta C_2} \cdot \Delta \tan \delta_2$$ | |

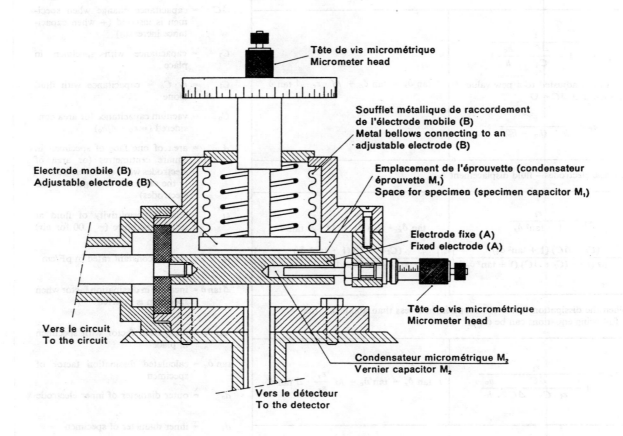

Fig. 1. — Système de condensateurs micrométriques pour diélectriques solides.
Micrometer-capacitor assembly for solid dielectrics.

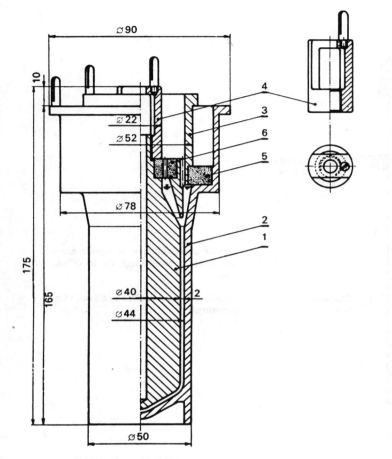

1 – électrode interne
inner electrode

2 – électrode externe
outer electrode

3 – anneau de garde
guard-ring

4 – poignée de levage
lifting handle

5 – anneau de quartz ou de borosilicate
borosilicate or quartz washer

6 – anneau de quartz ou de borosilicate
borosilicate or quartz washer

*Dimensions en millimètres*                    *Dimensions in millimetres*

FIG. 2. — Exemple d'une cellule à trois bornes pour mesures sur liquides.
Example of a three-terminal cell for measurements on liquids.

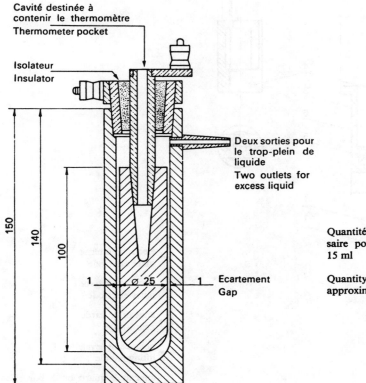

Cavité destinée à
contenir le thermomètre
Thermometer pocket

Isolateur
Insulator

Deux sorties pour
le trop-plein de
liquide
Two outlets for
excess liquid

Ecartement
Gap

Quantité approximative de liquide néces-
saire pour le remplissage de la cellule:
15 ml

Quantity of liquid required to fill cell 15 ml
approximately

*Dimensions en millimètres*

*Dimensions in millimetres*

FIG. 3. — Exemple d'une cellule à deux bornes pour mesures sur liquides.
Example of a two-terminal cell for measurements on liquids.

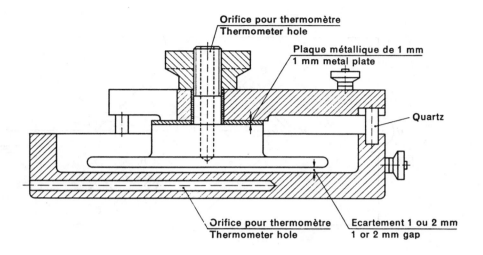

FIG. 4. — Cellule à deux bornes à électrodes planes pour mesures sur liquides.
Two-terminal cell with plane electrodes for measurements on liquids.

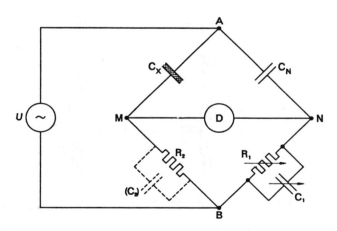

FIG. 5. — Pont de Schering, schéma de principe.
Schering bridge, circuit diagram.

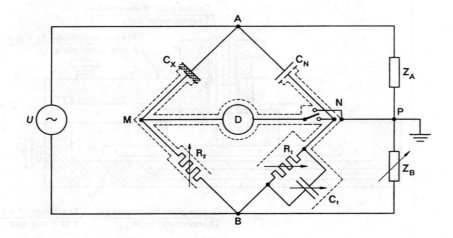

FIG. 6. — Pont de Schering avec dispositif de Wagner.
**Schering bridge with Wagner earth circuit.**

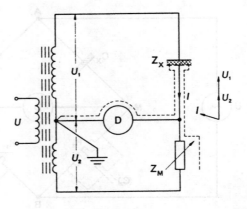

FIG. 7. — Pont à transformateur, schéma de principe.
**Transformer bridge, circuit diagram.**

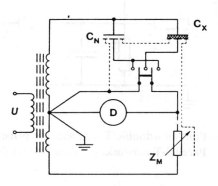

**Fig. 8.** — Pont à transformateur, tarage à charge constante.
Transformer bridge, constant load calibration.

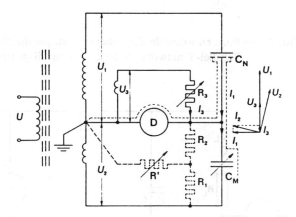

En pointillé: constitution d'une résistance élevée en parallèle sur $C_M$ (cas de l'avance de $I_2$ sur $I_1$)

Dotted line: formation of a high resistance in parallel with $C_M$ (when $I_2$ leads with respect to $I_1$)

**Fig. 9.** — Pont à transformateur, compensation lorsque $U_2$ est en retard par rapport à $U_1$ (enroulement $U_3$).
Transformer bridge, compensation when $U_2$ is lagging with respect to $U_1$ (winding $U_3$).

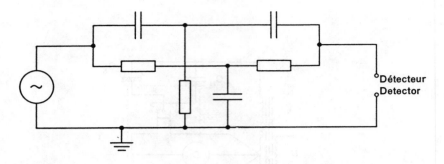

FIG. 10. — Pont en «double T», schéma de principe du circuit.
Parallel-T network, principal circuit diagram.

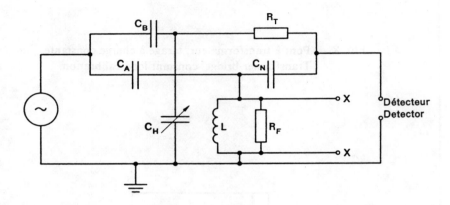

FIG. 11. — Pont en «double T», schéma pratique du circuit.
Parallel-T network, practical circuit diagram.

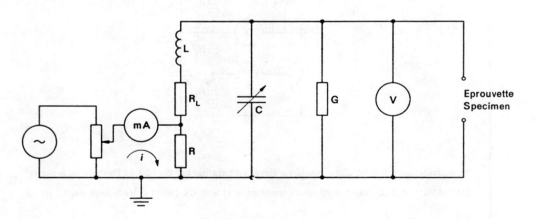

FIG. 12. — Méthode de résonance, schéma du circuit.
Resonant-rise method, circuit diagram.